T0255622

[illegible]

Lecture Notes in Mathematics

Edited by J.-M. Morel, F. Takens and B. Teissier

Editorial Policy for Multi-Author Publications: Summer Schools / Intensive Courses

1. Lecture Notes aim to report new developments in all areas of mathematics and their applications – quickly, informally and at a high level. Mathematical texts analysing new developments in modelling and numerical simulation are welcome. Manuscripts should be reasonably self-contained and rounded off. Thus they may, and often will, present not only results of the author but also related work by other people. They should provide sufficient motivation, examples and applications. There should also be an introduction making the text comprehensible to a wider audience. This clearly distinguishes Lecture Notes from journal articles or technical reports which normally are very concise. Articles intended for a journal but too long to be accepted by most journals, usually do not have this „lecture notes" character.

2. In general SUMMER SCHOOLS and other similar INTENSIVE COURSES are held to present mathematical topics that are close to the frontiers of recent research to an audience at the beginning or intermediate graduate level, who may want to continue with this area of work, for a thesis or later. This makes demands on the didactic aspects of the presentation. Because the subjects of such schools are advanced, there often exists no textbook, and so ideally, the publication resulting from such a school could be a first approximation to such a textbook.

   Usually several authors are involved in the writing, so it is not always simple to obtain a unified approach to the presentation.

   For prospective publication in LNM, the resulting manuscript should not be just a collection of course notes, each of which has been developed by an individual author with little or no co-ordination with the others, and with little or no common concept. The subject matter should dictate the structure of the book, and the authorship of each part or chapter should take secondary importance. Of course the choice of authors is crucial to the quality of the material at the school and in the book, and the intention here is not to belittle their impact, but simply to say that the book should be planned to be written by these authors jointly, and not just assembled as a result of what these authors happen to submit.

   This represents considerable preparatory work (as it is imperative to ensure that the authors know these criteria before they invest work on a manuscript), and also considerable editing work afterwards, to get the book into final shape. Still it is the form that holds the most promise of a successful book that will be used by its intended audience, rather than yet another volume of proceedings for the library shelf.

3. Manuscripts should be submitted (preferably in duplicate) either to Springer's mathematics editorial in Heidelberg, or to one of the series editors (with a copy to Springer). Volume editors are expected to arrange for the refereeing, to the usual scientific standards, of the individual contributions. If the resulting reports can be forwarded to us (series editors or Springer) this is very helpful. If no reports are forwarded or if other questions remain unclear in respect of homogeneity etc, the series editors may wish to consult external referees for an overall evaluation of the volume. A final decision to publish can be made only on the basis of the complete manuscript; however a preliminary decision can be based on a pre-final or incomplete manuscript. The strict minimum amount of material that will be considered should include a detailed outline describing the planned contents of each chapter.

   Volume editors and authors should be aware that incomplete or insufficiently close to final manuscripts almost always result in longer evaluation times. They should also be aware that parallel submission of their manuscript to another publisher while under consideration for LNM will in general lead to immediate rejection.

Continued on inside back-cover

Lecture Notes in Mathematics 1875

Editors:
J.-M. Morel, Cachan
F. Takens, Groningen
B. Teissier, Paris

Subseries:
Ecole d'Eté de Probabilités de Saint-Flour

J. Pitman

Combinatorial Stochastic Processes

Ecole d'Eté de Probabilités de Saint-Flour XXXII – 2002

Editor: Jean Picard

Author

Jim Pitman
Department of Statistics
University of California, Berkeley
367 Evans Hall
Berkeley, CA 94720-3860
USA
e-mail: pitman@stat.Berkeley.edu

Editor

Jean Picard
Laboratoire de Mathématiques Appliquées
UMR CNRS 6620
Université Blaise Pascal (Clermont-Ferrand)
63177 Aubière Cedex
France
e-mail: jean.picard@math.univ-bpclermont.fr

Cover: Blaise Pascal (1623–1662)

Library of Congress Control Number: 2006921042

Mathematics Subject Classification (2000): 05Axx, 60C05, 60J65, 60G09, 60J80

ISSN print edition: 0075-8434
ISSN electronic edition: 1617-9692
ISSN Ecole d'Eté de Probabilités de St. Flour, print edition: 0721-5363
ISBN-10 3-540-30990-X Springer Berlin Heidelberg New York
ISBN-13 978-3-540-30990-1 Springer Berlin Heidelberg New York
DOI 10.1007/b11601500

This work is subject to copyright. All rights are reserved, whether the whole or part of the material is concerned, specifically the rights of translation, reprinting, reuse of illustrations, recitation, broadcasting, reproduction on microfilm or in any other way, and storage in data banks. Duplication of this publication or parts thereof is permitted only under the provisions of the German Copyright Law of September 9, 1965, in its current version, and permission for use must always be obtained from Springer. Violations are liable for prosecution under the German Copyright Law.

Springer is a part of Springer Science+Business Media
springer.com
© Springer-Verlag Berlin Heidelberg 2006
Printed in The Netherlands

The use of general descriptive names, registered names, trademarks, etc. in this publication does not imply, even in the absence of a specific statement, that such names are exempt from the relevant protective laws and regulations and therefore free for general use.

Typesetting: by the authors and TechBookse using a Springer LaTeX package

Cover design: *design & production* GmbH, Heidelberg

Printed on acid-free paper SPIN: 11601500 41/TechBooks 5 4 3 2 1 0

Foreword

Three series of lectures were given at the 32nd Probability Summer School in Saint-Flour (July 7–24, 2002), by the Professors Pitman, Tsirelson and Werner. The courses of Professors Tsirelson ("Scaling limit, noise, stability") and Werner ("Random planar curves and Schramm-Loewner evolutions") have been published in a previous issue of *Lectures Notes in Mathematics* (volume 1840). This volume contains the course "Combinatorial stochastic processes" of Professor Pitman. We cordially thank the author for his performance in Saint-Flour and for these notes.

76 participants have attended this school. 33 of them have given a short lecture. The lists of participants and of short lectures are enclosed at the end of the volume.

The Saint-Flour Probability Summer School was founded in 1971. Here are the references of Springer volumes which have been published prior to this one. All numbers refer to the *Lecture Notes in Mathematics* series, except S-50 which refers to volume 50 of the *Lecture Notes in Statistics* series.

1971: vol 307	1980: vol 929	1990: vol 1527	1998: vol 1738
1973: vol 390	1981: vol 976	1991: vol 1541	1999: vol 1781
1974: vol 480	1982: vol 1097	1992: vol 1581	2000: vol 1816
1975: vol 539	1983: vol 1117	1993: vol 1608	2001: vol 1837 & 1851
1976: vol 598	1984: vol 1180	1994: vol 1648	2002: vol 1840
1977: vol 678	1985/86/87: vol 1362 & S-50	1995: vol 1690	2003: vol 1869
1978: vol 774	1988: vol 1427	1996: vol 1665	
1979: vol 876	1989: vol 1464	1997: vol 1717	

Further details can be found on the summer school web site
`http://math.univ-bpclermont.fr/stflour/`

Université Blaise Pascal
September 2005

Jean Picard

Contents

Preliminaries

0.0. Preface

This is a collection of expository articles about various topics at the interface between enumerative combinatorics and stochastic processes. These articles expand on a course of lectures given at the Ecole d'Eté de Probabilités de St. Flour in July 2002. The articles are also called 'chapters'. Each chapter is fairly self-contained, so readers with adequate background can start reading any chapter, with occasional consultation of earlier chapters as necessary. Following this Chapter 0, there are 10 chapters, each divided into *sections*. Most sections conclude with some *Exercises*. Those for which I don't know solutions are called *Problems*.

Acknowledgments Much of the research reviewed here was done jointly with David Aldous. Much credit is due to him, especially for the big picture of continuum approximations to large combinatorial structures. Thanks also to my other collaborators in this work, especially Jean Bertoin, Michael Camarri, Steven Evans, Sasha Gnedin, Ben Hansen, Jacques Neveu, Mihael Perman, Ravi Sheth, Marc Yor and Jim Young. A preliminary version of these notes was developed in Spring 2002 with the help of a dedicated class of ten graduate students in Berkeley: Noam Berger, David Blei, Rahul Jain, Şerban Nacu, Gabor Pete, Lea Popovic, Alan Hammond, Antar Bandyopadhyay, Manjunath Krishnapur and Grégory Miermont. The last four deserve special thanks for their contributions as research assistants. Thanks to the many people who have read versions of these notes and made suggestions and corrections, especially David Aldous, Jean Bertoin, Aubrey Clayton, Shankar Bhamidi, Rui Dong, Steven Evans, Sasha Gnedin, Bénédicte Haas, Jean-François Le Gall, Neil O'Connell, Mihael Perman, Lea Popovic, Jason Schweinsberg. Special thanks to Marc Yor and Matthias Winkel for their great help in preparing the final version of these notes for publication. Thanks also to Jean Picard for his organizational efforts in making arrangements for the St. Flour Summer School. This work was supported in part by NSF Grants DMS-0071448 and DMS-0405779.

0.1. Introduction

The main theme of this course is the study of various combinatorial models of random partitions and random trees, and the asymptotics of these models related to continuous parameter stochastic processes. A basic feature of models for random partitions is that the sum of the parts is usually constant. So the sizes of the parts cannot be independent. But the structure of many natural models for random partitions can be reduced by suitable conditioning or scaling to classical probabilistic results involving sums of independent random variables. Limit models for combinatorially defined random partitions are consequently related to the two fundamental limit processes of classical probability theory: Brownian motion and Poisson processes. The theory of Brownian motion and related stochastic processes has been greatly enriched by the recognition that some fundamental properties of these processes are best understood in terms of how various random partitions and random trees are embedded in their paths. This has led to rapid developments, particularly in the theory of continuum random trees, continuous state branching processes, and Markovian superprocesses, which go far beyond the scope of this course. Following is a list of the main topics to be treated:

- models for random combinatorial structures, such as trees, forests, permutations, mappings, and partitions;
- probabilistic interpretations of various combinatorial notions e.g. Bell polynomials, Stirling numbers, polynomials of binomial type, Lagrange inversion;
- Kingman's theory of exchangeable random partitions and random discrete distributions;
- connections between random combinatorial structures and processes with independent increments: Poisson-Dirichlet limits;
- random partitions derived from subordinators;
- asymptotics of random trees, graphs and mappings related to excursions of Brownian motion;
- continuum random trees embedded in Brownian motion;
- Brownian local times and squares of Bessel processes;
- various processes of fragmentation and coagulation, including Kingman's coalescent, the additive and multiplicative coalescents

Next, an incomplete list and topics of current interest, with inadequate references. These topics are close to those just listed, and certainly part of the realm of combinatorial stochastic processes, but not treated here:

- probability on trees and networks, as presented in [292];
- random integer partitions [159, 104], random Young tableaux, growth of Young diagrams, connections with representation theory and symmetric functions [245, 420, 421, 239];
- longest increasing subsequence of a permutation, connections with random matrices [28];

- random partitions related to uniformly chosen invertible matrices over a finite field, as studied by Fulman [160];
- random maps, coalescing saddles, singularity analysis, and Airy phenomena, [81];
- random planar lattices and integrated superbrownian excursion [94].

The reader of these notes is assumed to be familiar with the basic theory of probability and stochastic processes, at the level of Billingsley [64] or Durrett [122], including continuous time stochastic processes, especially Brownian motion and Poisson processes. For background on some more specialized topics (local times, Bessel processes, excursions, SDE's) the reader is referred to Revuz-Yor [384]. The rest of this Chapter 0 reviews some basic facts from this probabilistic background for ease of later reference. This material is organized as follows:

0.2. Brownian motion and related processes This section provides some minimal description of the background expected of the reader to follow some of the more advanced sections of the text. This includes the definition and basic properties of Brownian motion $B := (B_t, t \geq 0)$, and of some important processes derived from B by operations of scaling and conditioning. These processes include the Brownian bridge, Brownian meander and Brownian excursion. The basic facts of Itô's excursion theory for Brownian motion are also recorded.

0.3. Subordinators This section reviews a few basic facts about increasing Lévy processes in general, and some important facts about gamma and stable processes in particular.

0.2. Brownian motion and related processes

Let $S_n := X_1 + \cdots + X_n$ where the X_i are independent random variables with mean 0 and variance 1, and let S_t for real t be defined by linear interpolation between integer values. According to *Donsker's theorem* [64, 65, 122, 384]

$$(S_{nt}/\sqrt{n}, 0 \leq t \leq 1) \xrightarrow{d} (B_t, 0 \leq t \leq 1) \tag{0.1}$$

in the usual sense of convergence in distribution of random elements of $C[0,1]$, where $(B_t, t \geq 0)$ is a *standard Brownian motion* meaning that B is a process with continuous paths and stationary independent Gaussian increments, with $B_t \overset{d}{=} \sqrt{t} B_1$ where B_1 is standard Gaussian.

Brownian bridge Assuming now that the X_i are integer valued, some conditioned forms of Donsker's theorem can be presented as follows. Let $o(\sqrt{n})$ denote any sequence of possible values of S_n with $o(\sqrt{n})/\sqrt{n} \to 0$ as $n \to \infty$. Then [128]

$$(S_{nt}/\sqrt{n}, 0 \leq t \leq 1 \,|\, S_n = o(\sqrt{n})) \xrightarrow{d} (B_t^{\text{br}}, 0 \leq t \leq 1) \tag{0.2}$$

where B^{br} is the *standard Brownian bridge*, that is, the centered Gaussian process obtained by conditioning $(B_t, 0 \leq t \leq 1)$ on $B_1 = 0$. Some well known descriptions of the distribution of B^{br} are [384, Ch. III, Ex (3.10)]

$$(B_t^{\mathrm{br}}, 0 \leq t \leq 1) \stackrel{d}{=} (B_t - tB_1, 0 \leq t \leq 1) \stackrel{d}{=} ((1-t)B_{t/(1-t)}, 0 \leq t \leq 1) \quad (0.3)$$

where $\stackrel{d}{=}$ denotes equality of distributions on the path space $C[0,1]$, and the rightmost process is defined to be 0 for $t = 1$.

Brownian meander and excursion Let $T_- := \inf\{n : S_n < 0\}$. Then as $n \to \infty$

$$(S_{nt}/\sqrt{n}, 0 \leq t \leq 1 \,|\, T_- > n) \stackrel{d}{\to} (B_t^{\mathrm{me}}, 0 \leq t \leq 1) \quad (0.4)$$

where B^{me} is the *standard Brownian meander* [205, 71], and as $n \to \infty$ through possible values of T_-

$$(S_{nt}/\sqrt{n}, 0 \leq t \leq 1 \,|\, T_- = n) \stackrel{d}{\to} (B_t^{\mathrm{ex}}, 0 \leq t \leq 1) \quad (0.5)$$

where B_t^{ex} is the *standard Brownian excursion* [225, 102]. Informally,

$$B^{\mathrm{me}} \stackrel{d}{=} (B \,|\, B_t > 0 \text{ for all } 0 < t < 1)$$
$$B^{\mathrm{ex}} \stackrel{d}{=} (B \,|\, B_t > 0 \text{ for all } 0 < t < 1, B_1 = 0)$$

where $\stackrel{d}{=}$ denotes equality in distribution. These definitions of conditioned Brownian motions have been made rigorous in a number of ways: for instance by the method of Doob h-transforms [255, 394, 155], and as weak limits as $\varepsilon \downarrow 0$ of the distribution of B given suitable events A_ε, as in [124, 69], for instance

$$(B \,|\, \underline{B}(0,1) > -\varepsilon) \stackrel{d}{\to} B^{\mathrm{me}} \text{ as } \varepsilon \downarrow 0 \quad (0.6)$$

$$(B^{\mathrm{br}} \,|\, \underline{B^{\mathrm{br}}}(0,1) > -\varepsilon) \stackrel{d}{\to} B^{\mathrm{ex}} \text{ as } \varepsilon \downarrow 0 \quad (0.7)$$

where $\underline{X}(s,t)$ denotes the infimum of a process X over the interval (s,t).

Brownian scaling For a process $X := (X_t, t \in J)$ parameterized by an interval J, and $I = [G_I, D_I]$ a subinterval of J with length $\lambda_I := D_I - G_I > 0$, we denote by $X[I]$ or $X[G_I, D_I]$ the *fragment of* X *on* I, that is the process

$$X[I]_u := X_{G_I + u} \qquad (0 \leq u \leq \lambda_I). \quad (0.8)$$

We denote by $X_*[I]$ or $X_*[G_I, D_I]$ the *standardized fragment of* X *on* I, defined by the *Brownian scaling operation*

$$X_*[I]_u := \frac{X_{G_I + u\lambda_I} - X_{G_I}}{\sqrt{\lambda_I}} \qquad (0 \leq u \leq 1). \quad (0.9)$$

For $T > 0$ let $G_T := \sup\{s : s \leq T, B_s = 0\}$ be the last zero of B before time T and $D_T := \inf\{s : s > T, B_s = 0\}$ be the first zero of B after time

T. Let $|B| := (|B_t|, t \geq 0)$, called *reflecting Brownian motion.* It is well known [211, 98, 384] that for each fixed $T > 0$, there are the following identities in distribution derived by *Brownian scaling*:

$$B_*[0,T] \stackrel{d}{=} B[0,1]; \qquad B_*[0,G_T] \stackrel{d}{=} B^{\mathrm{br}} \tag{0.10}$$

$$|B|_*[G_T,T] \stackrel{d}{=} B^{\mathrm{me}}; \qquad |B|_*[G_T,D_T] \stackrel{d}{=} B^{\mathrm{ex}}. \tag{0.11}$$

It is also known that $B^{\mathrm{br}}, B^{\mathrm{me}}$ and B^{ex} can be constructed by various other operations on the paths of B, and transformed from one to another by further operations [53].

For $0 < t < \infty$ let $B^{\mathrm{br},t}$ be a *Brownian bridge of length* t, which may be regarded as a random element of $C[0,t]$ or of $C[0,\infty]$, as convenient:

$$B^{\mathrm{br},t}(s) := \sqrt{t}B^{\mathrm{br}}((s/t)\wedge 1) \qquad (s \geq 0). \tag{0.12}$$

Let $B^{\mathrm{me},t}$ denote a *Brownian meander of length* t, and $B^{\mathrm{ex},t}$ be a *Brownian excursion of length* t, defined similarly to (0.12) with B^{me} or B^{ex} instead of B^{br}.

Brownian excursions and the three-dimensional Bessel process The following theorem summarizes some important relations between Brownian excursions and a particular time-homogeneous diffusion process R_3 on $[0,\infty)$, commonly known as the *three-dimensional Bessel process* BES(3), due to the representation

$$(R_3(t), t \geq 0) \stackrel{d}{=} \left(\sqrt{\sum_{i=1}^{3}(B_i(t))^2}, t \geq 0\right) \tag{0.13}$$

where the B_i are three independent standard Brownian motions. It should be understood however that this particular representation of R_3 is a relatively unimportant coincidence in distribution. What is more important, and can be understood entirely in terms of the random walk approximations (0.1) and (0.5) of Brownian motion and Brownian excursion, is that there exists a time-homogeneous diffusion process R_3 on $[0,\infty)$ with $R_3(0) = 0$, which has the same self-similarity property as B, meaning invariance under Brownian scaling, and which can be characterized in various ways, including (0.13), but most importantly as a Doob h-transform of Brownian motion.

Theorem 0.1. *For each fixed $t > 0$, the Brownian excursion $B^{\mathrm{ex},t}$ of length t is the BES(3) bridge from 0 to 0 over time t, meaning that*

$$(B^{\mathrm{ex},t}(s), 0 \leq s \leq t) \stackrel{d}{=} (R_3(s), 0 \leq s \leq t \,|\, R_3(t) = 0).$$

Moreover, as $t \to \infty$

$$B^{\mathrm{ex},t} \stackrel{d}{\to} R_3, \tag{0.14}$$

and R_3 can be characterized in two other ways as follows:

(i) [303, 436] *The process R_3 is a Brownian motion on $[0, \infty)$ started at 0 and conditioned never to return to 0, as defined by the Doob h-transform, for the harmonic function $h(x) = x$ of Brownian motion on $[0, \infty)$, with absorbtion at 0. That is, R_3 has continuous paths starting at 0, and for each $0 < a < b$ the stretch of R_3 between when it first hits a and first hits b is distributed like B with $B_0 = a$ conditioned to hit b before 0.*
(ii) [345]

$$R_3(t) = B(t) - 2\underline{B}(t) \qquad (t \geq 0) \tag{0.15}$$

where B is a standard Brownian motion with past minimum process

$$\underline{B}(t) := \underline{B}[0, t] = -\underline{R_3}[t, \infty).$$

Lévy's identity The identity in distribution (0.15) admits numerous variations and conditioned forms [345, 53, 55] by virtue of Lévy's identity of joint distributions of paths [384]

$$(B - \underline{B}, -\underline{B}) \stackrel{d}{=} (|B|, L) \tag{0.16}$$

where $L := (L_t, t \geq 0)$ is the local time process of B at 0, which may be defined almost surely as the occupation density

$$L_t = \lim_{\varepsilon \downarrow 0} \frac{1}{2\varepsilon} \int_0^t ds 1(|B_s| \leq \varepsilon).$$

For instance,

$$(R_3(t), t \geq 0) \stackrel{d}{=} (|B_t| + L_t, t \geq 0).$$

Lévy-Itô-Williams theory of Brownian excursions Due to (0.16), the process of excursions of $|B|$ away from 0 is equivalent in distribution to the process of excursions of B above $\underline{B}$. According to the Lévy-Itô description of this process, if $I_\ell := [T_{\ell-}, T_\ell]$ for $T_\ell := \inf\{t : B(t) < -\ell\}$, the points

$$\{(\ell, \mu(I_\ell), (B - \underline{B})[I_\ell]) : \ell > 0, \mu(I_\ell) > 0\}, \tag{0.17}$$

where μ is Lebesgue measure, are the points of a Poisson point process on $\mathbb{R}_{>0} \times \mathbb{R}_{>0} \times C[0, \infty)$ with intensity

$$d\ell \frac{dt}{\sqrt{2\pi}\, t^{3/2}} \mathbb{P}(B^{\mathrm{ex},t} \in d\omega). \tag{0.18}$$

On the other hand, according to Williams [437], if $M_\ell := \overline{B}[I_\ell] - \underline{B}[I_\ell]$ is the maximum height of the excursion of B over $\underline{B}$ on the interval I_ℓ, the points

$$\{(\ell, M_\ell, (B - \underline{B})[I_\ell]) : \ell > 0, \mu(I_\ell) > 0\}, \tag{0.19}$$

are the points of a Poisson point process on $\mathbb{R}_{>0} \times \mathbb{R}_{>0} \times C[0, \infty)$ with intensity

$$d\ell \frac{dm}{m^2} \mathbb{P}(B^{\mathrm{ex}\,|\,m} \in d\omega) \tag{0.20}$$

where $B^{\text{ex}\,|\,m}$ is a *Brownian excursion conditioned to have maximum* m. That is to say $B^{\text{ex}\,|\,m}$ is a process X with $X(0) = 0$ such that for each $m > 0$, and $H_x(X) := \inf\{t : t > 0, X(t) = x\}$, the processes $X[0, H_m(X)]$ and $m - X[H_m(X), H_0(X)]$ are two independent copies of $R_3[0, H_m(R_3)]$, and X is stopped at 0 at time $H_0(X)$. *Itô's law of Brownian excursions* is the σ-finite measure ν on $C[0,\infty)$ which can be presented in two different ways according to (0.18) and (0.20) as

$$\nu(\cdot) = \int_0^\infty \frac{dt}{\sqrt{2\pi} t^{3/2}} \mathbb{P}(B^{\text{ex},t} \in \cdot) = \int_0^\infty \frac{dm}{m^2} \mathbb{P}(B^{\text{ex}\,|\,m} \in \cdot) \tag{0.21}$$

where the first expression is a disintegration according to the lifetime of the excursion, and the second according to its maximum. The identity (0.21) has a number of interesting applications and generalizations [60, 367, 372].

BES(3) bridges Starting from three independent standard Brownian bridges $B_i^{\text{br}}, i = 1,2,3$, for $x, y \geq 0$ let

$$R_3^{x\to y}(u) := \sqrt{(x + (y-x)u + B_{1,u}^{\text{br}})^2 + (B_{2,u}^{\text{br}})^2 + (B_{3,u}^{\text{br}})^2} \qquad (0 \leq u \leq 1). \tag{0.22}$$

The random element $R_3^{x\to y}$ of $C[0,1]$ is the *BES(3) bridge from* x *to* y, in terms of which the laws of the standard excursion and meander are represented as

$$B^{\text{ex}} \stackrel{d}{=} R_3^{0\to 0} \text{ and } B^{\text{me}} \stackrel{d}{=} R_3^{0\to\rho} \tag{0.23}$$

where ρ is a random variable with the *Rayleigh density*

$$P(\rho \in dx)/dx = xe^{-\frac{1}{2}x^2} \qquad (x > 0) \tag{0.24}$$

and ρ is independent of the family of Bessel bridges $R_3^{0\to r}, r \geq 0$. Then by construction

$$B_1^{\text{me}} = \rho \stackrel{d}{=} \sqrt{2\Gamma_1} \tag{0.25}$$

where Γ_1 is a *standard exponential variable*, and

$$(B_1^{\text{me}} \,|\, B_1^{\text{me}} = r) \stackrel{d}{=} R_3^{0\to r}. \tag{0.26}$$

These descriptions are read from [435, 208]. See also [98, 53, 62, 384] for further background. By (0.22) and Itô's formula, the process $R_3^{x\to y}$ can be characterized for each $x, y \geq 0$ as the solution over $[0,1]$ of the Itô SDE

$$R_0 = x; \qquad dR_s = \left(\frac{1}{R_s} + \frac{(y - R_s)}{(1-s)}\right) ds + d\gamma_s \tag{0.27}$$

for a Brownian motion γ.

Exercises

0.2.1. [384] Show, using stochastic calculus, that the three dimensional Bessel process R_3 is characterized by description (i) of Theorem 0.1.

0.2.2. Check that $R_3^{x\to y}$ solves (0.27), and discuss the uniqueness issue.

0.2.3. [344, 270] Formulate and prove a discrete analog for simple symmetric random walk of the equivalence of the two descriptions of R_3 given in Theorem 0.1, along with a discrete analog of the following fact: if $R(t) := B(t) - 2\underline{B}(t)$ for a Brownian motion B then

$$\text{the conditional law of } \underline{B}(t) \text{ given } (R(s), 0 \le s \le t) \text{ is uniform on } [-R(t), 0]. \tag{0.28}$$

Deduce the Brownian results by embedding a simple symmetric random walk in the path of B.

0.2.4. (Williams' time reversal theorem)[436, 344, 270] Derive the identity in distribution

$$(R_3(t), 0 \le t \le K_x) \stackrel{d}{=} (x - B(H_x - t), 0 \le t \le H_x), \tag{0.29}$$

where K_x is the last hitting time of $x > 0$ by R_3, and where H_x the first hitting time of $x > 0$ by B, by first establishing a corresponding identity for paths of a suitably conditioned random walk with increments of ± 1, then passing to a Brownian limit.

0.2.5. [436, 270] Derive the identity in distribution

$$(R_3(t), 0 \le t \le H_x) \stackrel{d}{=} (x - R_3(H_x - t), 0 \le t \le H_x), \tag{0.30}$$

where H_x is the hitting time of $x > 0$ by R_3.

0.2.6. Fix $x > 0$ and for $0 < y < x$ let K_y be the last time before $H_x(R_3)$ that R_3 hits y, let $I_y := [K_{y-}, K_y]$, and let $R_3[I_y] - y$ be the excursion of R_3 over the interval I_y pulled down so that it starts and ends at 0. Let M_y be the maximum height of this excursion. Show that the points

$$\{(y, M_y, R_3[I_y] - y) : M_y > 0\}, \tag{0.31}$$

are the points of a Poisson point process on $[0, x] \times \mathbb{R}_{>0} \times C[0, \infty)$ with intensity measure of the form

$$f(y, m)\, dy\, dm\, \mathbb{P}(B^{\text{ex}\,|\,m} \in d\omega)$$

for some $f(y, m)$ to be computed explicitly, where $B^{\text{ex}\,|\,m}$ is a Brownian excursion of maximal height m. See [348] for related results.

Notes and comments

See [387, 270, 39, 384, 188] for different approaches to the basic path transformation (0.15) from B to R_3, its discrete analogs, and various extensions. In terms of $X := -B$ and $M := \overline{X} = -\underline{B}$, the transformation takes X to $2M - X$. For a generalization to exponential functionals, see Matsumoto and Yor [299]. This is also discussed in [331], where an alternative proof is given using reversibility and symmetry arguments, with an application to a certain directed polymer problem. A multidimensional extension is presented in [332], where a representation for Brownian motion conditioned never to exit a (type A) Weyl chamber is obtained using reversibility and symmetry properties of certain queueing networks. See also [331, 262] and the survey paper [330]. This representation theorem is closely connected to random matrices, Young tableaux, the Robinson-Schensted-Knuth correspondence, and symmetric functions theory [329, 328]. A similar representation theorem has been obtained in [75] in a more general symmetric spaces context, using quite different methods. These multidimensional versions of the transformation from X to $2M - X$ are intimately connected with combinatorial representation theory and Littelmann's path model [286].

0.3. Subordinators

A *subordinator* $(T_s, s \geq 0)$ is an increasing process with right continuous paths, stationary independent increments, and $T_0 = 0$. It is well known [40] that every such process can be represented as

$$T_t = ct + \sum_{0<s\leq t} \Delta_s \qquad (t \geq 0)$$

for some $c \geq 0$ where $\Delta_s := T_s - T_{s-}$ and $\{(s, \Delta_s) : s > 0, \Delta_s > 0\}$ is the set of points of a Poisson point process on $(0,\infty)^2$ with intensity measure $ds\Lambda(dx)$ for some measure Λ on $(0,\infty)$, called the *Lévy measure* of T_1 or of $(T_t, t \geq 0)$, such that the *Laplace exponent*

$$\Psi(u) = cu + \int_0^\infty (1 - e^{-ux})\Lambda(dx) \tag{0.32}$$

is finite for some (hence all) $u > 0$. The Laplace transform of the distribution of T_t is then given by the following special case of the *Lévy-Khintchine formula* [40]:

$$\mathbb{E}[e^{-uT_t}] = e^{-t\Psi(u)}. \tag{0.33}$$

The gamma process Let $(\Gamma_s, s \geq 0)$ denote a *standard gamma process*, that is the subordinator with marginal densities

$$\mathbb{P}(\Gamma_s \in dx)/dx = \frac{1}{\Gamma(s)}\, x^{s-1}\, e^{-x} \qquad (x > 0). \tag{0.34}$$

The Laplace exponent $\Psi(u)$ of the standard gamma process is

$$\Psi(u) = \log(1+u) = u - \frac{u^2}{2} + \frac{u^3}{3} - \cdots$$

and the Lévy measure is $\Lambda(dx) = x^{-1}e^{-x}dx$. A special feature of the gamma process is the multiplicative structure exposed by Exercise 0.3.1 and Exercise 0.3.2 . See also [416].

Stable subordinators Let $\mathbb{P}_\alpha$ govern a stable subordinator $(T_s, s \geq 0)$ with index $\alpha \in (0,1)$. So under $\mathbb{P}_\alpha$

$$T_s \stackrel{d}{=} s^{1/\alpha} T_1 \tag{0.35}$$

where

$$\mathbb{E}_\alpha[\exp(-\lambda T_1)] = \exp(-\lambda^\alpha) = \int_0^\infty e^{-\lambda x} f_\alpha(x)\, dx \tag{0.36}$$

with $f_\alpha(x)$ the stable(α) density of T_1, that is [377]

$$f_\alpha(t) = \frac{1}{\pi} \sum_{k=0}^\infty \frac{(-1)^{k+1}}{k!} \sin(\pi\alpha k) \frac{\Gamma(\alpha k+1)}{t^{\alpha k+1}}. \tag{0.37}$$

For $\alpha = \frac{1}{2}$ this reduces to the formula of Doetsch [112, pp. 401-402] and Lévy [284]

$$f_{\frac{1}{2}}(t) = \frac{t^{-\frac{3}{2}}}{2\sqrt{\pi}} e^{-\frac{1}{4t}} = \mathbb{P}(\tfrac{1}{2}B_1^{-2} \in dt)/dt \tag{0.38}$$

where B_1 is a standard Gaussian variable. For general α, the Lévy density of T_1 is well known to be

$$\rho_\alpha(x) := \frac{\alpha}{\Gamma(1-\alpha)} \frac{1}{x^{\alpha+1}} \qquad (x > 0) \tag{0.39}$$

Note the useful formula

$$\mathbb{E}_\alpha(T_1^{-\theta}) = \frac{\Gamma(\frac{\theta}{\alpha}+1)}{\Gamma(\theta+1)} \qquad (\theta > -\alpha) \tag{0.40}$$

which is read from (0.36) using $T_1^{-\theta} = \Gamma(\theta)^{-1} \int_0^\infty \lambda^{\theta-1} e^{-\lambda T_1} d\lambda$. Let $(S_t, t \geq 0)$ denote the continuous inverse of $(T_s, s \geq 0)$. For instance, $(S_t, t \geq 0)$ may be the local time process at 0 of some self-similar Markov process, such as a Brownian motion ($\alpha = \frac{1}{2}$) or a Bessel process of dimension $2 - 2\alpha \in (0,2)$. See [384, 41]. Easily from (0.35), under $\mathbb{P}_\alpha$ there is the identity in law

$$S_t / t^\alpha \stackrel{d}{=} S_1 \stackrel{d}{=} T_1^{-\alpha} \tag{0.41}$$

Thus the $\mathbb{P}_\alpha$ distribution of S_1 is the *Mittag-Leffler distribution* with Mellin transform

$$\mathbb{E}_\alpha(S_1^p) = \mathbb{E}_\alpha((T_1^{-\alpha})^p) = \frac{\Gamma(p+1)}{\Gamma(p\alpha+1)} \qquad (p > -1) \tag{0.42}$$

and density at $s > 0$

$$\mathbb{P}_\alpha(S_1 \in ds)/ds = g_\alpha(s) := \frac{f_\alpha(s^{-1/\alpha})}{\alpha s^{1+1/\alpha}} = \frac{1}{\pi\alpha}\sum_{k=0}^{\infty}\frac{(-1)^{k+1}}{k!}\Gamma(\alpha k+1)s^{k-1}\sin(\pi\alpha k) \tag{0.43}$$

See [314, 66] for background.

Exercises

0.3.1. (Beta-Gamma algebra) Let $(\Gamma_t, t \geq 0)$ be a standard gamma process. For $a, b > 0$ let

$$\beta_{a,b} := \Gamma_a/\Gamma_{a+b}. \tag{0.44}$$

Then $\beta_{a,b}$ has the *beta*(a,b) *distribution*

$$\mathbb{P}(\beta_{a,b} \in du) = \frac{\Gamma(a+b)}{\Gamma(a)\Gamma(b)}u^{a-1}(1-u)^{b-1}du \qquad (0 < u < 1) \tag{0.45}$$

and $\beta_{a,b}$ is independent of Γ_{a+b}. See [117] for a review of algebraic properties of beta and gamma distributions, and [87] for developments related to intertwining of Markov processes.

0.3.2. (Dirichlet Process) [153, 350] Let $(\Gamma_t, t \geq 0)$ be a standard gamma process, and for $\theta > 0$ set

$$F_\theta(u) := \Gamma_{u\theta}/\Gamma_\theta \qquad (0 \leq u \leq 1). \tag{0.46}$$

Call $F_\theta(\cdot)$, *the standard Dirichlet process with parameter* θ, or *Dirichlet*(θ) process for short. This process $F_\theta(\cdot)$ is increasing with exchangeable increments, and independent of Γ_θ. Note that $F_\theta(\cdot)$ is the cumulative distribution function of a random discrete probability distribution on $[0,1]$, which may also be denoted F_θ. For each partition of $[0,1]$ into m disjoint intervals $I_1, \ldots, I_m$ of lengths $a_1, \ldots, a_m$, with $\sum_{i=1}^m a_i = 1$, the random vector $(F_\theta(I_1), \ldots, F_\theta(I_m))$ has the *Dirichlet*$(\theta_1, \ldots, \theta_m)$ distribution with $\theta_i = \theta a_i$, with density

$$\frac{\Gamma(\theta_1 + \cdots + \theta_m)}{\Gamma(\theta_1)\cdots\Gamma(\theta_m)}p_1^{\theta_1 - 1}\cdots p_m^{\theta_m - 1}dp_1 \cdots dp_{m-1} \tag{0.47}$$

on the simplex $(p_1, \ldots, p_m)$ with $p_i \geq 0$ and $\sum_{i=1}^m p_i = 1$. Deduce a description of the laws of gamma bridges $(\Gamma_t, 0 \leq t \leq \theta \,|\, \Gamma_\theta = x)$ in terms of the standard Dirichlet process $F_\theta(\cdot)$ analogous to the well known description of Brownian bridges $(B_t, 0 \leq t \leq \theta \,|\, B_\theta = x)$ in terms of a standard Brownian bridge B^{br}.

1

Bell polynomials, composite structures and Gibbs partitions

This chapter provides an introduction to the elementary theory of Bell polynomials and their applications in probability and combinatorics.

1.1. Notation This section introduces some basic notation for factorial powers and power series.

1.2. Partitions and compositions The (n,k)th partial Bell polynomial

$$B_{n,k}(w_1, w_2, \ldots)$$

is introduced as a sum of products over all partitions of a set of n elements into k blocks. These polynomials arise naturally in the enumeration of composite structures, and in the *compositional* or *Faà di Bruno formula* for coefficients in the power series expansion for the composition of two functions. Various kinds of Stirling numbers appear as valuations of Bell polynomials for particular choices of the weights $w_1, w_2, \ldots$.

1.3. Moments and cumulants The classical formula for cumulants of a random variable as a polynomial function of its moments, and various related identities, provide applications of Bell polynomials.

1.4. Random sums Bell polynomials appear in the study of sums of a random number of independent and identically distributed non-negative random variables, as in the theory of branching processes, due to the well known expression for the generating function of such a random sum as the composition of two generating functions.

1.5. Gibbs partitions Bell polynomials appear again in a natural model for random partitions of a finite set, in which the probability of each partition is assigned proportionally to a product of weights w_j depending on the sizes j of blocks.

1.1. Notation

Factorial powers For $n = 0, 1, 2\ldots$, and arbitrary real x and α let $(x)_{n\uparrow\alpha}$ denote the nth *factorial power of x with increment α,* that is

$$(x)_{n\uparrow\alpha} := x(x+\alpha)\cdots(x+(n-1)\alpha) = \prod_{i=0}^{n-1}(x+i\alpha) = \alpha^n (x/\alpha)_{n\uparrow} \tag{1.1}$$

where $(x)_{n\uparrow} := (x)_{n\uparrow 1}$ and the last equality is valid only for $\alpha \neq 0$. Similarly, let

$$(x)_{n\downarrow\alpha} := (x)_{n\uparrow -\alpha} \tag{1.2}$$

be the nth *factorial power of x with decrement α* and $(x)_{n\downarrow} := (x)_{n\downarrow 1}$. Note that $(x)_{n\downarrow}$ for positive integer x is the number of permutations of x elements of length n, and that

$$(x)_{n\uparrow} = \Gamma(x+n)/\Gamma(x). \tag{1.3}$$

Recall the consequence of Stirling's formula that for each real r

$$\Gamma(x+r)/\Gamma(x) \sim x^r \text{as } x \to \infty. \tag{1.4}$$

Power series Notation such as

$$c_n = [x^n]f(x)$$

should be read as "c_n is the coefficient of x^n in $f(x)$", meaning

$$f(x) = \sum_n c_n x^n$$

where the power series might be convergent in some neighborhood of 0, or regarded formally [407]. Note that e.g.

$$\left[\frac{x^n}{n!}\right] f(x) = n![x^n]f(x)$$

1.2. Partitions and compositions

Let F be a finite set. A *partition of F into k blocks* is an *unordered* collection of non-empty disjoint sets $\{A_1, \ldots, A_k\}$ whose union is F. Let $\mathcal{P}_{[n]}^k$ denote the set of partitions of the set $[n] := \{1, \ldots, n\}$ into k blocks, and let $\mathcal{P}_{[n]} := \cup_{k=1}^n \mathcal{P}_{[n]}^k$, the set of all partitions of $[n]$. To be definite, the blocks A_i of a partition of $[n]$ are assumed to be listed in *order of appearance*, meaning the order of their least elements, except if otherwise specified. For instance, the blocks of the partition of $[6]$

$$\{3,4,5\}, \{6,1\}, \{2\}$$

in order of appearance are

$$\{1,6\}, \{2\}, \{3,4,5\}.$$

The sequence $(|A_1|,\ldots,|A_k|)$ of sizes of blocks of a partition of $[n]$ defines a *composition of* n, that is a sequence of positive integers with sum n. Let $\mathcal{C}_n$ denote the set of all compositions of n. An *integer composition* is an element of $\cup_{n=1}^{\infty}\mathcal{C}_n$. The multiset $\{|A_1|,\ldots,|A_k|\}$ of unordered sizes of blocks of a partition Π_n of $[n]$ defines a *partition of* n, customarily encoded by one of the following:

- the composition of n defined by the decreasing arrangement of block sizes of Π_n, say $(N^{\downarrow}_{n,1},\ldots,N^{\downarrow}_{n,|\Pi_n|})$ where $|\Pi_n|$ is the number of blocks of Π_n;
- the infinite decreasing sequence of non-negative integers $(N^{\downarrow}_{n,1}, N^{\downarrow}_{n,2},\ldots)$ defined by appending an infinite string of zeros to $(N^{\downarrow}_{n,1},\ldots,N^{\downarrow}_{n,|\Pi|_n})$, so $N^{\downarrow}_{n,i}$ is the size of the ith largest block of Π_n if $|\Pi_n|\geq i$, and 0 otherwise,
- the sequence of non-negative integer counts $(|\Pi_n|_j, 1\leq j\leq n)$, where $|\Pi_n|_j$ is the number of blocks of Π_n of size j, with

$$\sum_j |\Pi_n|_j = |\Pi_n| \text{ and } \sum_j j|\Pi_n|_j = n. \tag{1.5}$$

Thus the set $\mathcal{P}_n$ of all partitions of n is bijectively identified with each of the following three sets of sequences of non-negative integers:

$$\bigcup_{k=1}^{n}\{(n_j)_{1\leq j\leq k} : n_1\geq n_2\geq\ldots\geq n_k\geq 1 \text{ and } \sum_j n_j = n\}$$

or

$$\{(n_j)_{1\leq j<\infty} : n_1\geq n_2\geq\ldots\geq 0 \text{ and } \sum_j n_j = n\}$$

or

$$\{(m_i)_{1\leq i\leq n} : \sum_i i m_i = n\}.$$

with the bijection from either (n_j) to (m_i) defined by $m_i = \sum_j 1(n_j = i)$.

Composite structures Let $v_\bullet := (v_1, v_2,\ldots)$ and $w_\bullet := (w_1, w_2,\ldots)$ be two sequences of non-negative integers. Let V be some species of combinatorial structures [37, 38], so for each finite set F_n with $|F_n| = n$ elements there is some construction of a set $V(F_n)$ of V*-structures on* F_n, such that the number of V-structures on a set of n elements is $|V(F_n)| = v_n$. For instance $V(F_n)$ might be $F_n\times F_n$, or $F_n^{F_n}$, or permutations from F_n to F_n, or rooted trees labeled F_n, corresponding to the sequences $v_n = n^2$, or n^n, or $n!$, or n^{n-1} respectively.

Let W be another species of combinatorial structures, such that the number of W-structures on a set of j elements is w_j. Let $(V\circ W)(F_n)$ denote the *composite structure* on F_n defined as the set of all ways to partition F_n into blocks $\{A_1,\ldots,A_k\}$ for some $1\leq k\leq n$, assign this collection of blocks a V-structure, and assign each block A_i a W-structure. Then for each set F_n with n elements, the number of such composite structures is evidently

$$|(V \circ W)(F_n)| = B_n(v_\bullet, w_\bullet) := \sum_{k=1}^{n} v_k B_{n,k}(w_\bullet), \tag{1.6}$$

where

$$B_{n,k}(w_\bullet) := \sum_{\{A_1,\ldots,A_k\} \in \mathcal{P}_{[n]}^k} \prod_{i=1}^{k} w_{|A_i|} \tag{1.7}$$

is the number of ways to partition F_n into k blocks and assign each block a W-structure.

The sum $B_{n,k}(w_\bullet)$ is a polynomial in variables $w_1, \ldots, w_{n-k+1}$, known as the (n,k)th *partial Bell polynomial* [100]. For a partition π_n of n into k parts with m_j parts equal to j for $1 \le j \le n$, the coefficient of $\prod_j w_j^{m_j}$ in $B_{n,k}(w_\bullet)$ is the number of partitions Π_n of $[n]$ corresponding to π_n. That is to say,

$$\left[\prod_j w_j^{m_j}\right] B_{n,k}(w_\bullet) = \frac{n!}{\prod_j (j!)^{m_j} m_j!} \qquad \left(\sum_j j m_j = n, \sum_j m_j = k\right) \tag{1.8}$$

as indicated for $1 \le k \le n \le 5$ in the following table:

Table 1.1. Some partial Bell polynomials

n	$B_{n,1}(w_\bullet)$	$B_{n,2}(w_\bullet)$	$B_{n,3}(w_\bullet)$	$B_{n,4}(w_\bullet)$	$B_{n,5}(w_\bullet)$
1	w_1				
2	w_2	w_1^2			
3	w_3	$3w_1w_2$	w_1^3		
4	w_4	$4w_1w_3 + 3w_2^2$	$6w_1^2w_2$	w_1^4	
5	w_5	$5w_1w_4 + 10w_2w_3$	$10w_1^2w_3 + 15w_1w_2^2$	$10w_1^3w_2$	w_1^5

Stirling numbers are obtained as evaluations of $B_{n,k}(w_\bullet)$ for particular $w_\bullet$, as discussed in the exercises of this section. The Bell polynomials and Stirling numbers have many interpretations and applications, some of them reviewed in the exercises of this section. See also [100]. Three different probabilistic interpretations discussed in the next three sections involve:

- formulae relating the moments and cumulants of a random variable X, particularly for X with infinitely divisible distribution;
- the probability function of a random sum $X_1 + \ldots + X_K$ of independent and identically distributed positive integer valued random variables X_i;
- the normalization constant in the definition of Gibbs distributions on partitions.

The second and third of these interpretations turn out to be closely related, and will be of fundamental importance throughout this course. The first interpretation can be related to the second in special cases. But this interpretation is rather different in nature, and not so closely connected to the main theme of the course.

Useful alternative expressions for $B_{n,k}(w_\bullet)$ and $B_n(v_\bullet, w_\bullet)$ can be given as follows. For each partition of $[n]$ into k disjoint non-empty blocks there are $k!$ different *ordered partitions of* $[n]$ into k such blocks. Corresponding to each composition $(n_1, \ldots, n_k)$ of n with k parts, there are

$$\binom{n}{n_1, \ldots, n_k} = n! \prod_{i=1}^{k} \frac{1}{n_i!}$$

different ordered partitions $(A_1, \ldots, A_k)$ of $[n]$ with $|A_i| = n_i$. So the definition (1.7) of $B_{n,k}(w_\bullet)$ as a sum of products over partitions of $[n]$ with k blocks implies

$$B_{n,k}(w_\bullet) = \frac{n!}{k!} \sum_{(n_1,\ldots,n_k)} \prod_{i=1}^{k} \frac{w_{n_i}}{n_i!} \tag{1.9}$$

where the sum is over all compositions of n into k parts. In view of this formula, it is natural to introduce the *exponential generating functions* associated with the weight sequences $v_\bullet$ and $w_\bullet$, say

$$v(\theta) := \sum_{k=1}^{\infty} v_k \frac{\theta^k}{k!} \text{ and } w(\xi) := \sum_{j=1}^{\infty} w_j \frac{\xi^j}{j!}$$

where the power series can either be assumed convergent in some neighborhood of 0, or regarded formally. Then (1.9) reads

$$B_{n,k}(w_\bullet) = \left[\frac{\xi^n}{n!}\right] \frac{w(\xi)^k}{k!} \tag{1.10}$$

and (1.6) yields the formula

$$B_n(v_\bullet, w_\bullet) = \left[\frac{\xi^n}{n!}\right] v(w(\xi)) \tag{1.11}$$

known as the *compositional* or *Faà di Bruno formula* [407],[100, 3.4]. Thus the combinatorial operation of composition of species of combinatorial structures corresponds to the analytic operation of composition of exponential generating functions. Note that (1.10) is the particular case of the compositional formula (1.11) when $v_\bullet = 1(\bullet = k)$, meaning $v_j = 1$ if $j = k$ and 0 else, for some $1 \leq k \leq n$. Another important case of (1.11) is the *exponential formula* [407]

$$B_n(x^\bullet, w_\bullet) = \left[\frac{\xi^n}{n!}\right] e^{xw(\xi)} \tag{1.12}$$

where $x^\bullet$ is the sequence whose kth term is x^k. For positive integer x, this exponential formula gives the number of ways to partition the set $[n]$ into an unspecified number of blocks, and assign each block of size j one of w_j possible structures and one of x possible colors.

Exercises

1.2.1. (Number of Compositions) [406, p. 14] The number of compositions of n with k parts is $\binom{n-1}{k-1}$, and the number of compositions of n is 2^{n-1}.

1.2.2. (Stirling numbers of the second kind) Let

$$S_{n,k} := B_{n,k}(1^\bullet) = \#\{\text{partitions of } [n] \text{ into } k \text{ blocks}\}, \tag{1.13}$$

where the substitution $w_\bullet = 1^\bullet$ means $w_n = 1^n \equiv 1$. The numbers $S_{n,k}$ are known as *Stirling numbers of the second kind.*

n	$S_{n,1}$	$S_{n,2}$	$S_{n,3}$	$S_{n,4}$	$S_{n,5}$
1	1				
2	1	1			
3	1	3	1		
4	1	7	6	1	
5	1	15	25	10	1

Show combinatorially that the $S_{n,k}$ are the connection coefficients determined by the identity of polynomials in x

$$x^n = \sum_{k=1}^{n} S_{n,k}\,(x)_{k\downarrow}. \tag{1.14}$$

1.2.3. (Stirling numbers of the first kind) Let

$$c_{n,k} := B_{n,k}((\bullet - 1)!) = \#\{\text{permutations of } [n] \text{ with } k \text{ cycles}\} \tag{1.15}$$

where the substitution $w_\bullet = (\bullet - 1)!$ means $w_n = (n-1)!$. Since $(n-1)!$ is the number of cyclic permutations of $[n]$, the second equality in (1.15) corresponds to the representation of a permutation of $[n]$ as the product of cyclic permutations acting on the blocks of some partition of $[n]$. The $c_{n,k}$ are known as *unsigned Stirling numbers of the first kind.*

n	$c_{n,1}$	$c_{n,2}$	$c_{n,3}$	$c_{n,4}$	$c_{n,5}$
1	1				
2	1	1			
3	2	3	1		
4	6	11	6	1	
5	24	50	35	10	1

Show combinatorially that

$$(x)_{n\uparrow} = \sum_{k=1}^{n} c_{n,k}\,x^k \text{ and } (x)_{n\downarrow} = \sum_{k=1}^{n} s_{n,k}x^k \tag{1.16}$$

where the $s_{n,k} = (-1)^{n-k}c_{n,k} = B_{n,k}((-1)^{\bullet-1}(\bullet - 1)!)$ are the *Stirling numbers of the first kind.* Check that the matrix of Stirling numbers of the first kind is the inverse of the matrix of Stirling numbers of the second kind.

1.2.4. (Matrix representation of composition) Jabotinsky [100]. Regard the numbers $B_{n,k}(w_\bullet)$ for fixed $w_\bullet$ as an infinite matrix indexed by $n, k \geq 1$. For sequences $v_\bullet$ and $w_\bullet$ with exponential generating functions $v(\xi)$ and $w(\xi)$, let $(v \circ w)_\bullet$ denote the sequence whose exponential generating function is $v(w(\xi))$. Then the matrix associated with the sequence $(v \circ w)_\bullet$ is the product of the matrices associated with $w_\bullet$ and $v_\bullet$ respectively. In particular, for $w_\bullet$ with $w_1 \neq 0$, and $w_\bullet^{-1}$ the sequence whose exponential generating function w^{-1} is the compositional inverse of w, so $w^{-1}(w(\xi)) = w(w^{-1}(\xi)) = \xi$, the matrix $\mathbf{B}(w_\bullet^{-1})$ is the matrix inverse of $\mathbf{B}(w_\bullet)$.

1.2.5. (Polynomials of binomial type). Given some fixed weight sequence $w_\bullet$, define a sequence of polynomials $B_n(x)$ by $B_0(x) := 1$ and for $n \geq 1$ $B_n(x) := B_n(x^\bullet, w_\bullet)$ as in (1.12). The sequence of polynomials $B_n(x)$ is of *binomial type*, meaning that

$$\sum_{j=0}^{n} \binom{n}{j} B_j(x) B_{n-j}(y) = B_n(x+y). \tag{1.17}$$

Conversely, it is known [391, 390] that if $B_n(x)$ is a sequence of polynomials of binomial type such that $B_n(x)$ is of degree n, then $B_n(x) = B_n(x^\bullet, w_\bullet)$ as in (1.12) for some weight sequence $w_\bullet$. Note that then $w_j = [x]B_j(x)$.

1.2.6. (Change of basis) Each sequence of polynomials of binomial type $B_n(x)$ with B_n of degree n defines a basis for the space of polynomials in x. The matrix of *connection coefficients* involved in changing from one basis to another can be described in a number of different ways [391]. For instance, given two sequences of polynomials of binomial type, say $B_n(x^\bullet, u_\bullet)$ and $B_n(x^\bullet, v_\bullet)$, for some weight sequences $u_\bullet$ and $v_\bullet$, with $v_1 \neq 0$,

$$B_n(x^\bullet, u_\bullet) = \sum_{j=0}^{n} B_{n,j}(w_\bullet) B_j(x^\bullet, v_\bullet) \tag{1.18}$$

where

$$w(\xi) := v^{-1}(u(\xi)) \text{ is the unique solution of } u(\xi) = v(w(\xi)).$$

for u and v the exponential generating functions associated with $u_\bullet$ and $v_\bullet$.

1.2.7. (Generalized Stirling numbers) Toscano [415], Riordan [385, p. 46], Charalambides and Singh [89], Hsu and Shiue [203]. For arbitrary distinct reals α and β, show that the connection coefficients $S_{n,k}^{\alpha,\beta}$ defined by

$$(x)_{n\downarrow\alpha} = \sum_{k=0}^{n} S_{n,k}^{\alpha,\beta} \, (x)_{k\downarrow\beta} \tag{1.19}$$

are

$$S_{n,k}^{\alpha,\beta} = B_{n,k}((\beta-\alpha)_{\bullet-1\downarrow\alpha}) = \frac{n!}{k!}[\xi^n](w^{\alpha,\beta}(\xi))^k \tag{1.20}$$

where

$$w^{\alpha,\beta}(\xi) := \sum_{j=1}^{\infty} (\beta-\alpha)_{j-1\downarrow\alpha} \frac{\xi^j}{j!} = \begin{cases} \beta^{-1}((1+\alpha\xi)^{\beta/\alpha}-1) & \text{if } \alpha \neq 0, \beta \neq 0 \\ \beta^{-1}(e^{\beta\xi}-1) & \text{if } \alpha = 0 \\ \alpha^{-1}\log(1+\alpha\xi) & \text{if } \beta = 0. \end{cases} \tag{1.21}$$

n	$S_{n,1}^{\alpha,\beta}$	$S_{n,2}^{\alpha,\beta}$	$S_{n,3}^{\alpha,\beta}$	$S_{n,4}^{\alpha,\beta}$
1	1			
2	$(\beta-\alpha)$	1		
3	$(\beta-\alpha)(\beta-2\alpha)$	$3(\beta-\alpha)$	1	
4	$(\beta-\alpha)(\beta-2\alpha)(\beta-3\alpha)$	$4(\beta-\alpha)(\beta-2\alpha)+3(\beta-\alpha)^2$	$6(\beta-\alpha)$	1

Alternatively

$$S_{n,k}^{\alpha,\beta} = \sum_{j=k}^{n} s_{n,j} S_{j,k} \alpha^{n-j} \beta^{j-k} \tag{1.22}$$

where $s_{n,j} := S_{n,j}^{1,0}$ is a Stirling number of the first kind and $S_{j,k} := S_{j,k}^{0,1}$ is a Stirling number of the second kind.

1.3. Moments and cumulants

Let $(X_t, t \geq 0)$ be a real-valued *Lévy process*, that is a process with stationary independent increments, started at $X_0 = 0$, with sample paths which are cadlag (right continuous with left limits) [40]. According to the exponential formula of probability theory, i.e. the Lévy-Khintchine formula, if we assume that X_t has a convergent moment generating function in some neighborhood of 0 then

$$\mathbb{E}[e^{\theta X_t}] = \exp(t\Psi(\theta)) \tag{1.23}$$

for a characteristic exponent Ψ which can be represented as

$$\Psi(\theta) = \sum_{n=1}^{\infty} \kappa_n \frac{\theta^n}{n!} \tag{1.24}$$

where $\kappa_1 = \mathbb{E}(X_1)$, κ_2 is the variance of X_1, and

$$\kappa_n = \int_{\mathbb{R}} x^n \Lambda(dx) \quad (n = 3, 4, \ldots). \tag{1.25}$$

where Λ is the Lévy measure of X_1. Compare (1.23) with the exponential formula of combinatorics (1.12) to see that the coefficient of $\theta^n/n!$ in (1.23) is

$$\mathbb{E}(X_t^n) = B_n(t^\bullet, \kappa_\bullet). \tag{1.26}$$

Thus the moments of X_t define a sequence of polynomials in t which is the sequence of polynomials of binomial type associated with the sequence $\kappa_\bullet$ of *cumulants* of X_1. Two special cases are worthy of note.

Gaussian case If X_1 is standard Gaussian, the sequence of cumulants of X_1 is $\kappa_\bullet = 1(\bullet = 2)$. It follows from (1.26) and the combinatorial meaning of $B_{n,k}$ that the nth moment μ_n of X_1 is the number of *matchings* of $[n]$, meaning the number of partitions of $[n]$ into $n/2$ pairs. Thus

$$\mathbb{E}(X_t^n)/t^{n/2} = \begin{cases} 0 & \text{if } n \text{ is odd} \\ 1 \times 3 \times \cdots (n-1) & \text{if } n \text{ is even.} \end{cases} \tag{1.27}$$

Exercise 1.3.4 and Exercise 1.3.5 offer some generalizations.

Poisson case If $X_1 = N_1$ is Poisson with mean 1, the sequence of cumulants is $\kappa_\bullet = 1^\bullet$. The positive integer moments of N_t are therefore given by

$$\mathbb{E}(N_t^n) = \sum_{m=0}^{\infty} \frac{e^{-t} t^m m^n}{m!} = \sum_{k=1}^{n} B_{n,k}(1^\bullet) t^k \qquad (n = 1, 2, \ldots) \tag{1.28}$$

where the $B_{n,k}(1^\bullet)$ are the Stirling numbers of the second kind. These polynomials in t are known as *exponential polynomials.* In particular, the nth moment of the Poisson(1) distribution of N_1 is the nth *Bell number*

$$B_n(1^\bullet, 1^\bullet) := \sum_{k=1}^{n} B_{n,k}(1^\bullet) = \left[\frac{\xi^n}{n!}\right] \exp(e^\xi - 1)$$

which is the number of partitions of $[n]$. The first six Bell numbers are

1, 2, 5, 15, 52, 203.

Now (1.28) for $t = 1$ gives the famous *Dobiński formula* [111]

$$B_n(1^\bullet, 1^\bullet) = e^{-1} \sum_{m=1}^{\infty} \frac{m^n}{m!}. \tag{1.29}$$

As noted by Comtet [100], for each n the infinite sum in (1.29) can be evaluated as the least integer greater than the sum of the first $2n$ terms.

Exercises

1.3.1. (Moments and Cumulants) [290, 100] Let X_1 be any random variable with a moment generating function which is convergent in some neighborhood of 0. Let $\mu_n := E(X_1^n)$ and let the cumulants κ_n of X_1 be defined by the expansion (1.24) of $\Psi(\theta) := \log \mathbb{E}[e^{\theta X_1}]$. Show that the moment and cumulant sequences $\mu_\bullet$ and $\kappa_\bullet$ determine each other by the formulae

$$\mu_n = \sum_{k=1}^{n} B_{n,k}(\kappa_\bullet) \text{ and } \kappa_n = \sum_{k=1}^{n} (-1)^{k-1}(k-1)! B_{n,k}(\mu_\bullet) \tag{1.30}$$

for $n = 1, 2, \ldots$. These formulae allow the first n cumulants of X to be defined for any X with $\mathbb{E}[|X|^n] < \infty$, and many of the following exercises can be adapted to this case.

1.3.2. (Thiele's recursion) [186, p. 144, (4.2)], [316, p.74, Th. 2], [107, Th. 2.3.6]. Two sequences $\mu_\bullet$ and $\kappa_\bullet$ are related by (1.30) if and only if

$$\mu_n = \sum_{i=0}^{n-1} \binom{n-1}{i} \mu_i \kappa_{n-i} \quad (n = 1, 2, \ldots) \tag{1.31}$$

where $\mu_0 = 0$.

1.3.3. (Moment polynomials) [77], [187], [319, p. 80], [146] [147, Prop. 2.1.4]. For $t = 1, 2, \ldots$ let $S_t = \sum_{i=1}^{t} X_i$ where the X_i are independent copies of X with moment sequence $\mu_\bullet$ and cumulant sequence $\kappa_\bullet$. Then

$$\mathbb{E}[S_t^n] = \sum_{k=1}^{n} B_{n,k}(\kappa_\bullet) t^k = \sum_{k=1}^{n} B_{n,k}(\mu_\bullet)(t)_{k\downarrow}. \tag{1.32}$$

For $n = 0, 1, \ldots$, and x real, let

$$\mu_n(x, t) := \mathbb{E}[(x + S_t)^n] = \sum_{k=0}^{n} \binom{n}{k} x^k \mathbb{E}[S_t^{n-k}] \quad (t = 0, 1, 2, \ldots) \tag{1.33}$$

where $S_0 := 0$, and let

$$H_n(x, t) := \mu_n(x, -t) \tag{1.34}$$

where the right side is defined for each x by polynomial continuation of $\mu_n(x, t)$ in (1.33). Then

$$(H_n(S_t, t), t = 0, 1, 2, \ldots) \text{ is an } (\mathcal{F}_t)\text{-martingale} \tag{1.35}$$

where $\mathcal{F}_t$ is the σ-field generated by $(S_u, u = 0, 1, \ldots, t)$.

1.3.4. (Matchings and Stirling numbers). Check that for X_1 standard Gaussian (1.32) for even $n = 2q$ gives

$$\mathbb{E}(S_t^{2q}) = t^q \mu_{2q} = \sum_{k=1}^{q} B_{2q,k}(\mu_\bullet)(t)_{k\downarrow}. \tag{1.36}$$

Compare with the definition (1.16) of Stirling numbers of the second kind to see that

$$B_{2q,k}(\mu_\bullet) = \mu_{2q} B_{q,k}(1^\bullet). \tag{1.37}$$

Give a combinatorial proof of (1.37) and deduce more generally that

$$B_{2q,k}(\mu_\bullet x_{\bullet/2}) = \mu_{2q} B_{2q,k}(x_\bullet) \tag{1.38}$$

for an arbitrary sequence $x_\bullet$, where the nth term of $\mu_\bullet x_{\bullet/2}$ is $\mu_q x_q$ if $n = 2q$ is even, and 0 else.

1.3.5. (Feynman diagrams) [214, Theorem 1.28] [297, Lemma 4.5]. Check the following generalization of (1.27): if $X_1, \ldots, X_n$ are centered jointly Gaussian variables, then

$$\mathbb{E}(X_1 \cdots X_n) = \sum \prod_k E(X_{i_k} X_{j_k}) \tag{1.39}$$

where the sum is over all partitions of $[n]$ into $n/2$ pairs $\{\{i_k, j_k\}, 1 \le k \le n/2\}$. See [297] for applications to local times of Markov processes.

1.3.6. (Poisson moments)[353] Deduce (1.28) from (1.16) and the more elementary formula $\mathbb{E}[(N_t)_{n\downarrow}] = t^n$.

Notes and comments

Moment calculations for commutative and non-commutative Gaussian random variables in terms of partitions, matchings etc. are described in [196]. There, for instance, is a discussion of the fact that the Catalan numbers are the moments of the semicircle law, which appears in Wigner's limit theorem for the empirical distribution of the eigenvalues of a random Hermitian matrix. Combinatorial representations for the moments of superprocesses, in terms of expansions over forests, were given by Dynkin [129], where a connection is made with similar calculations arising in quantum field theory. This is further explained with pictures in Etheridge [136]. These ideas are further applied in [139, 140, 388].

1.4. Random sums

Recall that if X, X_1, $X_2, \ldots$ are independent and identically distributed non-negative integer valued random variables with probability generating function

$$G_X(z) := \mathbb{E}[z^X] = \sum_{n=0}^{\infty} \mathbb{P}(X = n) z^n,$$

and K is a non-negative integer valued random variable independent of $X_1, X_2, \ldots$ with probability generating function G_K, and $S_K := X_1 + \cdots + X_K$, then, by conditioning on K, the probability generating function of S_K is found to be the composition of G_K and G_X:

$$G_{S_K}(z) = G_K(G_X(z)). \tag{1.40}$$

Comparison of this formula with the compositional formula (1.11) for $B_n(v_\bullet, w_\bullet)$ in terms of the exponential generating functions $v(z) = \sum_{n=0}^{\infty} v_n z^n/n!$ and $w(\xi) = \sum_{n=1}^{\infty} w_n \xi^n/n!$, suggests the following construction. (It is convenient here to allow v_0 to be non-zero, which makes no difference in (1.11)). Let $\xi > 0$ be such that $v(w(\xi)) < \infty$. Let $\mathbb{P}_{\xi, v_\bullet, w_\bullet}$ be a probability distribution which makes X_i independent and identically distributed with the power series distribution

$$\mathbb{P}_{\xi, v_\bullet, w_\bullet}(X = n) = \frac{w_n \xi^n}{n! w(\xi)} \text{ for } n = 1, 2, \ldots, \text{ so } G_X(z) = \frac{w(z\xi)}{w(\xi)} \tag{1.41}$$

and K independent of the X_i with the power series distribution

$$\mathbb{P}_{\xi,v_\bullet,w_\bullet}(K=k) = \frac{v_k w(\xi)^k}{k! v(w(\xi))} \text{ for } k=0,1,2,\dots \text{ so } G_K(y) = \frac{v(yw(\xi))}{v(w(\xi))}. \quad (1.42)$$

Let $S_K := X_1 + \cdots + X_K$. Then from (1.40) and (1.11),

$$\mathbb{P}_{\xi,v_\bullet,w_\bullet}(S_K=n) = \frac{\xi^n}{n! v(w(\xi))} B_n(v_\bullet, w_\bullet) \quad (1.43)$$

or

$$B_n(v_\bullet, w_\bullet) = \frac{n! v(w(\xi))}{\xi^n} \mathbb{P}_{\xi,v_\bullet,w_\bullet}(S_K=n) \quad (1.44)$$

This probabilistic representation of $B_n(v_\bullet, w_\bullet)$ was given in increasing generality by Holst [201], Kolchin [260], and Kerov [240]. Rényi's formula for the Bell numbers $B_n(1^\bullet, 1^\bullet)$ in Exercise 1.4.1 is a variant of (1.44) for $v_\bullet = w_\bullet = 1^\bullet$. Holst [201] gave (1.44) for $v_\bullet$ and $w_\bullet$ with values in $\{0,1\}$, when $B_n(v_\bullet, w_\bullet)$ is the number of partitions of $[n]$ into some number of blocks k with $v_k = 1$ and each block of size j with $w_j = 1$. As observed by Rényi and Holst, for suitable $v_\bullet$ and $w_\bullet$ the probabilistic representation (1.44) allows large n asymptotics of $B_n(v_\bullet, w_\bullet)$ to be derived from local limit approximations to the distribution of sums of independent random variables. This method is closely related to classical saddle point approximations: see notes and comments at the end of Section 1.5.

Exercises

1.4.1. (Rényi's formula for the Bell numbers)[382, p. 11]. Let $(N_t, t \geq 0)$ and $(M_t, t \geq 0)$ be two independent standard Poisson processes. Then for $n = 1, 2, \dots$ the number of partitions of $[n]$ is

$$B_n(1^\bullet, 1^\bullet) = n!\, e^{e-1}\, \mathbb{P}(N_{M_e} = n). \quad (1.45)$$

1.4.2. (Asymptotic formula for the Bell numbers)[288, 1.9]. Deduce from (1.45) the asymptotic equivalence

$$B_n(1^\bullet, 1^\bullet) \sim \frac{1}{\sqrt{n}} \lambda(n)^{n+1/2} e^{\lambda(n)-n-1} \qquad \text{as } n \to \infty, \quad (1.46)$$

where $\lambda(n)\log(\lambda(n)) = n$.

1.5. Gibbs partitions

Suppose as in Section 1.2 that $(V \circ W)([n])$ is the set of all composite $V \circ W$-structures built over $[n]$, for some species of combinatorial structures V and W. Let a composite structure be picked uniformly at random from $(V \circ W)([n])$, and let Π_n denote the random partition of $[n]$ generated by blocks of this random composite structure. Recall that v_j and w_j denote the number of V- and W-structures respectively on a set of j elements. Then for each particular partition $\{A_1, \ldots, A_k\}$ of $[n]$ it is clear that

$$\mathbb{P}(\Pi_n = \{A_1, \ldots, A_k\}) = p(|A_1|, \ldots, |A_k|; v_\bullet, w_\bullet) \tag{1.47}$$

where for each composition $(n_1, \ldots, n_k)$ of n

$$p(n_1, \ldots, n_k; v_\bullet, w_\bullet) := \frac{v_k \prod_{i=1}^k w_{n_i}}{B_n(v_\bullet, w_\bullet)} \tag{1.48}$$

with the normalization constant $B_n(v_\bullet, w_\bullet) := \sum_{k=1}^n v_k B_{n,k}(w_\bullet)$ as in (1.6) and (1.11), assumed strictly positive. More generally, given two non-negative sequences $v_\bullet := (v_1, v_2, \ldots)$ and $w_\bullet := (w_1, w_2, \ldots)$, call Π_n a *Gibbs*$_{[n]}(v_\bullet, w_\bullet)$ *partition* if the distribution of Π_n on $\mathcal{P}_{[n]}$ is given by (1.47)-(1.48). Note that due to the normalization in (1.48), there is the following redundancy in the parameterization of Gibbs$_{[n]}(v_\bullet, w_\bullet)$ partitions: for arbitrary positive constants a, b and c,

$$\text{Gibbs}_{[n]}(ab^\bullet v_\bullet, c^\bullet w_\bullet) = \text{Gibbs}_{[n]}(v_\bullet, bw_\bullet). \tag{1.49}$$

That is to say, the Gibbs$_{[n]}(v_\bullet, w_\bullet)$ distribution is unaffected by multiplying $v_\bullet$ by a constant factor a, or multiplying $w_\bullet$ by a geometric factor $c^\bullet$, while multiplying $v_\bullet$ by the geometric factor $b^\bullet$ is equivalent to multiplication of $w_\bullet$ by the constant factor b.

The block sizes in exchangeable order The following theorem provides a fundamental representation of Gibbs partitions.

Theorem 1.2. ***(Kolchin's representation of Gibbs partitions)*** [260], [240] *Let $(N^{\text{ex}}_{n,1}, \ldots, N^{\text{ex}}_{n,|\Pi|_n})$ be the random composition of n defined by putting the block sizes of a Gibbs$_{[n]}(v_\bullet, w_\bullet)$ partition Π_n in an* exchangeable random order, *meaning that given k blocks, the order of the blocks is randomized by a uniform random permutation of $[k]$. Then*

$$(N^{\text{ex}}_{n,1}, \ldots, N^{\text{ex}}_{n,|\Pi|_n}) \stackrel{d}{=} (X_1, \ldots, X_K) \text{ under } \mathbb{P}_{\xi, v_\bullet, w_\bullet} \text{ given } X_1 + \cdots + X_K = n \tag{1.50}$$

where $\mathbb{P}_{\xi, v_\bullet, w_\bullet}$ governs independent and identically distributed random variables $X_1, X_2, \ldots$ with $E(z^{X_i}) = w(z\xi)/w(\xi)$ and K is independent of these variables with $E(y^K) = v(yw(\xi))/v(w(\xi))$ as in (1.41) and (1.42).

Proof. It is easily seen that the manipulation of sums leading to (1.9) can be interpreted probabilistically as follows:

$$\mathbb{P}((N^{\text{ex}}_{n,1}, \ldots, N^{\text{ex}}_{n,|\Pi|_n}) = (n_1, \ldots, n_k)) = \frac{n!\, v_k}{k! B_n(v_\bullet, w_\bullet)} \prod_{i=1}^k \frac{w_{n_i}}{n_i!} \tag{1.51}$$

for all compositions $(n_1, \ldots, n_k)$ of n. Compare with formula (1.43) and the conclusion is evident. □

Note that for fixed $v_\bullet$ and $w_\bullet$, the $\mathbb{P}_{\xi,v_\bullet,w_\bullet}$ distribution of the random integer composition $(X_1, \ldots, X_K)$ depends on the parameter ξ, but the $\mathbb{P}_{\xi,v_\bullet,w_\bullet}$ conditional distribution of $(X_1, \ldots, X_K)$ given $S_K = n$ does not. In statistical terms, with $v_\bullet$ and $w_\bullet$ regarded as fixed and known, the sum S_K is a sufficient statistic for ξ. Note also that for any fixed n, the distribution of Π_n depends only on the weights v_j and w_j for $j \leq n$, so the condition $v(w(\xi)) < \infty$ can always be arranged by setting $v_j = w_j = 0$ for $j > n$.

The partition of n Recall that the random partition of n induced by a random partition Π_n of $[n]$ is encoded by the random vector $(|\Pi_n|_j, 1 \leq j \leq n)$ where $|\Pi_n|_j$ is the number of blocks of Π_n of size j. Using (1.8), the distribution of the partition of n induced by a $\text{Gibbs}_{[n]}(v_\bullet, w_\bullet)$ partition Π_n is given by

$$\mathbb{P}(|\Pi_n|_j = m_j, 1 \leq j \leq n) = \frac{n!\, v_k}{B_n(v_\bullet, w_\bullet)} \prod_{j=1}^{n} \left(\frac{w_j}{j!}\right)^{m_j} \frac{1}{m_j!} \tag{1.52}$$

where $\sum_{j=1}^n m_j = k$ and $\sum_{j=1}^n j m_j = n$. In particular, for a $\text{Gibbs}_{[n]}(1^\bullet, w_\bullet)$ partition

$$(|\Pi_n|_j, 1 \leq j \leq n) \stackrel{d}{=} \left(M_j, 1 \leq j \leq n \,\middle|\, \sum_{j=1}^{n} j M_j = n \right) \tag{1.53}$$

where the M_j are independent Poisson variables with parameters $(w_j \xi^j / j!)$ for arbitrary $\xi > 0$. This can also be read from (1.50). For $v_\bullet = 1^\bullet$ the random variable K has Poisson $(w(\xi))$ distribution. Hence, by the classical Poissonization of the multinomial distribution, the number M_j of i such that $i \leq K$ and $X_i = j$ has a Poisson $(w_j \xi^j / j!)$ distribution, and $S_K = \sum_j j M_j$ is compound Poisson. See also Exercise 1.5.1 . Arratia, Barbour and Tavaré [27] make the identity in distribution (1.53) the starting point for a detailed analysis of the asymptotic behaviour of the counts $(|\Pi_n|_j, 1 \leq j \leq n)$ of a $\text{Gibbs}_{[n]}(1^\bullet, w_\bullet)$ partition as $n \to \infty$ for $w_\bullet$ in the *logarithmic class*, meaning that $j w_j / j! \to \theta$ as $j \to \infty$ for some $\theta > 0$. One of their main results is presented later in Chapter 2.

Physical interpretation Suppose that n particles labeled by elements of the set $[n]$ are partitioned into *clusters* in such a way that each particle belongs to a unique cluster. Formally, the collection of clusters is represented by a partition of $[n]$. Suppose further that each cluster of size j can be in any one of w_j different *internal states* for some sequence of non-negative integers $w_\bullet = (w_j)$. Let the *configuration* of the system of n particles be the partition of the set of n particles into clusters, together with the assignment of an internal state to each cluster. For each partition π of $[n]$ with k blocks of sizes $n_1, \ldots, n_k$, there are $\prod_{i=1}^k w_{n_i}$ different configurations with that partition π. So $B_{n,k}(w_\bullet)$ gives *the number of configurations with k clusters.* For $v_\bullet = 1(\bullet = k)$ the sequence with kth component 1 and all other components 0, the $\text{Gibbs}(v_\bullet, w_\bullet)$ partition of $[n]$ corresponds to assuming that all configurations with k clusters are equally likely.

This distribution on the set $\mathcal{P}_{[n]}^k$ of partitions of $[n]$ with k blocks, is known in the physics literature as a *microcanonical Gibbs state*. It may also be called here the *Gibbs*($w_\bullet$) *distribution on* $\mathcal{P}_{[n]}^k$. A general weight sequence $v_\bullet$ randomizes k, to allow any probabilistic mixture over k of these microcanonical states. For fixed $w_\bullet$ and n, the set of all Gibbs($v_\bullet, w_\bullet$) distributions on partitions of $[n]$, as $v_\bullet$ varies, is an $(n-1)$-dimensional simplex whose set of extreme points is the collection of n different microcanonical states. Whittle [432, 433, 434] showed how the Gibbs distribution (1.52) on partitions of n arises as the reversible equilibrium distribution in a Markov process with state space $\mathcal{P}_n$, where parts of various sizes can split or merge at appropriate rates. In this setting, the Poisson variables M_j represent equilibrium counts in a corresponding unconstrained system where the total size is also subject to variation. See also [123] for further studies of equilibrium models for processes of coagulation and fragmentation.

Example 1.3. Uniform random set partitions. Let Π_n be a uniformly distributed random partition of $[n]$. Then Π_n is a random $(V \circ V)$-structure on $[n]$ for V the species of non-empty sets. Thus Π_n has the Gibbs$(1^\bullet, 1^\bullet)$ distribution on $\mathcal{P}_{[n]}$. Note that $\mathbb{P}(|\Pi_n| = k) = B_{n,k}(1^\bullet)/B_n(1^\bullet, 1^\bullet)$ but there is no simple formula, either for the Stirling numbers of the second kind $B_{n,k}(1^\bullet)$, or for the Bell numbers $B_n(1^\bullet, 1^\bullet)$. Exercise 1.5.5 gives a normal approximation for $|\Pi_n|$. The independent and identically distributed variables X_i in Kolchin's representation are Poisson variables conditioned not to be 0. See [159, 174, 424] and papers cited there for further probabilistic analysis of Π_n for large n.

Example 1.4. Random permutations. Let $W(F)$ be the set of all permutations of F with a single cycle. Then $w_n = (n-1)!$, so

$$w(\xi) = \sum_{n=1}^{\infty} (n-1)! \frac{\xi^n}{n!} = -\log(1-\xi)$$

and

$$e^{\theta w(\xi)} = e^{-\theta \log(1-\xi)} = (1-\xi)^{-\theta} = \sum_{n=0}^{\infty} (\theta)_{n\uparrow 1} \frac{\xi^n}{n!}.$$

So

$$B_n(1^\bullet, \theta(\bullet - 1)!) = (\theta)_{n\uparrow 1}. \tag{1.54}$$

In particular, for $\theta = 1$, $B_n(1^\bullet, (\bullet - 1)!) = (1)_{n\uparrow 1} = n!$ is just the number of permutations of $[n]$. Since each permutation corresponds bijectively to a partition of $[n]$ and an assignment of a cycle to each block of the partition, the random partition Π_n of $[n]$ generated by the cycles of a uniform random permutation of $[n]$ is a Gibbs$_{[n]}(1^\bullet, (\bullet - 1)!)$ partition. While there is no simple formula for the unsigned Stirling numbers $B_{n,k}((\bullet - 1)!)$ which determine the distribution of $|\Pi_n|$, asymptotic normality of this distribution is easily shown (Exercise 1.5.4). Similarly, for $\theta = 1, 2, \ldots$ the number in (1.54) is the number of different ways to pick a permutation of $[n]$ and assign each cycle of the permutation one of θ possible colors. If each of these ways is assumed equally likely, the resulting random partition of $[n]$ is a $\text{Gibbs}_{[n]}(1^\bullet, \theta(\bullet - 1)!)$ partition. For any $\theta > 0$, the X_i in Kolchin's representation have *logarithmic series distribution*

$$p_j = \frac{1}{-\log(1-b)} \frac{b^j}{j} \qquad (j = 1, 2, \ldots)$$

where $0 < b < 1$ is a positive parameter. This example is developed further in Chapter 3.

Example 1.5. Cutting a rooted random segment. Suppose that the internal state of a cluster C of size j is one of $w_j = j!$ linear orderings of the set C. Identify each cluster as a directed graph in which there is a directed edge from a to b if and only if a is the immediate predecessor of b in the linear ordering. Call such a graph a *rooted segment.* Then $B_{n,k}(\bullet!)$ is the number of directed graphs labeled by $[n]$ with k such segments as its connected components. In the previous two examples, with $w_j = 1^j$ and $w_j = (j-1)!$, the $B_{n,k}(w_\bullet)$ were Stirling numbers for which there is no simple formula. Since $j! = (\beta - \alpha)_{j-1\downarrow\alpha}$ for $\alpha = -1$ and $\beta = 1$, formula (1.20) shows that the Bell matrix $B_{n,k}(\bullet!)$ is the array of generalized Stirling numbers

$$B_{n,k}(\bullet!) = S_{n,k}^{-1,1} = \binom{n-1}{k-1} \frac{n!}{k!} \tag{1.55}$$

known as *Lah numbers* [100, p. 135], though these numbers were already considered by Toscano [415]. The Gibbs model in this instance is a variation of Flory's model for a linear polymerization process. It is easily shown in this case that a sequence of random partitions $(\Pi_{n,k}, 1 \le k \le n)$ such that $\Pi_{n,k}$ has the microcanonical Gibbs distribution on clusters with k components may be obtained as follows. Let G_1 be a uniformly distributed random rooted segment labeled by $[n]$. Let G_k be derived from G_1 by deletion of a set of $k-1$ edges picked uniformly at random from the set of $n-1$ edges of G_1, and let $\Pi_{n,k}$ be the partition induced by the components of G_k. If the $n-1$ edges of G_1 are deleted sequentially, one by one, cf. Figure 1.1, the random sequence $(\Pi_{n,1}, \Pi_{n,2}, \ldots, \Pi_{n,n})$ is a *fragmenting sequence*, meaning that $\Pi_{n,j}$ is coarser than $\Pi_{n,k}$ for $j < k$, such that $\Pi_{n,k}$ has the microcanonical Gibbs distribution on $\mathcal{P}_{[n]}^k$ derived from the weight sequence $w_j = j!$. The time-reversed sequence $(\Pi_{n,n}, \Pi_{n,n-1}, \ldots, \Pi_{n,1})$ is then a discrete time Markov chain governed by the rules of *Kingman's coalescent* [30, 253]: conditionally given Π_k with k components, Π_{k-1} is equally likely to be any one of the $\binom{k}{2}$ different partitions of $[n]$ obtained by merging two of the components of Π_k. Equivalently, the sequence $(\Pi_{n,1}, \Pi_{n,2}, \ldots, \Pi_{n,n})$ has uniform distribution over the set $\mathcal{R}_n$ of all fragmenting sequences of partitions of $[n]$ such that the kth term of the sequence has k components. The consequent enumeration $\#\mathcal{R}_n = n!(n-1)!/2^{n-1}$ was found by Erdös et al. [135]. That $\Pi_{n,k}$ determined by this model has the microcanonical Gibbs($\bullet$!) distribution on $\mathcal{P}_{[n]}^k$ was shown by Bayewitz et. al. [30] and Kingman [253]. See also Chapter 5 regarding Kingman's coalescent with continuous time parameter.

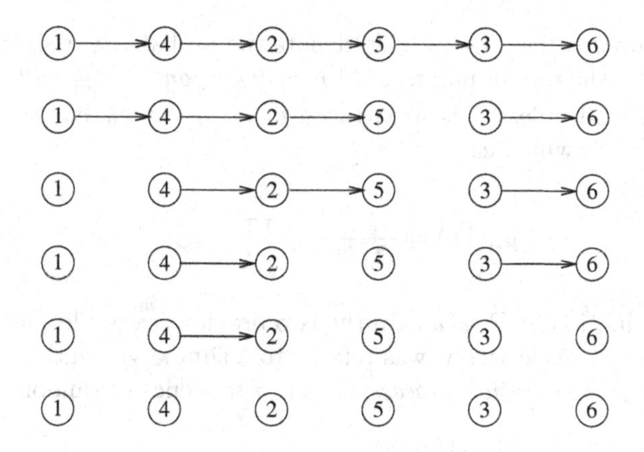

Figure 1.1: Cutting a rooted random segment.

Example 1.6. Cutting a rooted random tree. Suppose the internal state of a cluster C of size j is one of the $w_j = j^{j-1}$ rooted trees labeled by C. Then $B_{n,k}(\bullet^{\bullet-1})$ is the number of forests of k rooted trees labeled $[n]$. This time again there is a simple construction of the microcanonical Gibbs states by sequential deletion of random edges, hence a simple formula for $B_{n,k}$.

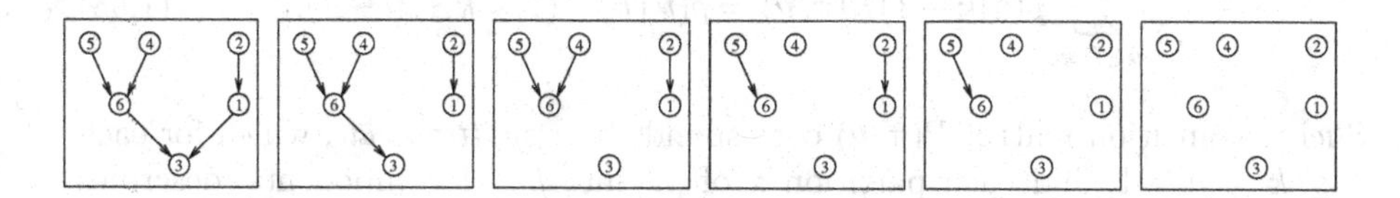

Figure 1.2: Cutting a rooted random tree with 5 edges.

By a reprise of the previous argument [356],

$$B_{n,k}(\bullet^{\bullet-1}) = \binom{n}{k} kn^{n-k-1} \tag{1.56}$$

which is an equivalent of Cayley's formula kn^{n-k-1} for the number of rooted trees labeled by $[n]$ whose set of roots is $[k]$. The Gibbs model in this instance corresponds to assuming that all forests of k rooted trees labeled by $[n]$ are equally likely. This model turns up naturally in the theory of random graphs and has been studied and applied in several other contexts. The coalescent obtained by reversing the process of deletion of edges is the *additive coalescent* studied in [356]. The structure of large random trees is one of the main themes of this course, to be taken up in Chapter 6. This leads in Chapter 7 to the notion of continuum trees embedded in Brownian paths, then in Chapter 10 to a representation in terms of continuum trees of the large n asymptotics of the additive coalescent.

Gibbs fragmentations Let $p(\lambda \,|\, k)$ denote the probability assigned to a partition λ of $[n]$ by the *microcanonical Gibbs distribution on* $\mathcal{P}^k_{[n]}$ *with weights* $w_\bullet$, that is $p(\lambda \,|\, k) = 0$ unless λ is a partition of of $[n]$ into k blocks of sizes say $n_i(\lambda)$, $1 \le i \le k$, in which case

$$p(\lambda \,|\, k) = \frac{1}{B_{n,k}(w_\bullet)} \prod_{i=1}^{k} w_{n_i(\lambda)} \tag{1.57}$$

The simple evaluation of $B_{n,k}(w_\bullet)$ in the two previous examples, for $w_n = (n-1)!$ and $w_n = n^{n-1}$ respectively, was related to a simple sequential construction of a *Gibbs*($w_\bullet$) *fragmentation process*, that is a sequence of random partitions

$$(\Pi_{n,1}, \Pi_{n,2}, \ldots, \Pi_{n,n})$$

such that $\Pi_{n,k}$ has the Gibbs($w_\bullet$) distribution on $\mathcal{P}^k_{[n]}$, and for each $1 \le k \le n-1$ the partition $\Pi_{n,k}$ is a refinement of $\Pi_{n,k-1}$ obtained by splitting some block $\Pi_{n,k-1}$ in two. This leads to the question of which weight sequences $(w_1, \ldots, w_{n-1})$ are such that there exists a Gibbs($w_\bullet$) fragmentation process. Is there exists such a process, then one can also be constructed as a Markov chain with some transition matrix $P(\pi, \nu)$ indexed by $\mathcal{P}_{[n]}$ such that $P(\pi, \nu) > 0$ only if ν is a refinement of π, and

$$\sum_{\nu \in \mathcal{P}_{[n]}} p(\pi | k-1) P(\pi, \nu) = p(\nu \,|\, k) \quad (1 \le k \le n-1). \tag{1.58}$$

Such a transition matrix $P(\pi, \nu)$ corresponds to a *splitting rule*, which for each $1 \le k \le n-1$, and each partition π of $[n]$ into $k-1$ components, describes the probability that π splits into a partition ν of $[n]$ into k components. Given that $\Pi_{k-1} = \{A'_1, \ldots, A'_{k-1}\}$ say, the only possible values of Π_k are partitions $\{A_1, \ldots, A_k\}$ such that two of the A_j, say A_1 and A_2, form a partition of one of the A'_i, and the remaining A_j are identical to the remaining A'_i. The initial splitting rule starting with $\pi_1 = \{1, \ldots, n\}$ is described by the Gibbs formula $p(\cdot \,|\, 2)$ determined by the weight sequence $(w_1, \ldots, w_{n-1})$. The simplest way to continue is to use the following

Recursive Gibbs Rule: whenever a component is split, given that the component currently has size m, it is split according to the Gibbs formula $p(n_1, n_2 \,|\, 2)$ for n_1 and n_2 with $n_1 + n_2 = n$.

To complete the description of a splitting rule, it is also necessary to specify for each partition $\pi_{k-1} = \{A'_1, \ldots, A'_{k-1}\}$ the probability that the next component to be split is A'_i, for each $1 \le i \le k-1$. Here the simplest possible assumption seems to be the following:

Linear Selection Rule: Given $\pi_{k-1} = \{A'_1, \ldots, A'_{k-1}\}$, split A'_i with probability proportional to $\#A_i - 1$.

While this selection rule is somewhat arbitrary, it is natural to investigate its implications for the following reasons. Firstly, components of size 1 cannot be split, so the probability of picking a component to split must depend on size. This probability must be 0 for a component of size 1, and 1 for a component of size $n-k+2$. The simplest way to achieve this is by linear interpolation. Secondly, both the segment splitting model and the tree splitting model described in Examples 1.5 and 1.6 follow this rule. In each of these examples a component of size m is derived from a graph component with $m-1$ edges, so the linear selection rule corresponds to picking an edge uniformly at random from the set of all edges in the random graph whose components define Π_{k-1}. Given two natural combinatorial examples with the same selection rule, it is natural ask what other models there might be following the same rule. At the level of random partitions of $[n]$, this question is answered by the following proposition. It also seems reasonable to expect that the conclusions of the proposition will remain valid under weaker assumptions on the selection rule.

Proposition 1.7. *Fix $n \geq 4$, let $(w_j, 1 \leq j \leq n-1)$ be a sequence of positive weights with with $w_1 = 1$, and let let $(\Pi_k, 1 \leq k \leq n)$ be a $\mathcal{P}_{[n]}$-valued fragmentation process defined by the recursive Gibbs splitting rule derived from these weights, with the linear selection rule. Then the following statements are equivalent:*

(i) For each $1 \leq k \leq n$ the random partition Π_k has the Gibbs distribution on $\mathcal{P}_{[n]}^k$ derived from $(w_j, 1 \leq j \leq n-1)$;

(ii) The weight sequence is of the form

$$w_j = \prod_{m=2}^{j} (mc + jb) \tag{1.59}$$

for every $1 \leq j \leq n-1$ for some constants c and b.

(iii) For each $2 \leq k \leq n$, given that Π_k has k components of sizes $n_1, \cdots, n_k$, the partition Π_{k-1} is derived from Π_k by merging the ith and jth of these components with probability proportional to $2c + b(n_i + n_j)$ for some constants c and b.

The constants c and b appearing in (ii) and (iii) are only unique up to constant factors. To be more precise, if either of conditions (ii) or (iii) holds for some (b, c), then so does the other condition with the same (b, c). Hendriks et al. [195] showed that the construction (iii) of a $\mathcal{P}_{[n]}$-valued coalescent process $(\Pi_n, \Pi_{n-1}, \ldots, \Pi_1)$ corresponds to the discrete skeleton of the continuous time Marcus-Lushnikov model with state space $\mathcal{P}_n$ and merger rate $2c + b(x + y)$ between every pair of components of sizes (x, y), and that the distribution of Π_k is then determined as in (i) and (ii). Note the implication of Proposition 1.7 that if $(\Pi_n, \Pi_{n-1}, \ldots, \Pi_1)$ is so constructed as a coalescent process, then the reversed process $(\Pi_1, \Pi_2, \ldots, \Pi_n)$ is a Gibbs fragmentation process governed by the recursive Gibbs splitting rule with weights (w_j) as in (ii) and linear selection probabilities.

Lying behind Proposition 1.7 is the following evaluation of the associated Bell polynomial:

Lemma 1.8. *For* $w_j = \prod_{m=2}^{j}(mc + jb), j = 2, 3, \ldots$

$$B_{n,k}(1, w_2, w_3, \ldots) = \binom{n-1}{k-1} \prod_{m=k+1}^{n} (mc + nb) \quad (1 \le k \le n). \tag{1.60}$$

This evaluation can be read from [195, (19)–(21)]. See *Gibbs distributions for random partitions generated by a fragmentation process* by Nathanael Berestycki and Jim Pitman (arXiv:math.PR/0512378) for the proof of Proposition 1.7.

Example 1.9. Random mappings. Let M_n be a uniformly distributed random mapping from $[n]$ to $[n]$, meaning that all n^n such maps are equally likely. Let Π_n the partition of $[n]$ induced by the tree components of the usual functional digraph of M_n. Then Π_n is the random partition of $[n]$ associated a random $(V \circ W)$-structure on $[n]$ for V the species of permutations and W the species of rooted labeled trees. So Π_n has the Gibbs$(\bullet!, \bullet^{\bullet-1})$ distribution on $\mathcal{P}_{[n]}$. Let $\hat{\Pi}_n$ denote the partition of $[n]$ induced by the connected components of the usual functional digraph of M_n, so each block of $\hat{\Pi}_n$ is the union of tree components in Π_n attached to some cycle of M_n. Then $\hat{\Pi}_n$ is the random partition of $[n]$ derived from a random $(V \circ W)$-structure on $[n]$ for V the species of non-empty sets and W the species of mappings whose digraphs are connected. So $\hat{\Pi}_n$ has a Gibbs$(1^\bullet, w_\bullet)$ distribution on $\mathcal{P}_{[n]}$ where w_j is the number of mappings on $[j]$ whose digraphs are connected. Classify by the number c of cyclic points of the mapping on $[j]$, and use (1.56), to see that

$$w_j = \sum_{c=1}^{j} (c-1)! \binom{j}{c} c j^{j-c-1} = \mathbb{P}(N_j < j)(j-1)! e^j \sim \tfrac{1}{2}(j-1)! e^j \text{ as } j \to \infty \tag{1.61}$$

where N_j is a Poisson(j) variable. This example is further developed in Chapter 9.

Exercises

1.5.1. (Compound Poisson) Stam [404]. Let $\Delta_1, \Delta_2, \ldots$ denote the successive jumps of a non-negative integer valued compound Poisson process $(X(t), t \ge 0)$ with jump intensities $\lambda_j, j = 1, 2, \ldots$, with $\lambda := \sum_j \lambda_j \in (0, \infty)$, and let $N(t)$ be the number of jumps of X in $[0, t]$, so that $X_t = \sum_{i=1}^{N_t} \Delta_i$. The Δ_i are independent and identically distributed with distribution $P(\Delta_i = j) = \lambda_j / \lambda$, independent of $N(t)$, hence

$$\mathbb{P}(X(t) = n \,|\, N(t) = k) = \lambda_n^{k*} / \lambda^k$$

where (λ_n^{k*}) is the k-fold convolution of the sequence (λ_n) with itself, with the convention $\lambda_n^{0*} = 1(n = 0)$. So for all $t \ge 0$ and $n = 0, 1, 2, \ldots$

$$\mathbb{P}[X(t) = n] = e^{-\lambda t} \sum_{k=0}^{n} \lambda_n^{k*} \frac{t^k}{k!} = \frac{e^{-\lambda t}}{n!} B_n(t^\bullet, w_\bullet) \text{ for } w_j := j! \lambda_j. \tag{1.62}$$

Moreover for each $t > 0$ and each $n = 1, 2, \ldots$,

$$(\Delta_1, \ldots, \Delta_{N(t)}) \text{ given } X(t) = n \tag{1.63}$$

has the exchangeable Gibbs distribution on compositions of n defined by (1.51) for $v_k \equiv 1$ and $w_j := tj!\lambda_j$. Let $N_j(t)$ be the number of $i \leq N(t)$ such that $\Delta_i = j$. Then the $N_j(t)$ are independent Poisson variables with means $\lambda_j t$. The random partition of n derived from the random composition (1.63) of n is identical to

$$(N_j(t), 1 \leq j \leq n) \text{ given } \sum_{j=1}^{\infty} jN_j(t) = n, \tag{1.64}$$

and the distribution of $(N_j(t), 1 \leq j \leq n)$ remains the same with conditioning on $\sum_{j=1}^{n} jN_j(t) = n$ instead of $\sum_{j=1}^{\infty} jN_j(t) = n$.

1.5.2. (Distribution of the number of blocks) For a Gibbs$(1^\bullet, w_\bullet)$ partition of $[n]$,

$$\mathbb{P}(|\Pi_n| = k) = \frac{B_{n,k}(w_\bullet)}{B_n(w_\bullet)} = \frac{n!}{k!} \frac{(w(\xi))^k}{\xi^n B_n(w_\bullet)} \mathbb{P}_{\xi,w_\bullet}(S_k = n) \tag{1.65}$$

for $1 \leq k \leq n$, where $\mathbb{P}_{\xi,w_\bullet}$ governs S_k as the sum of k independent variables X_i with the power series distribution (1.41), assuming that $\xi > 0$ is such that $w(\xi) < \infty$, and the complete Bell polynomial $B_n(1^\bullet, w_\bullet) := \sum_{k=1}^{n} B_{n,k}(w_\bullet)$ is determined via the exponential formula (1.12). Deduce from (1.65) the formula [99]

$$\mathbb{E}(|\Pi_n|) = \frac{1}{B_n(1^\bullet, w_\bullet)} \left[\frac{\xi^n}{n!}\right] w(\xi) e^{w(\xi)}. \tag{1.66}$$

Kolchin [260, §1.6] and other authors [317], [99] have exploited the representation (1.65) to deduce asymptotic normality of $|\Pi_n|$ for large n, under appropriate assumptions on $w_\bullet$, from the asymptotic normality of the $\mathbb{P}_{\xi,w_\bullet}$ asymptotic distribution of S_k for large k and well chosen ξ, which is typically determined by a local limit theorem. See also [27] for similar results obtained by other techniques.

1.5.3. (Normal approximation for combinatorial sequences: Harper's method)

(i) (Lévy) Let $(a_0, a_1, \ldots a_n)$ be a sequence of nonnegative real numbers, with generating polynomial $A(z) := \sum_{k=0}^{n} a_k z^k, z \in \mathbb{C}$, such that $A(1) > 0$. Show that A has only real zeros if and only if there exist independent Bernoulli trials $X_1, X_2, \ldots, X_n$ with $\mathbb{P}(X_i = 1) = p_i \in (0, 1]$, $1 \leq i \leq n$, such that $\mathbb{P}(X_1 + X_2 + \cdots + X_n = k) = a_k / A(1)$, $\forall\, 0 \leq k \leq n$. Then the roots α_i of A are related to the p_i by $\alpha_i = -(1 - p_i)/p_i$.

(ii) (Harper, [190]) Let $\{a_{n,k}\}_{k=0}^{n}$ be a sequence of nonnegative real numbers. Suppose that $H_n(z) := \sum_{k=0}^{n} a_{n,k} z^k, z \in \mathbb{C}$ with $H_n(1) > 0$ has only real roots, say $\alpha_{n,i} = -(1 - p_{n,i})/p_{n,i}$. Suppose K_n is a random variable with distribution

$$P_n(k) := \mathbb{P}(K_n = k) = a_{n,k}/H_n(1) \quad (0 \leq k \leq n).$$

Then
$$\frac{K_n - \mu_n}{\sigma_n} \xrightarrow{d} N(0,1) \quad \text{if and only if} \quad \sigma_n \to \infty,$$
where $\mu_n := \mathbb{E}(K_n) = \sum_{i=1}^n p_{n,i}$, and $\sigma_n^2 := \text{Var}(K_n) = \sum_{i=1}^n p_{n,i}(1 - p_{n,i})$. See [352] and papers cited there for numerous applications. Two basic examples are provided by the next two exercises. Harper [190] also proved a local limit theorem for such a sequence provided the central limit theorem holds. Hence both kinds of Stirling numbers admit local normal approximations.

1.5.4. (Number of cycles of a uniform random permutation) Let $a_{n,k} = B_{n,k}((\bullet - 1)!)$ be the Stirling numbers of the first kind, note from (1.14) that
$$H_n(z) = z(z+1)(z+2)\cdots(z+n-1)$$
Deduce that if K_n is the number of cycles from a uniformly chosen permutation of $[n]$ then the Central Limit Theorem holds, $\mathbb{E}(K_n) - \log n = O(1)$, $\text{Var}(K_n) \sim \log n$, and hence
$$\frac{K_n - \log n}{\sqrt{\log n}} \xrightarrow{d} N(0,1).$$

1.5.5. (Number of blocks of a uniform random partition) Let $a_{n,k} = B_{n,k}(1^\bullet)$ be the Stirling numbers of the second kind. Let K_n be the number of blocks of a uniformly chosen partition of $[n]$.

(a) Show that $B_{n+1,k}(1^\bullet) = B_{n,k-1}(1^\bullet) + k\, B_{n,k}(1^\bullet)$.
(b) Using (a) deduce that
$$e^z H_{n+1}(z) = z\frac{d}{dz}\left(e^z H_n(z)\right).$$
(c) Apply induction to show that for all $n \geq 1$, H_n has only real zeros.
(d) Use the recursion in (a) again to show
$$\mu_n := \mathbb{E}[K_n] = \frac{B_{n+1}}{B_n} - 1, \quad \text{and}$$
$$\sigma_n^2 := \text{Var}(K_n) = \frac{B_{n+2}}{B_n} - \left(\frac{B_{n+1}}{B_n}\right)^2 - 1,$$
where $B_n := H_n(1) = B_n(1^\bullet, 1^\bullet)$ is the nth Bell number.
(e) Deduce the Central Limit Theorem.

1.5.6. (Problem: existence of Gibbs fragmentations) For given $w_\bullet$, describe the set of n for which such a fragmentation process exists. In particular, for which $w_\bullet$ does such a process exist for all n? Even the following particular case for $w_j = (j-1)!$ does not seem easy to resolve:

1.5.7. (Problem: cyclic fragmentations) Does there exist for each n a $\mathcal{P}_n$-valued fragmentation process $(\Pi_{n,k}, 1 \leq k \leq n)$ such that $\Pi_{n,k}$ is distributed like the partition generated by cycles of a uniform random permutation of $[n]$ conditioned to have k cycles?

1.5.8. Show that for $w_\bullet = 1^\bullet$, for all sufficiently large n there does not exist a Gibbs$(w_\bullet)$ fragmentation process of $[n]$. [Hint: $\Pi_{n,k}$ has the same distribution as the partition generated by n independent random variables with uniform distribution on $[k]$, conditioned on the event that all k values appear].

1.5.9. (Problem) For exactly which n does there exist a Gibbs$(1^\bullet)$ fragmentation process of $[n]$? What is the largest such n?

Notes and comments

See also Bender [31, 32] and Canfield [83] for more general analytic methods to obtain central and local limit theorems for combinatorial sequences. Canfield [83] gives nice sufficient conditions for central and local limit theorems for the coefficients of polynomials of binomial type. Similar results may also be derived using classical analytic techniques like the saddle point approximation [333] and Hayman's criterion [194].

2

Exchangeable random partitions

This chapter is a review of basic ideas from Kingman's theory of exchangeable random partitions [253], as further developed in [14, 347, 350]. This theory turns out to be of interest in a number of contexts, for instance in the study of population genetics, Bayesian statistics, and models for processes of coagulation and fragmentation. The chapter is arranged as follows.

2.1. Finite partitions This section introduces the *exchangeable partition probability function (EPPF)* associated with an exchangeable random partition Π_n of the set $[n] := \{1, \ldots, n\}$. This symmetric function of compositions $(n_1, \cdots, n_k)$ of n gives the probability that Π_n equals any *particular* partition of $[n]$ into k subsets of sizes $n_1, n_2, \ldots, n_k$, where $n_i \geq 1$ and $\Sigma_i n_i = n$. Basic examples are provided by *Gibbs partitions* for which the EPPF assumes a product form.

2.2. Infinite partitions A random partition Π_∞ of the set $\mathbb{N}$ of positive integers is called exchangeable if its restriction Π_n to $[n]$ is exchangeable for every n. The distribution of Π_∞ is determined by an EPPF which is a function of compositions of positive integers subject to an addition rule expressing the consistency of the partitions Π_n as n varies. Kingman [250] established a one-to-one correspondence between distributions of such exchangeable random partitions of $\mathbb{N}$ and distributions for a sequence of nonnegative random variables $P_1^\downarrow, P_2^\downarrow, \ldots$ with $P_1^\downarrow \geq P_2^\downarrow \geq \ldots$ and $\sum_k P_k^\downarrow \leq 1$. In Kingman's *paintbox representation*, the blocks of Π_∞ are the equivalence classes generated by the random equivalence relation $\sim$ on positive integers, constructed as follows from *ranked frequencies* $(P_k^\downarrow)$ and a sequence of independent random variables U_i with uniform distribution on $[0, 1]$, where (U_i) and $(P_k^\downarrow)$ are independent: $i \sim j$ iff either $i = j$ or both U_i and U_j fall in I_k for some k, where the I_k are some disjoint random sub-intervals of $[0, 1]$ of lengths $P_k^\downarrow$. Each $P_k^\downarrow$ with $P_k^\downarrow > 0$ is then the asymptotic frequency of some corresponding block of Π_∞, and if $\sum_k P_k^\downarrow < 1$ there is also a remaining subset of $\mathbb{N}$ with asymptotic frequency $1 - \sum_k P_k^\downarrow$, each of whose elements is a singleton block of Π_∞.

2.3. Structural distributions A basic property of every exchangeable random partition Π_∞ of $\mathbb{N}$ is that each block of Π_∞ has a limiting relative frequency almost surely. The *structural distribution* associated with Π_∞ is the probability distribution on $[0,1]$ of the asymptotic frequency of the block of Π_∞ that contains a particular positive integer, say 1. In terms of Kingman's representation, this is the distribution of a size-biased pick from the associated sequence of random frequencies $(P_k^\downarrow)$. Many important features of exchangeable random partitions and associated random discrete distributions, such as the mean number of frequencies in a given interval, can be expressed in terms of the structural distribution.

2.4. Convergence Convergence in distribution of a sequence of exchangeable random partitions Π_n of $[n]$ as $n \to \infty$ can be expressed in several equivalent ways: in terms of induced partitions of $[m]$ for fixed m, in terms of ranked or size-biased frequencies, and in terms of an associated process with exchangeable increments.

2.5. Limits of Gibbs partitions Limits of Gibbs partitions lead to exchangeable random partitions of $\mathbb{N}$ with ranked frequencies $(P_i^\downarrow, i \geq 1)$ distributed according to some mixture over s of the conditional distribution of ranked jumps of some subordinator $(T_u, 0 \leq u \leq s)$ given $T_s = 1$. Two important special cases arise when T is a gamma process, or a stable subordinator of index $\alpha \in (0,1)$. The study of such limit distributions is pursued further in Chapter 4.

2.1. Finite partitions

A random partition Π_n of $[n]$ is called *exchangeable* if its distribution is invariant under the natural action on partitions of $[n]$ by the symmetric group of permutations of $[n]$. Equivalently, for each partition $\{A_1, \ldots, A_k\}$ of $[n]$,

$$\mathbb{P}(\Pi_n = \{A_1, \ldots, A_k\}) = p(|A_1|, \ldots, |A_k|)$$

for some symmetric function p of compositions $(n_1, \ldots, n_k)$ of n. This function p is called the *exchangeable partition probability function (EPPF)* of Π_n. For instance, given two positive sequences $v_\bullet = (v_1, v_2, \ldots)$ and $w_\bullet = (w_1, w_2, \ldots)$, the formula

$$p(n_1, \ldots, n_k; v_\bullet, w_\bullet) := \frac{v_k \prod_{i=1}^k w_{n_i}}{B_n(v_\bullet, w_\bullet)} \tag{2.1}$$

where $B_n(v_\bullet, w_\bullet)$ is a normalization constant, defines the EPPF of a Gibbs partition determined by $v_\bullet$ and $w_\bullet$ as discussed in Section 1.5 . In most applications, it is the sizes of blocks of an exchangeable random partition Π_n which are of primary interest. The next three paragraphs present three different ways to encode these block sizes as a random composition of $[n]$, and show how the distributions of these encodings are determined by the EPPF p.

Decreasing order Let $(N_{n,1}^\downarrow, \ldots, N_{n,K_n}^\downarrow)$ denote the *partition of n induced by Π_n*, that is the random composition of n defined by the sizes of blocks of

Π_n with blocks in decreasing order of size. Then for each partition of n with component sizes (n_i) in decreasing order,

$$\mathbb{P}((N^{\downarrow}_{n,1}, \ldots, N^{\downarrow}_{n,K_n}) = (n_1, \ldots, n_k)) = \frac{n!}{\prod_{i=1}^{n} (i!)^{m_i} m_i!} p(n_1, \ldots, n_k) \qquad (2.2)$$

where

$$m_i := \sum_{\ell=1}^{k} 1(n_\ell = i) \qquad (2.3)$$

is the number of components of size i in the given partition of n, and the combinatorial factor is the number of partitions of $[n]$ corresponding to the given partition of n. Let $|\Pi_n|_j$ denote the number of blocks of Π_n of size j. Due to the bijection between partitions of n and possible vectors of counts $(m_i, 1 \le i \le n)$, for (m_i) a vector of non-negative integers subject to $\sum_i m_i = k$ and $\sum_i i m_i = n$, the probability

$$\mathbb{P}(|\Pi_n|_i = m_i \text{ for } 1 \le i \le n), \qquad (2.4)$$

that is the probability that Π_n has m_i blocks of size i for each $1 \le i \le n$, is identical to the the probability in (2.2) for $(n_1, \ldots, n_k)$ the decreasing sequence subject to (2.3).

Size-biased order of least elements Let $(\tilde{N}_{n,1}, \ldots, \tilde{N}_{n,K_n})$ denote the random composition of n defined by the sizes of blocks of Π_n with blocks in order of appearance. Then for all compositions $(n_1, \ldots, n_k)$ of n into k parts,

$$\mathbb{P}((\tilde{N}_{n,1}, \ldots, \tilde{N}_{n,K_n}) = (n_1, \ldots, n_k)) \qquad (2.5)$$

$$= \frac{n!}{n_k(n_k + n_{k-1}) \cdots (n_k + \cdots + n_1) \prod_{i=1}^{k} (n_i - 1)!} p(n_1, \ldots, n_k) \qquad (2.6)$$

where the combinatorial factor is the number of partitions of $[n]$ with the prescribed block sizes in order of appearance [115]. Note that $(\tilde{N}_{n,1}, \ldots, \tilde{N}_{n,K_n})$ is a *size-biased random permutation* of $(N^{\downarrow}_{n,1}, \ldots, N^{\downarrow}_{n,K_n})$, meaning that given the decreasing rearrangement, the blocks appear in the random order in which they would be discovered in a process of simple random sampling without replacement.

Exchangeable random order It is often convenient to consider the block sizes of a random partition of $[n]$ in *exchangeable random order*, meaning that conditionally given $\Pi_n = \{A_1, \ldots, A_k\}$, random variables $(N^{ex}_{n,1}, \ldots, N^{ex}_{n,k})$ are constructed as $N^{ex}_{n,i} = |A_{\sigma(i)}|$ where σ is a uniformly distributed random permutation of $[k]$. Then

$$\mathbb{P}((N^{ex}_{n,1}, \ldots, N^{ex}_{n,K_n}) = (n_1, \ldots, n_k)) = \binom{n}{n_1, \ldots, n_k} \frac{1}{k!} p(n_1, \ldots, n_k). \qquad (2.7)$$

To see this, recall that $p(n_1, \ldots, n_k)$ is the probability of any particular partition of $[n]$ with block sizes $(n_1, \ldots, n_k)$ in some order. Dividing by $k!$ gives the probability of obtaining a particular ordered partition of $[n]$ after randomizing the order of the blocks, and the multinomial coefficent is the number of such ordered partitions consistent with $(n_1, \ldots, n_k)$.

Partitions generated by sampling without replacement. Let

$$\Pi(x_1, \ldots, x_n)$$

denote the partition of $[n]$ generated by a sequence $x_1, \ldots, x_n$. That is the partition whose blocks are the equivalence classes for the random equivalence relation $i \sim j$ iff $x_i = x_j$. If $(X_1, \ldots, X_n)$ is a sequence of exchangeable random variables, then $\Pi(X_1, \ldots, X_n)$ is an exchangeable random partition of $[n]$. Moreover, the most general possible distribution of an exchangeable random partition of $[n]$ is obtained this way. To be more precise, there is the following basic result. See Figure 2.1 for a less formal statement.

Proposition 2.1. [14] *Let Π_n be an exchangeable random partition of $[n]$, and let $\pi_n := (N_{n,i}^{\downarrow}, 1 \le i \le K_n)$ be the corresponding partition of n defined by the decreasing rearrangement of block sizes of Π_n. Then the joint law of Π_n and π_n is that of $\Pi(X_1, \ldots, X_n)$ and π_n, where conditionally given π_n the sequence $(X_1, \ldots, X_n)$ is defined by simple random sampling without replacement from a list $x_1, \ldots, x_n$ with $N_{n,i}^{\downarrow}$ values equal to i for each $1 \le i \le K_n$, say $(X_1, \ldots, X_n) = (x_{\sigma(1)}, \ldots, x_{\sigma(n)})$ where σ is a uniform random permutation of $[n]$.*

Exercises

2.1.1. Prove Proposition 2.1.

2.1.2. Corresponding to each probability distribution Q on the set $\mathcal{P}_n$ of partitions of n, there is a unique distribution of an exchangeable partition Π_n of $[n]$ which induces a partition π_n of n with distribution Q: given π_n, let Π_n have uniform distribution on the set of all partitions of $[n]$ whose block sizes are consistent with π_n.

2.1.3. A function p defined on the set of compositions of n is the EPPF of some exchangeable random partition Π_n of $[n]$ if and only if p is non-negative, symmetric, and

$$\sum_{k=1}^{n} \sum_{(n_1, \ldots, n_k)} \binom{n}{n_1, \ldots, n_k} \frac{1}{k!} p(n_1, \ldots, n_k) = 1,$$

where the second sum is over all compositions $(n_1, \ldots, n_k)$ of n with k parts.

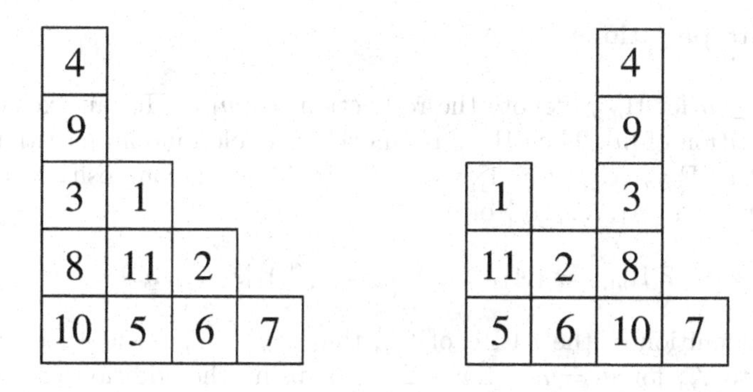

Block sizes in decreasing order of size : $(5,3,2,1)$

Block sizes in the size-biased order of the least element : $(3,2,5,1)$

Figure 2.1: A random partition Π_{11} of $[11]$. To state Proposition 2.1 less formally: if Π_n is exchangeable, then given that the block sizes of Π_n in decreasing order define some pattern of boxes, as above left for $n = 11$, known as a Ferrer's diagram, corresponding to a partition of the integer n, the partition of $[n]$ is recovered by filling the boxes with numbers sampled from $[n]$ without replacement, then taking the partition generated by the columns of boxes, to get e.g. $\Pi_n = \{\{4,9,3,8,10\},\{1,11,5\},\{2,6\},\{7\}\}$ as above.

2.1.4. (Serban Nacu [318]) . Let X_i be the indicator of the event that i is the least element of some block of an exchangeable random partition Π_n of $[n]$. Show that the joint law of the $(X_i, 1 \le i \le n)$ determines the law of Π_n.

2.1.5. (Problem) Characterize all possible laws of strings of 0's and 1's which can arise as in the previous exercise. Variants of this problem, with side conditions on the laws, are easier but still of some interest. Compare with Exercise 3.2.4 .

2.1.6. The EPPF of an exchangeable random partition Π_n of $[n]$ is $p(n_1, \dots, n_k) := \mathbb{P}(\Pi_n = \Pi)$ for each particular partition $\Pi = \{A_1, \dots, A_k\}$ of $[n]$ with $|A_i| = n_i$ for all $1 \le i \le n$. Let $q(n_1, \dots, n_k)$ be the common value of $\mathbb{P}(\Pi_n \ge \Pi)$ for each such Π, where $\Pi_n \ge \Pi$ means that Π_n is *coarser than* Π, i.e. each block of Π_n is some union of blocks of Π. Each of the functions p and q determines the other via the formula

$$q(n_1, \dots, n_k) = \sum_{j=1}^{k} \sum_{\{B_1,\dots,B_j\}} p(n_{B_1}, \dots, n_{B_j}) \tag{2.8}$$

where the second sum is over partitions $\{B_1, \dots, B_j\}$ of $[k]$, and $n_B := \sum_{i \in B} n_i$.

2.2. Infinite partitions

For $1 \leq m \leq n$ let $\Pi_{m,n}$ denote the restriction to $[m]$ of Π_n, an exchangeable random partition of $[n]$. Then $\Pi_{m,n}$ is an exchangeable random partition of $[m]$ with some EPPF $p_n : \mathcal{C}_m \to [0,1]$, where $\mathcal{C}_m$ is the set of compositions of m. So for each partition $\{A_1, \dots, A_k\}$ of $[m]$

$$\mathbb{P}(\Pi_{m,n} = \{A_1, \dots, A_k\}) = p_n(|A_1|, \dots, |A_k|)$$

where the definition of the EPPF of Π_n, that is $p_n : \mathcal{C}_n \to [0,1]$, is extended recursively to $\mathcal{C}_m$ for $m = n-1, n-2, \dots, 1$, using the addition rule of probability. Thus the function $p = p_n$ satisfies the following *addition rule*: for each composition $(n_1, \dots, n_k)$ of $m < n$

$$p(n_1, \dots, n_k) = \sum_{j=1}^{k} p(\dots, n_j + 1, \dots) + p(n_1, \dots, n_k, 1) \tag{2.9}$$

where $(\dots, n_j+1, \dots)$ is derived from $(n_1, \dots, n_k)$ by substituting $n_j + 1$ for n_j. For instance,

$$1 = p(1) = p(2) + p(1,1) \tag{2.10}$$

and

$$p(2) = p(3) + p(2,1); \quad p(1,1) = p(1,2) + p(2,1) + p(1,1,1) \tag{2.11}$$

where $p(1,2) = p(2,1)$ by symmetry of the EPPF.

Consistency [253, 14, 347] Call a sequence of exchangeable random partitions (Π_n) *consistent in distribution* if Π_m has the same distribution as $\Pi_{m,n}$ for every $m < n$. Equivalently, there is a symmetric function p defined on the set of all integer compositions (an *infinite EPPF*) such that $p(1) = 1$, the addition rule (2.9) holds for all integer compositions $(n_1, \dots, n_k)$, and the restriction of p to $\mathcal{C}_n$ is the EPPF of Π_n. Such (Π_n) can then be constructed so that $\Pi_m = \Pi_{m,n}$ almost surely for every $m < n$. The sequence of random partitions $\Pi_\infty := (\Pi_n)$ is then called an *exchangeable random partition of* $\mathbb{N}$, or an *infinite exchangeable random partition.* Such a Π_∞ can be regarded as a random element of the set $\mathcal{P}_\mathbb{N}$ of partitions of $\mathbb{N}$, equipped with the σ-field generated by the restriction maps from $\mathcal{P}_\mathbb{N}$ to $\mathcal{P}_{[n]}$ for all n. One motivation for the study of exchangeable partitions of $\mathbb{N}$ is that if (Π_n) is any sequence of exchangeable partitions of $[n]$ for $n = 1, 2, \dots$ which has a limit in distribution in the sense that $\Pi_{m,n} \xrightarrow{d} \Pi_{m,\infty}$ for each m as $n \to \infty$, then the sequence of limit partitions $(\Pi_{m,\infty}, m = 1, 2, \dots)$ is consistent in distribution, hence constructible as an exchangeable partition of $\mathbb{N}$. This notion of weak convergence of random partitions is further developed in Section 2.4.

Partitions generated by random sampling Let (X_n) be an infinite exchangeable sequence of real random variables. According to de Finetti's theorem, (X_n) is obtained by *sampling from some random probability distribution* F. That is to say there is a random probability distribution F on the line, such that conditionally given F the X_i are i.i.d. according to F. To be more explicit, if

$$F_n(x) := \frac{1}{n}\sum_{i=1}^{n} 1(X_i \leq x)$$

is the empirical distribution of the first n values of the sequence, then by combination of de Finetti's theorem [122, p. 269] and the Glivenko-Cantelli theorem [122, p. 59]

$$F(x) = \lim_n F_n(x) \text{ uniformly in } x \text{ almost surely.} \tag{2.12}$$

Let Π_∞ be the exchangeable random partition of $\mathbb{N}$ generated by (X_n), meaning that the restriction Π_n of Π_∞ to $[n]$ is the partition generated by $(X_1, \ldots, X_n)$, as defined above Proposition 2.1. The distribution of $\Pi_\infty := (\Pi_n)$ is determined by the distribution of $(P_i^\downarrow, i \geq 1)$, where $P_i^\downarrow$ is the magnitude of the ith largest atom of F. Note that $1 - \sum_i P_i^\downarrow$ is the magnitude of the continuous component of F, which might be strictly positive, and that almost surely each i such that X_i is not an atom of F contributes a singleton component $\{i\}$ to Π_∞. To summarize this setup, say Π_∞ is *generated by sampling from a random distribution with ranked atoms* $(P_i^\downarrow, i \geq 1)$. According to the following theorem, every infinite exchangeable partition has the same distribution as one generated this way. This is the infinite analog of Proposition 2.1, according to which every finite exchangeable random partition can be generated by a process of random sampling without replacement from some random population.

Theorem 2.2. (Kingman's representation [251, 253]) *Let* $\Pi_\infty := (\Pi_n)$ *be an exchangeable random partition of* $\mathbb{N}$, *and let* $(N_{n,i}^\downarrow, i \geq 1)$ *be the decreasing rearrangement of block sizes of* Π_n, *with* $N_{n,i}^\downarrow = 0$ *if* Π_n *has fewer than* i *blocks. Then* $N_{n,i}^\downarrow/n$ *has an almost sure limit* $P_i^\downarrow$ *as* $n \to \infty$ *for each* i. *Moreover the conditional distribution of* Π_∞ *given* $(P_i^\downarrow, i \geq 1)$ *is as if* Π_∞ *were generated by random sampling from a random distribution with ranked atoms* $(P_i^\downarrow, i \geq 1)$.

Proof. (Sketch, following Aldous [14, p. 88]) Without loss of generality, it can be supposed that on the same probability space as Π_∞ there is an independent sequence of i.i.d. uniform $[0, 1]$ variables U_j. Let $X_n = U_j$ if n falls in the jth class of Π_∞ to appear. Then $(X_n, n = 1, 2, \ldots)$ is exchangeable. Hence Π_∞ is generated by random sampling from F which is the uniform almost sure limit of

$$F_n(u) := \frac{1}{n}\sum_{m=1}^{n} 1(X_m \leq u) = \sum_{i=1}^{\infty} \frac{N_{n,i}^\downarrow}{n} 1(\hat{U}_{n,i} \leq u)$$

for some Π_n-dependent rearrangement $\hat{U}_{n,i}$ of the U_j. By the almost sure uniformity (2.12) of convergence of F_n to F, the size $N^{\downarrow}_{n,i}/n$ of the ith largest atom of F_n has almost sure limit $P^{\downarrow}_i$ which is the size of the ith largest atom of F. □

Theorem 2.2 sets up a bijection (*Kingman's correspondence*) between probability distributions for an infinite exchangeable random partition, as specified by an infinite EPPF, and probability distributions of $(P^{\downarrow}_i)$ on the set

$$\mathcal{P}^{\downarrow}_{[0,1]} := \{(p_1, p_2, \ldots) : p_1 \geq p_2 \geq \cdots \geq 0 \text{ and } \sum_{i=1}^{\infty} p_i \leq 1\} \tag{2.13}$$

of ranked sub-probability distributions on $\mathbb{N}$.

Note that the set of all infinite EPPF's $p : \cup_{n=1}^{\infty} \mathcal{C}_n \to [0,1]$, with the topology of pointwise convergence, is compact.

Theorem 2.3. (Continuity of Kingman's correspondence [250, §5], [252, p. 45]) *Pointwise convergence of EPPF's is equivalent to weak convergence of finite dimensional distributions of the corresponding ranked frequencies.*

A similar result holds for the frequencies of blocks in order of appearance. See Theorem 3.1. Assuming for simplicity that Π_∞ has *proper frequencies*, meaning that $\sum_i P^{\downarrow}_i = 1$ a.s., Kingman's correspondence can be made more explicit as follows. Let (P_i) denote any rearrangement of the ranked frequencies $(P^{\downarrow}_i)$, which can even be a random rearrangement. Then

$$p(n_1, \ldots, n_k) = \sum_{(j_1,\ldots,j_k)} \mathbb{E}\left[\prod_{i=1}^{k} P_{j_i}^{n_i}\right] \tag{2.14}$$

where $(j_1, \ldots, j_k)$ ranges over all ordered k-tuples of distinct positive integers. This is easily seen from Kingman's representation for $(P_i) = (P^{\downarrow}_i)$. The formula holds also for any rearrangement of these frequencies, because the right side is the expectation of a function of $(P_1, P_2, \ldots)$ which is invariant under finite or infinite permutations of its arguments. In particular (P_i) could be the sequence $(\tilde{P}_i)$ of limit frequencies of classes of (Π_n) in order of appearance, which is a size-biased random permutation of $(P^{\downarrow}_i)$. A much simpler formula in this case is described later in Theorem 3.1.

Exercises

The first two exercises recall some forms of Pólya's urn scheme [151, VII.4], which allow explicit sequential constructions of exchangeable sequences and random partitions. See [300],[350] for more in this vein.

2.2.1. (Beta-binomial) Fix $a, b > 0$. Let $S_n := X_1 + \cdots + X_n$, where the X_i have values 0 or 1. Check that

$$\mathbb{P}(X_{n+1} = 1 \,|\, X_1, \ldots, X_n) = \frac{a + S_n}{a + b + n} \tag{2.15}$$

for all $n \geq 0$ if and only if the X_i are exchangeable and the almost sure limit of S_n/n has the beta(a, b) distribution.

2.2.2. (Dirichlet-multinomial) Fix $\theta_1, \ldots, \theta_m > 0$. Let $(X_n, n = 1, 2, \ldots)$ be a process with values in $\{1, \ldots, m\}$. If for each $n \geq 0$, given $(X_1, \ldots, X_n)$ with n_i values equal to i for each $1 \leq i \leq m$, where $n_1 + \cdots + n_m = n$,

$$X_{n+1} = i \text{ with probability } \frac{\theta_i + n_i}{\theta_1 + \cdots + \theta_m + n}$$

then (X_n) is exchangeable with asymptotic frequencies P_i with the *Dirichlet* $(\theta_1, \ldots, \theta_m)$ distribution (0.47), and conversely.

2.2.3. (Sampling from exchangeable frequencies) Let $p(n_1, \ldots, n_k)$ be the EPPF corresponding to some sequence of random ranked frequencies $(P_1^\downarrow, \ldots, P_m^\downarrow)$ with $\sum_{i=1}^m P_i^\downarrow = 1$ for some $m < \infty$. Let $(P_1, \ldots, P_m)$ be the exchangeable sequence with $\sum_{i=1}^m P_i = 1$ obtained by putting these ranked frequencies in exchangeable random order. Then

$$p(n_1, \ldots, n_k) = (m)_{k\downarrow} \mathbb{E}\left[\prod_{i=1}^k P_i^{n_i}\right].$$

2.2.4. (Coupon Collecting) If $P_i^\downarrow = 1/m$ for $1 \leq i \leq m$ then

$$p(n_1, \ldots, n_k) = (m)_{k\downarrow}/m^n \text{ where } n := \sum_{i=1}^k n_i. \tag{2.16}$$

2.2.5. (Sampling from exchangeable Dirichlet frequencies) [428] If $(P_1, \ldots, P_m)$ has the symmetric Dirichlet distribution (0.47) with parameters $\theta_1 = \cdots = \theta_m = \kappa > 0$, then

$$p(n_1, \ldots, n_k) = (m)_{k\downarrow} \frac{\prod_{i=1}^k (\kappa)_{n_i\uparrow}}{(m\kappa)_{n\uparrow}}. \tag{2.17}$$

Note that the coupon collector's partition (2.16) is recovered in the limit as $\kappa \to \infty$.

2.2.6. (The Blackwell-MacQueen urn scheme) [68]. Fix $\theta > 0$. Let (X_n) with values in $[0, 1]$ be governed by the following prediction rule: $n \geq 0$,

$$\mathbb{P}(X_{n+1} \in \cdot \,|\, X_1, \ldots, X_n) = \frac{\theta\,\lambda(\cdot) + \sum_{i=1}^n 1(X_i \in \cdot)}{\theta + n} \tag{2.18}$$

where $\lambda(\cdot)$ is Lebesgue measure on $[0, 1]$. Then (X_n) is exchangeable, distributed as a sample from a Dirichlet(θ) process F_θ as in (0.46).

2.2.7. (The Ewens sampling formula) [144, 26, 145] As $m \to \infty$ and $\kappa \to 0$ with $m\kappa \to \theta$, the EPPF in (2.17) converges to the EPPF

$$p_{0,\theta}(n_1, \dots, n_k) = \frac{\theta^k}{(\theta)_{n\uparrow}} \prod_{i=1}^{k} (n_i - 1)! \tag{2.19}$$

Such a partition is generated by $X_1, \dots, X_n$ governed by the Blackwell-MacQueen urn scheme (2.18). The corresponding partition of n has distribution

$$\mathbb{P}(|\Pi_n|_i = m_i, 1 \le i \le n) = \frac{n!\, \theta^k}{(\theta)_{n\uparrow}} \prod_{i=1}^{n} \frac{1}{i^{m_i} m_i!} \tag{2.20}$$

for (m_i) as in (2.4). The corresponding ranked frequencies are the ranked jumps of the Dirichlet(θ) process. The frequencies in order of appearance are described in Theorem 3.2.

2.2.8. (Continuity of Kingman's correspondence) Prove Theorem 2.3.

Notes and comments

The theory of exchangeable random partitions described here, following [14] and [347], is essentially equivalent to Kingman's theories of *partition structures* [250, 251] and of *exchangeable random equivalence relations* [253]. The theory is simplified by describing the consistent sequence of distributions of partitions of $[n]$ by its EPPF, rather than by the corresponding sequence of distributions of integer partitions, which is what Kingman called a partition structure. Donnelly and Joyce [114] and Gnedin [172] developed a parallel theory of *composition structures*, whose extreme points are represented by open subsets of $[0, 1]$. See also [198, 199] for alternate approaches.

In the work of Kerov and Vershik on multiplicative branchings [241, 242, 244], each extreme infinite exchangeable partition corresponds to a real-valued character of the algebra of symmetric functions, with certain positivity conditions. See also Aldous [14] regarding exchangeable arrays, and Kallenberg [231] for paintbox representations of random partitions with general symmetries.

2.3. Structural distributions

Let (P_i) be a random discrete probability distribution with size-biased permutation $(\tilde{P}_j)$. So in particular

$$\tilde{P}_1 = P_{\sigma(1)} \text{ where } \mathbb{P}(\sigma(1) = i \,|\, P_1, P_2, \dots) = P_i \qquad (i = 1, 2, \dots). \tag{2.21}$$

The random variable $\tilde{P}_1$ may be called a *size-biased pick* from (P_i). Let $\tilde{\nu}$ denote the distribution of $\tilde{P}_1$ on $(0, 1]$. Following the terminology of Engen [132], $\tilde{\nu}$ is called the *structural distribution* associated with the random discrete distribution (P_i). Note that if a random partition Π_∞ is derived by sampling from

(P_i), then the size-biased permutation $(\tilde{P}_j)$ can be constructed as the sequence of class frequencies of Π_∞ in order of appearance. Then $\tilde{P}_1$ is the frequency of the class of Π_∞ that contains 1. It follows from (2.21) that for an arbitrary non-negative measurable function g,

$$\int \tilde{\nu}(dp)g(p) = \mathbb{E}[g(\tilde{P}_1)] = \mathbb{E}\left[\sum_i P_i g(P_i)\right]. \tag{2.22}$$

Hence, taking $g(p) = f(p)/p$, for arbitrary non-negative measurable function f there is the formula

$$\mathbb{E}\left[\sum_i f(P_i)\right] = \mathbb{E}\left[\frac{f(\tilde{P}_1)}{\tilde{P}_1}\right] = \int_0^1 \frac{f(p)}{p}\,\tilde{\nu}(dp). \tag{2.23}$$

Formula (2.23) shows that the structural distribution $\tilde{\nu}$ encodes much information about the entire sequence of random frequencies. Taking f in (2.23) to be the indicator of a subset B of $(0,1]$, formula (2.23) shows that the point process with a point at each $P_j \in (0,1]$ has mean intensity measure $p^{-1}\tilde{\nu}(dp)$. If $(P_i) = (P_i^\downarrow)$ is in decreasing order, for $x > \frac{1}{2}$ there can be at most one $P_i^\downarrow > x$, so the structural distribution $\tilde{\nu}$ determines the distribution ν of $P_1^\downarrow = \max_i \tilde{P}_i$ on $(\frac{1}{2}, 1]$ via the formula

$$\mathbb{P}(P_1^\downarrow > x) = \nu(x,1] = \int_{(x,1]} p^{-1}\tilde{\nu}(dp) \qquad (x > \tfrac{1}{2}). \tag{2.24}$$

Typically, formulas for $\mathbb{P}(P_1^\downarrow > x)$ get progressively more complicated on the intervals $(\frac{1}{3}, \frac{1}{2}]$, $(\frac{1}{4}, \frac{1}{3}]$, $\cdots$. See e.g. [339, 371]. Note that by (2.14) for $k = 1$ and $n_1 = n$ and (2.23)

$$p(n) = \mathbb{E}\left[\sum_i P_i^n\right] = \mathbb{E}[\,\tilde{P}_1^{n-1}\,] = \mu(n-1) \qquad (n = 1, 2, \cdots), \tag{2.25}$$

where $\mu(q)$ is the qth moment of the distribution $\tilde{\nu}$ of $\tilde{P}_1$ on $(0,1]$. From (2.10), (2.11), and (2.25) the following values of the EPPF of an infinite exchangeable random partition Π_∞ are also determined by the first two moments of the structural distribution:

$$p(1,1) = 1 - \mu(1); \;\; p(2,1) = \mu(1) - \mu(2); \;\; p(1,1,1) = 1 - 3\mu(1) + 2\mu(2). \tag{2.26}$$

So the distribution of Π_3 on partitions of the set $\{1,2,3\}$ is determined by the first two moments of $\tilde{P}_1$. The distribution of Π_n is not determined for all n by the structural distribution (Exercise 2.3.4). But moments of the structural distribution play a key role in the description of a number of particular models for random partitions. See for instance [362, 170].

Exercises

2.3.1. (Improper frequencies) Show how to modify the results of this section to be valid also for exchangeable random partitions of the positive integers with improper frequencies. Show that formula (2.14) is false in the improper case. Find the patch for that formula, which is not so pretty. See for instance Kerov [244, equation (1.3.1)].

2.3.2. (Mean number of blocks) Engen [132]. For an infinite exchangeable partition (Π_n) with $\tilde{P}_1$ the frequency of the block containing 1,

$$\mathbb{E}(|\Pi_n|) = \mathbb{E}[k_n(\tilde{P}_1)], \tag{2.27}$$

where $k_n(v) := (1 - (1 - v)^n)/v$ is a polynomial of degree $n - 1$.

2.3.3. (Proper frequencies) [350] For an infinite exchangeable partition (Π_n) with frequencies $\tilde{P}_i$, the frequencies are proper, meaning $\sum_i \tilde{P}_i = 1$ almost surely, iff $\mathbb{P}(\tilde{P}_1 > 0) = 1$, and also iff $|\Pi_n|/n \to 0$ almost surely.

2.3.4. (The structural distribution does not determine the distribution of the infinite partition) Provide an appropriate example.

2.3.5. (Problem: characterization of structural distributions) What is a necessary and sufficient condition for a probability distribution F on $[0, 1]$ to be a structural distribution? For some necessary and some sufficient conditions see [368].

2.4. Convergence

There are many natural combinatorial constructions of exchangeable random partitions Π_n of $[n]$ which are not consistent in distribution as n varies, so not immediately associated with an infinite exchangeable partition Π_∞. However, it is often the case that a sequence of combinatorially defined exchangeable partitions (Π_n) *converges in distribution as* $n \to \infty$ meaning that

$$\Pi_{m,n} \xrightarrow{d} \Pi_{m,\infty} \text{ for each fixed } m \text{ as } n \to \infty, \tag{2.28}$$

where $\Pi_{m,n}$ is the restriction to $[m]$ of Π_n, and $(\Pi_{m,\infty}, m = 1, 2 \ldots)$ is some sequence of limit random partitions. Let $p_n(n_1, \ldots, n_k)$ denote the EPPF of Π_n, defined as a function of compositions $(n_1, \ldots, n_k)$ of m for every $m \leq n$, as discussed in Section 2.2. Then (2.28) means that for all integer compositions $(n_1, \ldots, n_k)$ of an arbitrary fixed m,

$$p_n(n_1, \ldots, n_k) \to p(n_1, \ldots, n_k) \text{ as } n \to \infty \tag{2.29}$$

for some limit function p. It is easily seen that any such limit p must be an infinite EPPF, meaning that the sequence of random partitions $\Pi_{m,\infty}$ in (2.28) can be constructed consistently to make an infinite exchangeable random partition

$\Pi_\infty := (\Pi_{m,\infty}, m = 1, 2, \ldots)$ whose EPPF is p. Let $(\tilde{P}_i)$ and $(P_i^\downarrow)$ denote the class frequencies of Π_∞, in order of appearance, and ranked order respectively. And let $(N_{n,i}, i \geq 1)$ and $(N_{n,i}^\downarrow, i \geq 1)$ denote the sizes of blocks of Π_n, in order of appearance and ranked order respectively, with padding by zeros to make infinite sequences. It follows from the continuity of Kingman's correspondence (Theorem 2.3) together with Proposition 2.1, and the obvious coupling between sampling with and without replacement for a sample of fixed size as the population size tends to ∞, that this notion (2.28)–(2.29) of convergence in distribution of Π_n to Π_∞ is further equivalent to

$$(N_{n,i}/n)_{i\geq 1} \stackrel{d}{\to} (\tilde{P}_i)_{i\geq 1} \tag{2.30}$$

meaning weak convergence of finite dimensional distributions, and similarly equivalent to

$$(N_{n,i}^\downarrow/n)_{i\geq 1} \stackrel{d}{\to} (P_i^\downarrow)_{i\geq 1} \tag{2.31}$$

in the same sense [173]. Let (U_i) be a sequence of independent and identically distributed uniform $(0,1)$ variables independent of the Π_n. Another equivalent condition is that for each fixed $u \in [0,1]$

$$\sum_{i=1}^{\infty} (N_{n,i}/n) 1(U_i \leq u) \stackrel{d}{\to} F(u) \tag{2.32}$$

for some random variable $F(u)$. According to Kallenberg's theory of processes with exchangeable increments [226], a limit process $(F(u), 0 \leq u \leq 1)$ can then be constructed as an increasing right-continuous process with exchangeable increments, with $F(0) = 0$ and $F(1) = 1$ a.s., and the convergence (2.32) then holds jointly as u varies, and in the sense of convergence in distribution on the Skorohod space $D[0,1]$. To be more explicit,

$$F(u) = \sum_{i=1}^{\infty} P_i 1(U_i \leq u) + (1 - \Sigma_{i=1}^\infty P_i) u \tag{2.33}$$

where the U_i with uniform distribution on $[0,1]$ are independent of the P_i, and either $(P_i) = (\tilde{P}_i)$ or $(P_i) = (P_i^\downarrow)$. The limit partition Π_∞ can then be generated by sampling from any random distribution such as F whose ranked atoms are distributed like $(P_i^\downarrow)$. The restriction $\Pi_{m,\infty}$ of Π_∞ to $[m]$ can then be generated for all $m = 1, 2, \ldots$ by sampling from F, meaning that i and j with $i, j \leq m$ lie in the same block of $\Pi_{m,\infty}$ iff $X_i = X_j$ where the X_i are random variables which conditionally given F are independent and identically distributed according to F:

$$\mathbb{P}(X_i \leq u \,|\, F) = F(u) \qquad (0 \leq u \leq 1).$$

This connects Kingman's theory of exchangeable random partitions to the theory of Bayesian statistical inference [350]. See also James [213, 212] for recent work in this vein.

2.5. Limits of Gibbs partitions

As an immediate consequence of (1.50), the decreasing arrangement of relative sizes of blocks of a $\text{Gibbs}_{[n]}(v_\bullet, w_\bullet)$ partition Π_n, say

$$(N_{n,1}^{\downarrow}/n, \ldots, N_{n,|\Pi_n|}^{\downarrow}/n) \tag{2.34}$$

has the same distribution as the decreasing sequence of order statistics of

$$(X_1/n, \ldots, X_K/n) \text{ given } S_K/n = 1$$

where the X_i have distribution (1.41) and K with distribution (1.42) is independent of the X_i, for some arbitrary $\xi > 0$ with $v(w(\xi)) < \infty$. Since the distribution of a $\text{Gibbs}_{[n]}(v_\bullet, w_\bullet)$ partition depends only on the v_j and w_j for $1 \leq j \leq n$, in this representation for fixed n the condition $v(w(\xi)) < \infty$ can always be arranged by setting $v_j = w_j = 0$ for $j > n$. It is well known [151, XVII.7] [230]that if $X_{n,1}, \ldots, X_{n,k_n}$ is for each n a sequence of independent and identically distributed variables of some non-random length k_n, with $k_n \to \infty$ as $n \to \infty$, then under appropriate conditions

$$\sum_{i=1}^{k_n} X_{n,i} \xrightarrow{d} T := \sum_{i=1}^{\infty} J_i^{\downarrow}$$

where $J_1^{\downarrow} \geq J_2^{\downarrow} \geq \ldots \geq 0$ are the points of a Poisson point process on $(0, \infty)$ with intensity measure $\Lambda(dx)$, for some Lévy measure Λ on $(0, \infty)$ with

$$\Psi(\lambda) := \int_0^{\infty} (1 - e^{-\lambda x}) \Lambda(dx) < \infty \tag{2.35}$$

for all $\lambda > 0$. Then

$$\Lambda(x, \infty) = \lim_{n \to \infty} k_n \mathbb{P}(X_{n,1} > x)$$

for all continuity points x of Λ, and the Laplace transform of T is given by the Lévy-Khintchine formula

$$\mathbb{E}(e^{-\lambda T}) = \exp(-\Psi(\lambda)).$$

It is also known that if such a sum $\sum_{i=1}^{k_n} X_{n,i}$ has T as its limit in distribution as $n \to \infty$, then the convergence in distribution of $\sum_{i=1}^{k_n} X_{n,i}$ to T holds jointly with convergence in distribution of the k largest order statistics of the $X_{n,i}$, $1 \leq i \leq k_n$ to the k largest points $J_1^{\downarrow}, \ldots, J_k^{\downarrow}$ of the Poisson process.

It is therefore to be anticipated that if a sequence of $\text{Gibbs}_{[n]}(v_\bullet, w_\bullet)$ partitions converges as $n \to \infty$ to some infinite partition Π_∞, where either $v_\bullet = v_\bullet^{(n)}$ or $w_\bullet = w_\bullet^{(n)}$ might be allowed to depend on n, and $v_\bullet^{(n)}$ is chosen to ensure that the distribution of the number of components K_n of Π_n grows to ∞ in a deterministic manner, say $K_n/k_n \to s > 0$ for some normalizing constants

k_n, then the distribution of ranked frequencies $(P_i^\downarrow)$ of Π_∞ obtained from the convergence of finite-dimensional distributions

$$(N_{n,i}^\downarrow/n)_{i\geq 1} \stackrel{d}{\to} (P_i^\downarrow)_{i\geq 1} \text{ with } K_n/k_n \to s \tag{2.36}$$

should be representable as

$$(P_i^\downarrow)_{i\geq 1} \stackrel{d}{=} ((J_{s,i}^\downarrow)_{i\geq 1} \,|\, T_s = 1) \tag{2.37}$$

for the ranked points $J_{s,i}^\downarrow$ of a Poisson point process with intensity $s\Lambda$, with $\sum_i J_{s,i}^\downarrow = T_s$. This Poisson process may be constructed as the jumps of $(T_u, 0 \leq u \leq s)$, where $(T_u, u \geq 0)$ is a subordinator with no drift and Lévy measure Λ. Then for $v_\bullet^{(n)}$ chosen so that K_n/k_n converges in distribution to S for some strictly positive random variable S, the limit law of $(P_i^\downarrow)$ in (2.36) should be

$$\int_0^\infty \mathbb{P}((J_{s,i}^\downarrow) \in \cdot \,|\, T_s = 1)\mathbb{P}(S \in ds). \tag{2.38}$$

To make rigorous sense of this, it is first necessary to give a rigorous meaning to the law of $(J_{s,i}^\downarrow)$ given $T_s = 1$, for instance by showing that for fixed s the law of $(J_{s,i}^\downarrow)$ given $T_s = t$ can be constructed to be weakly continuous in t. Second, to justify weak convergence of conditional probability distributions it is necessary to establish an appropriate local limit theorem.

This program has been carried out in two cases of combinatorial significance. One case, treated in detail in [27], covers the class of logarithmic combinatorial assemblies:

Theorem 2.4. [189, 27] *Let $w_\bullet = (w_j)$ be a sequence of weights with*

$$w_j \sim \theta(j-1)!y^j \quad as \ j \to \infty$$

for some $\theta > 0$ and $y > 0$. Let Π_n be a Gibbs$_{[n]}(v_\bullet^{(n)}, w_\bullet)$ partition, either for $v_\bullet^{(n)} \equiv 1^\bullet$, or more generally for any array of weights $v_\bullet^{(n)}$ such that $|\Pi_n|/\log n$ converges in probability to θ as $n \to \infty$. Then Π_n converges in distribution to Π_∞ as $n \to \infty$, where Π_∞ is a $(0,\theta)$-partition with EPPF (2.19), whose Poisson-Dirichlet$(0,\theta)$ frequencies are the ranked jumps of a gamma process $(T_u, 0 \leq u \leq \theta)$ given $T_\theta = 1$.

Sketch of proof. The case when $v_\bullet^{(n)} \equiv 1^\bullet$ can be read from the work of [27], where it is shown that in this case $|\Pi_n|/\log n$ converges in probability to θ as $n \to \infty$. The extension to more general $v_\bullet^{(n)}$ is quite straightforward. □

Two cases of Theorem 2.4 of special interest, discussed further in following chapters, are

- Π_n generated by the cycles of a uniform random permutation of $[n]$, when $w_j = (j-1)!, y = 1, \theta = 1$;

- Π_n generated by the basins of a uniform random mapping of $[n]$, with $w_j = (j-1)! \sum_{i=0}^{j-1} j^i/i!$ as in (1.61), $y = e, \theta = \frac{1}{2}$.

See [27] for many more examples. Note that mixtures over θ of $(0, \theta)$ partitions could arise by suitable choice of $v_\bullet^{(n)}$ so that $|\Pi_n|/\log n$ had a non-degenerate limit distribution, but this phenomenon does not seem to arise naturally in combinatorial examples.

Another case, treated by Pavlov [335, 336, 337], and Aldous-Pitman [17] covers a large number of examples involving random forests, where the limit involves the stable subordinator of index $\frac{1}{2}$. A more general result, where the limit partition is derived from a stable subordinator of index α for $\alpha \in (0,1)$, can be formulated as follows:

Theorem 2.5. *Let $w_\bullet = (w_j)$ be a sequence of weights with exponential generating function $w(\xi) := \sum_{j=1}^\infty \xi^j w_j/j!$ such that $w(\xi) = 1$ for some $\xi > 0$. Let $(p_j, j = 1, 2, \ldots)$ be the probability distribution defined by $p_j = \xi^j w_j/j!$ for ξ with $w(\xi) = 1$, and suppose that*

$$\sum_{j=i}^{\infty} p_j \sim \frac{c\, i^{-\alpha}}{\Gamma(1-\alpha)} \quad \text{as } i \to \infty \tag{2.39}$$

for some $\alpha \in (0,1)$. Let Π_n be a Gibbs$_{[n]}(v_\bullet^{(n)}, w_\bullet)$ partition, for any array of weights $v_\bullet^{(n)}$ such that $|\Pi_n|/n^\alpha$ converges in probability to s as $n \to \infty$. Then Π_n converges in distribution to Π_∞ as $n \to \infty$, where Π_∞ has ranked frequencies distributed like the

$$\text{ranked jumps of } (T_u, 0 \le u \le cs) \text{ given } T_{cs} = 1, \tag{2.40}$$

where $(T_u, u \ge 0)$ is the stable subordinator of index α with $\mathbb{E}\exp(-\lambda T_u) = \exp(-u\lambda^\alpha)$.

Sketch of proof. This was argued in some detail in [17] for the particular weight sequence $w_j = j^{j-1}$, corresponding to blocks with an internal structure specified by a rooted labeled tree. Then $\xi = e^{-1}$, $\alpha = \frac{1}{2}$, and the limiting partition can also be described in terms of the lengths of excursions of a Brownian motion or Brownian bridge, as discussed in Section 4.4 . The proof of the result stated above follows the same lines, appealing to the well known criterion for convergence to a stable law, and the local limit theorem of Ibragimov-Linnik [204]. □

In Section 4.3 the limiting partition Π_∞ appearing in Theorem 2.5 is called an $(\alpha|cs)$ partition. Mixtures of these distributions, obtained by randomizing s for fixed α, arise naturally in a number of different ways, as shown in Chapter 4.

Exercises

2.5.1. (Problem: Characterizing all weak limits of Gibbs partitions) Intuitively, the above discussion suggests that the only possible weak limits of Gibbs partitions are partitions whose ranked frequencies are mixtures over s of the law of ranked jumps of some subordinator $(T_u, 0 \leq u \leq s)$ given $T_s = 1$, allowing also the possibility of conditioning on the number of jumps in the compound Poisson case. Show that if the conditioning is well defined by some regularity of the distribution of T_s for all s, then such a partition can be achieved as a limit of Gibbs partitions, allowing both $v_\bullet$ and $w_\bullet$ to depend on n. But due to the difficulty in giving meaning to the conditioning when T_s does not have a density, it is not clear how to formulate a rigorous result. Can that be done? Does it make any difference whether or not $w_\bullet$ is allowed to depend on n?

3

Sequential constructions of random partitions

This chapter introduces a basic sequential construction of random partitions, motivated at first by consideration of uniform random permutations of $[n]$ which are consistent in a certain sense as n varies. This leads to consideration of a particular two-parameter family of exchangeable random partition structures, which can be characterized in various ways, and which is closely related to gamma and stable subordinators.

3.1. The Chinese restaurant process This process defines a sequence of random permutations σ_n of the set $[n] := \{1, \ldots, n\}$ such that the random partitions Π_n generated by cycles of σ_n are consistent as n varies. The most general exchangeable random partition of positive integers can be obtained this way.

3.2. The two-parameter model This section treats a particularly tractable family of random partitions of $\mathbb{N}$, parameterized by a pair of real numbers (α, θ) subject to appropriate constraints. The distribution $\mathbb{P}_{\alpha,\theta}$ of an (α, θ) partition is characterized by the product form of its partition probabilities, and by the induced distribution of its frequencies in size-biased random order. This distribution is known in the literature as GEM(α, θ), after Griffiths-Engen-McCloskey, while the corresponding distribution of ranked frequencies is known as the Poisson-Dirichlet distribution PD(α, θ). These distributions and associated random partitions arise in numerous contexts, such as population genetics, number theory, Bayesian nonparametric statistics, and the theory of excursions of Brownian motion and Bessel processes.

3.3. Asymptotics This section treats various asymptotic properties of partitions of $\mathbb{N}$, with special emphasis on the two-parameter family, whose asymptotic properties are radically different according to whether α is positive, negative, or zero. In particular, the number K_n of blocks of Π_n is bounded if $\alpha < 0$, grows like $\theta \log n$ if $\alpha = 0 < \theta$, and grows like a random multiple of n^α if $0 < \alpha < 1$, where the distribution of the multiplier depends on θ. For fixed $\alpha \in (0, 1)$ the probability laws of (α, θ) partitions turn out to be mutually absolutely continuous as θ varies with $\theta > -\alpha$, and the Radon-Nikodym density is described.

3.4. A branching process construction of the two-parameter model
This section offers a construction of the two-parameter model in terms of a branching process in continuous time.

3.1. The Chinese restaurant process

Consistent random permutations Consider a sequence of random permutations $(\sigma_n, n = 1, 2, \cdots)$ such that

(i) σ_n is a uniformly distributed random permutation of $[n]$ for each n;

(ii) for each n, if σ_n is written as a product of cycles, then σ_{n-1} is derived from σ_n by deletion of element n from its cycle.

For example, using standard cycle notation for permutations,

if $\sigma_5 = (134)(25)$ then $\sigma_4 = (134)(2)$;
if $\sigma_5 = (134)(2)(5)$ then $\sigma_4 = (134)(2)$.

It is easily seen that these requirements determine a unique distribution for the sequence (σ_n), which can be described as follows.

An initially empty restaurant has an unlimited number of circular tables numbered $1, 2, \ldots$, each capable of seating an unlimited number of customers. Customers numbered $1, 2, \cdots$ arrive one by one and are seated at the tables according to the following:

Simple random seating plan Person 1 sits at table 1. For $n \geq 1$ suppose that n customers have already entered the restaurant, and are seated in some arrangement, with at least one customer at each of the tables j for $1 \leq j \leq k$ say, where k is the number of tables occupied by the first n customers to arrive. Let customer $n + 1$ choose with equal probability to sit at any of the following $n + 1$ places: to the left of customer j for some $1 \leq j \leq n$, or alone at table $k + 1$. Define $\sigma_n : [n] \to [n]$ as follows. If after n customers have entered the restaurant, customers i and j are seated at the same table, with i to the left of j, then $\sigma_n(i) = j$, and if customer i is seated alone at some table then $\sigma_n(i) = i$. The sequence (σ_n) then has features (i) and (ii) above by a simple induction.

Many asymptotic properties of uniform random permutations can be read immediately from this construction. For instance, the number of occupied tables after n customers have been seated is

$$K_n = \#\{\text{ cycles of } \sigma_n\} = Z_1 + \cdots + Z_n \tag{3.1}$$

where the Z_j is the indicator of the event that the jth customer is seated at a new table. By construction, the Z_j are independent Bernoulli$(1/j)$ variables, hence,

$$\frac{K_n}{\log n} \to 1 \text{ almost surely}, \qquad \frac{K_n - \log n}{(\log n)^{1/2}} \stackrel{d}{\to} B_1 \tag{3.2}$$

where B_1 is a standard Gaussian variable. This and other results about random permutations now recalled are well known.

Let Π_n be the partition of $[n]$ generated by the cycles of σ_n. Then Π_n is an exchangeable random partition of $[n]$, and the Π_n are consistent as n varies. Thus the sequence $\Pi_\infty := (\Pi_n)$ is an exchangeable random partition of $\mathbb{N}$. Let X_n be the indicator of the event that the $(n+1)$th customer sits at table 1. Then the sequence $(X_n)_{n\geq 1}$ is an exchangeable sequence which evolves by the dynamics of Pólya's urn scheme (2.15) with $a = b = 1$. Hence $S_n := X_1 + \cdots + X_n$ has uniform distribution on $\{0, 1, \ldots, n\}$. Equivalently, the size $S_n + 1$ of the cycle of σ_{n+1} containing 1 has uniform distribution on $\{1, \ldots, n+1\}$. The asymptotic frequency of the class of Π_∞ containing 1 is the almost sure limit of S_n/n, which evidently has uniform distribution on $[0,1]$.

The limit frequencies Let $(N_{n,1}, \ldots, N_{n,K_n})$ denote the sizes of blocks of Π_n, in order of least elements. In terms of the restaurant construction, $N_{n,i}$ is the number of customers seated at table i after n customers have been seated. From above, $N_{n,1}$ has uniform distribution on $[n]$. Similarly, given $N_{n,1} = n_1 < n$, $N_{n,2}$ has uniform distribution on $[n - n_1]$. And so on. Asymptotic behavior of this *discrete uniform stick-breaking scheme* is quite obvious: as $n \to \infty$, the relative frequencies $(N_{n,i}/n, i \geq 1)$ of the sizes of cycles of σ_n, which are in a size-biased random order, converge in distribution to the *continuous uniform stick-breaking sequence*

$$(\tilde{P}_1, \tilde{P}_2, \ldots) = (U_1,\ \overline{U}_1 U_2,\ \overline{U}_1 \overline{U}_2 U_3, \ldots)$$

where the U_i are independent uniform$[0,1]$ variables, and $\bar{U} := 1 - U$. By an obvious combinatorial argument, the corresponding infinite exchangeable partition probability function (EPPF), which gives for each n the probability that Π_n equals any *particular* partition of $[n]$ with n_i elements in the ith cycle, for some arbitrary ordering of cycles, is

$$p_{0,1}(n_1, \ldots, n_k) := \frac{1}{n!} \prod_{i=1}^{k} (n_i - 1)!. \tag{3.3}$$

Compare with (2.19) to see that this *continuous uniform stick-breaking sequence* $(\tilde{P}_1, \tilde{P}_2, \ldots)$ has the same distribution as a size-biased permutation of the jumps of the Dirichlet process with exchangeable increments

$$(\Gamma_u/\Gamma_1, 0 \leq u \leq 1)$$

where $(\Gamma_u, u \geq 0)$ is a gamma process. Since the limiting ranked frequencies $P_i^\downarrow$ are recovered from the $(\tilde{P}_j)$ by ranking, it follows that if Γ_1 is a standard exponential variable independent of the limiting ranked frequencies $P_i^\downarrow$ defined by the Chinese restaurant construction of random permutations, then

$$\Gamma_1 P_1^\downarrow > \Gamma_1 P_2^\downarrow > \Gamma_1 P_3^\downarrow > \cdots > 0$$

are the ranked points of a Poisson point process whose intensity measure $x^{-1}e^{-x}dx$ on $(0, \infty)$ is the Lévy measure of the gamma process. This allows calculation of moments of the $P_i^\downarrow$. For instance

$$\mathbb{E}\#\{i : \Gamma_1 P_i^\downarrow > y\} = E_1(y) := \int_y^\infty x^{-1} e^{-x} dx.$$

So as $n \to \infty$ the asymptotic mean fraction of elements in the longest cycle of a uniform random permutation of $[n]$ is

$$\mathbb{E}(P_1^{\downarrow}) = \mathbb{E}(\Gamma_1)\mathbb{E}(P_1^{\downarrow}) = \mathbb{E}(\Gamma_1 P_1^{\downarrow}) = \int_0^{\infty} (1 - e^{-E_1(x)})dx.$$

This technique of *random scaling* to simplify the probabilistic structure of random partitions has many other applications. See for instance [85, 372, 374, 24]. The distribution of $(P_1^{\downarrow}, P_2^{\downarrow}, \ldots)$, constructed here from random permutations using the Chinese restaurant process, is known as the *Poisson-Dirichlet distribution* with parameter 1. Some references: Shepp-Lloyd [400], Vershik-Shmidt [422, 423], Flajolet-Odlyzko [156], Arratia-Barbour-Tavaré [27].

Generalization The Chinese restaurant construction is easily generalized to allow construction of a sequence of random permutations σ_n of $[n]$ such that the associated sequence of random partitions $\Pi_\infty := (\Pi_n)$ is the most general possible exchangeable random partition of integers, as discussed in Section 2.2. Recall that the corresponding exchangeable partition probability function (EPPF) $p(n_1, \ldots, n_k)$ gives for each $(n_1, \ldots, n_k)$ the the probability that Π_n equals any specific partition of $[n]$ into sets of sizes $(n_1, \ldots, n_k)$. In terms of the Chinese restaurant, the permutation σ_n is thought of as a configuration of n customers seated at K_n tables, where K_n is the number of cycles of σ_n. For present purposes, we only care about the random partition Π_n induced by the cycles of σ_n. So for $1 \leq i \leq K_n$ the statement "customer $n+1$ is placed at occupied table i" means Π_{n+1} is the partition of $[n+1]$ whose restriction to $[n]$ is Π_n, with $n+1$ belonging to the ith class of Π_n. Similarly "customer $n+1$ is placed at a new table" means Π_{n+1} is the partition of $[n+1]$ whose restriction to $[n]$ is Π_n, with $\{n+1\}$ a singleton block. Given an infinite EPPF $p(n_1, \ldots, n_k)$, a corresponding exchangeable random partition of $\mathbb{N}$ (Π_n) can thus be constructed as follows.

Random seating plan for an exchangeable partition The first customer is seated at the first table, that is $\Pi_1 = \{1\}$. For $n \geq 1$, given the partition Π_n, regarded as a placement of the first n customers at tables of the Chinese restaurant, with k occupied tables, the next customer $n+1$ is

- placed at occupied table j with probability $p(\ldots, n_j + 1, \ldots)/p(n_1, \ldots, n_k)$
- placed at new table with probability $p(n_1, \ldots, n_k, 1)/p(n_1, \ldots, n_k)$

In particular, it is clear that a simple product form for the EPPF will correspond to a simple prescription of these conditional probabilities. But before discussing specific examples, it is worth making some more general observations. Any sequential seating plan for the Chinese restaurant, corresponding to a *prediction rule* for the conditional distribution of Π_{n+1} given Π_n for each n, whereby $n+1$ is either assigned to one of the existing blocks of Π_n or declared to be a singleton block of Π_{n+1}, can be used to construct a random partition $\Pi_\infty := (\Pi_n)$ of the positive integers. Most seating plans will fail to produce a Π_∞ that is exchangeable. But it is instructive to experiment with simple plans to see which ones do generate exchangeable partitions. According to Kingman's theory of exchangeable random partitions described in Section 2.2, a necessary condition for Π_∞ to be exchangeable is that for each i there exists an almost sure limiting frequency $\tilde{P}_i$ of customers seated at table i. More formally, this is the limit frequency of the ith block of Π_∞ when blocks are put in order of appearance. The simplest way to achieve this is to consider the following:

Random seating plan for a partially exchangeable partition Let $(P_i, i = 1, 2, \dots)$ be an arbitrary sequence of random variables with $P_i \geq 0$ and $\sum_i P_i \leq 1$. Given the entire sequence $(P_i, i = 1, 2, \dots)$ let the first customer be seated at the first table, and for $n \geq 1$, given the partition Π_n, regarded as a placement of the first n customers at tables of the Chinese restaurant, with k occupied tables, let the next customer $n + 1$ be

- placed at occupied table j with probability P_j
- placed at new table with probability $1 - \sum_{i=1}^{k} P_i$

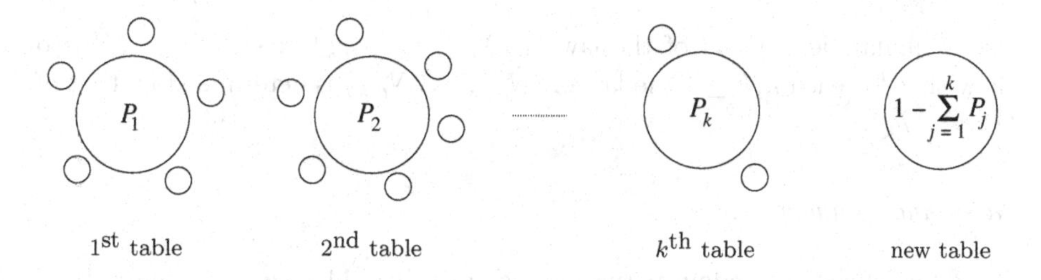

Figure 3.1: Chinese Restaurant Process with random seating plan.

By construction and the law of large numbers, for each i the limiting frequency of customers seated at table i exists and equals P_i. Moreover, by conditioning on the entire sequence P_i, the probability that Π_n equals any specific partition of $[n]$ into sets of sizes $(n_1, \dots, n_k)$, in order of least elements, is given by the formula

$$p(n_1, \cdots, n_k) = \mathbb{E}\left[\prod_{i=1}^{k} P_i^{n_i - 1} \prod_{i=1}^{k-1}\left(1 - \sum_{j=1}^{i} P_j\right)\right] \tag{3.4}$$

Such a random partition of $[n]$ is called *partially exchangeable* [347]. These considerations lead to the following variation of Kingman's representation:

Theorem 3.1. [347] *Let (P_i) be a sequence of random variables with $P_i \geq 0$ and $\sum_i P_i \leq 1$, and let $p(n_1, \dots, n_k)$ be defined by in formula* (3.4).
(i) *There exists an exchangeable random partition Π_∞ of $\mathbb{N}$ whose block frequencies in order of appearance $(\tilde{P}_i)$ are distributed like (P_i) if and only if the function $p(n_1, \dots, n_k)$ is a symmetric function of $(n_1, \dots, n_k)$ for each k.*
(ii) *If Π_∞ is such an exchangeable random partition of $\mathbb{N}$ with block frequencies $(\tilde{P}_i)$, then the EPPF of Π_∞ is $p(n_1, \dots, n_k)$ defined by* (3.4) *for $P_i = \tilde{P}_i$, and*

the conditional law of Π_∞ given $(\tilde{P}_i)$ is governed by the random seating plan for a partially exchangeable partition, described above.

Proof. The "if" part of (i) is read from the preceding argument. See [347] for the "only if" part of (i). Granted that, part (ii) follows easily. □

Exercises

3.1.1. Let $\Pi_\infty := (\Pi_n)$ be an infinite exchangeable (or partially exchangeable) random partition, with $\tilde{N}_{n,i}$ the number of elements of $[n]$ in the ith class of Π_∞ to appear, and $\tilde{P}_i := \lim_n \tilde{N}_{n,i}/n$. The conditional distribution of $\tilde{N}_{n,1} - 1$ given $\tilde{P}_1$ is binomial$(n-1, \tilde{P}_1)$, hence the distribution of $\tilde{N}_{n,1}$ is determined by that of $\tilde{P}_1$ via

$$\mathbb{P}(\tilde{N}_{n,1} = j) = \binom{n-1}{j-1} \mathbb{E}\left[\tilde{P}_1^{j-1}(1-\tilde{P}_1)^{n-j}\right] \quad (1 \le j \le n).$$

Use a similar description of the law of $(\tilde{N}_{n,1}, \ldots, \tilde{N}_{n,k})$ given $(\tilde{P}_1, \ldots, \tilde{P}_k)$ to show that for each $n, k \geq 1$ the law of $(\tilde{N}_{n,1}, \ldots, \tilde{N}_{n,k})$ is determined by that of $(\tilde{P}_1, \ldots, \tilde{P}_k)$.

Notes and comments

Basic references on random permutations are Feller [148] and Goncharov [177] from the 1940's. There is a nice bijection between the structure of records and cycles. For this and more see papers by Ignatov [207, 206], Rényi, Goldie [175], Stam [403, 405]. The fact that the cycle structure of uniform random permutations is consistent as n varies was pointed out by Greenwood [179]. Lester Dubins and I devised the Chinese Restaurant Process in the early 1980's as a way of constructing consistent random permutations and random partitions. The notion first appears in print in [14, (11.19)]. See also Joyce and Tavaré [224], and Arratia, Barbour and Tavaré [27] for many further results and references. The Chinese Restaurant Process and associated computations with random partitions have found applications in Bayesian statistics [109, 287, 210, 212], and in the theory of representations of the infinite symmetric group [243].

3.2. The two-parameter model

The EPPF's calculated in (2.17) and (2.19) suggest the following seating plan for the Chinese restaurant construction of a random partition of $\mathbb{N}$, say $\Pi_\infty := (\Pi_n)$, starting from $\Pi_1 := \{1\}$.

(α, θ) seating plan [347] Given at stage n there are k occupied tables, with n_i customers at the ith table, let the next customer be • placed at occupied table i with probability $(n_i - \alpha)/(n + \theta)$,

• placed at new table with probability $(\theta + k\alpha)/(n + \theta)$.

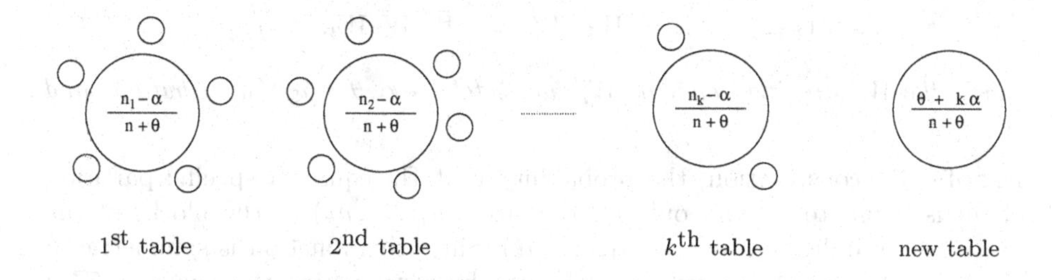

Figure 3.2: Chinese Restaurant Process with (α, θ) seating plan.

To satisfy the rules of probability it is necessary to suppose that

$$\begin{array}{ll} \bullet \text{ either} & \alpha = -\kappa < 0 \text{ and } \theta = m\kappa \text{ for some } m = 1, 2, \ldots \\ \bullet \text{ or} & 0 \le \alpha \le 1 \text{ and } \theta > -\alpha. \end{array} \tag{3.5}$$

Case ($\alpha = -\kappa < 0$ and $\theta = m\kappa$, for some $m = 1, 2, \ldots$) Compare the (α, θ) seating plan with Exercise 2.2.5 to see that in this case Π_∞ is distributed as if by sampling from a symmetric Dirichlet distribution with m parameters equal to κ. This can also be seen by comparison of (2.17) and (3.6) below.

Case ($\alpha = 0$ and $\theta > 0$) This is the weak limit of the previous case as $\kappa \to 0$ and $m\kappa \to \theta$. By consideration of this weak limit, or by the Blackwell-MacQueen urn scheme (2.18), such a Π_∞ is distributed as if by sampling from a Dirichlet process with parameter θ.

Case ($\alpha = 0$ and $\theta = 1$) This instance of the previous case corresponds to Π_∞ generated by the cycles of a consistent sequence of uniform random permutations, as in the previous section.

Case ($0 < \alpha < 1$ and $\theta > -\alpha$) This case turns out to be related to the stable subordinator of index α, as will be explained in detail in Section 4.2.

Theorem 3.2. [347] *For each pair of parameters (α, θ) subject to the constraints above, the Chinese restaurant with the (α, θ) seating plan generates an exchangeable random partition Π_∞ of $\mathbb{N}$. The corresponding EPPF is*

$$p_{\alpha,\theta}(n_1, \ldots, n_k) = \frac{(\theta + \alpha)_{k-1\uparrow\alpha} \prod_{i=1}^k (1 - \alpha)_{n_i - 1\uparrow 1}}{(\theta + 1)_{n-1\uparrow 1}} \tag{3.6}$$

where

$$(x)_{n\uparrow\alpha} := \prod_{i=0}^{n-1}(x + i\alpha). \tag{3.7}$$

The corresponding limit frequencies of classes, in size-biased order of least elements, can be represented as

$$(\tilde{P}_1, \tilde{P}_2, \ldots) = (W_1,\ \overline{W}_1 W_2,\ \overline{W}_1\overline{W}_2 W_3,\ \ldots) \tag{3.8}$$

where the W_i are independent, W_i has beta$(1-\alpha, \theta + i\alpha)$ distribution, and $\overline{W}_i := 1 - W_i$.

Proof. By construction, the probability that Π_n equals a specific partition of $[n]$ is found to depend only on the sizes $(n_1, \ldots, n_k)$ of the blocks of the partition, as indicated by $p_{\alpha,\theta}(n_1, \ldots, n_k)$. Since this function is symmetric in $(n_1, \ldots, n_k)$, each Π_n is exchangeable, and by construction the sequence (Π_n) is consistent. So $\Pi := (\Pi_n)$ is an exchangeable random partition of $\mathbb{N}$. The joint law of the W_i can be identified either using formula (3.4), or by repeated application of the beta-binomial relation described around (2.15) . □

Definition 3.3. (GEM and PD distributions) For (α, θ) subject to the constraints (3.5), call the distribution of size-biased frequencies $(\tilde{P}_j)$, defined by the residual allocation model (3.8), the *Griffiths-Engen-McCloskey distribution with parameters* (α, θ), abbreviated GEM(α, θ). Call the corresponding distribution on $\mathcal{P}^{\downarrow}_{[0,1]}$ of ranked frequencies $(P^{\downarrow}_i)$ of an (α, θ) partition, obtained by ranking $(\tilde{P}_j)$ with GEM(α, θ) distribution, the *Poisson-Dirichlet distribution with parameters* (α, θ), abbreviated PD(α, θ).

Explicit but complicated formulae are known for the joint density of the first j coordinates of a PD(α, θ) distributed sequence [371], but these formulae are of somewhat limited use.

Characterizations of the two-parameter scheme. The closure of the two-parameter family of models consists of the original two-parameter family subject to the constraints on (α, θ) discussed above, plus the following models:

- the degenerate case with Π_n the partition of singletons for all n; this arises for $\alpha = 1$ and as the weak limit of (α, θ) partitions as $\theta \to \infty$ for any fixed $\alpha \in [0, 1)$.
- for each $m = 1, 2, \ldots$ the coupon collectors partition (2.16) defined by m frequencies identically equal to $1/m$; this is the weak limit of $(-\kappa, m\kappa)$ partitions as $\kappa \to \infty$ for fixed m.
- for each $0 \le p \le 1$ the mixture with weights p and $1 - p$ of the one block partition and the partition into singletons. As observed by Kerov [240, (1.10)], this limit is obtained as $\alpha \to 1$ and $\theta \to -1$ with $(1-\alpha)/(1+\theta) \to p$. The cases $p = 0$ and $p = 1$ arise also as indicated just above.

Theorem 3.4. [240, 349] *Suppose that an exchangeable random partition* Π_∞ *of* $\mathbb{N}$ *has block frequencies* $\tilde{P}_j$ *(in order of least elements) such that* $0 < \tilde{P}_1 < 1$ *almost surely, and either*
(i) *The restriction* Π_n *of* Π_∞ *to* $[n]$ *is a* $Gibbs_{[n]}(v_\bullet, w_\bullet)$ *partition, meaning its EPPF is of the product form* (1.48), *for some pair of non-negative sequences* $v_\bullet$ *and* $w_\bullet$, *or*
(ii) *the frequencies* $\tilde{P}_j$ *are of the product form* (3.8) *for some independent random variables* W_i.
Then the distribution of Π_∞ *is either that determined by* (α, θ) *model for some* (α, θ), *or that of a coupon collectors partition, for some* $m = 2, 3, \ldots$.

Proof. Assuming (i), the form of the EPPF is forced by elementary arguments using addition rules of an EPPF [240]. Assuming (ii), the form of the distribution of the W_i is forced by symmetry of the EPPF and the formula (3.4). See [349] for details. □

See also Zabell [441] for closely related characterizations by the simple form of the prediction rule for (Π_n) defined by the (α, θ) seating plan. Note in particular the following consequence of the previous theorem:

Corollary 3.5. McCloskey [302]. *An exchangeable random partition* Π_∞ *of* $\mathbb{N}$ *has block frequencies* $\tilde{P}_j$ *(in order of least elements) of the product form* (3.8) *for some sequence of independent and identically distributed random variables* W_i *with* $0 < W_i < 1$ *if and only if the common distribution of the* W_i *is beta*$(1, \theta)$ *for some* $\theta > 0$, *in which case* Π_∞ *is generated by the* $(0, \theta)$ *model.*

This result of McCloskey is easily transformed into another characterization of the $(0, \theta)$ model due to Kingman. The following formulation is adapted from Aldous [14, p. 89].

Corollary 3.6. *Let* Π_∞ *be an exchangeable random partition of* $\mathbb{N}$. *The distribution of* Π_∞ *is governed by the* $(0, \theta)$ *model iff for each pair of integers* i *and* j, *the probability that* i *and* j *belong to the same component of* Π_∞ *is* $1/(1+\theta)$, *and* Π_∞ *has the following further property: for each pair of non-empty disjoint finite sets of positive integers* A *and* B, *the event that* A *is a block of the restriction of* Π_∞ *to* $A \cup B$ *is independent of the restriction of* Π_∞ *to* B.

Proof. That the $(0, \theta)$ model satisfies the independence condition is evident from the form of its EPPF. Conversely, in terms of the general Chinese restaurant construction, the exchangeability of Π_∞ plus the independence condition means that the process of seating customers at tables $2, 3, \ldots$, watched only when customers are placed at one of these tables, can be regarded in an obvious way as a copy of the original process of seating customers at tables $1, 2, 3, \ldots$, and that this copy of the original process is independent of the sequence of times at which customers are seated at table 1. It follows that if the block frequencies $(\tilde{P}_j)$ of Π_∞ are represented in the product form (3.8), then the asymptotic frequency $\tilde{P}_1 = W_1$ of customers arriving at table 1 is independent of the sequence $(W_2, W_3, \ldots)$ governing the relative frequencies of arrivals at tables $2, 3, \ldots$, and that $(W_2, W_3, \ldots) \stackrel{d}{=} (W_1, W_2, \ldots)$. So the W_i are i.i.d. and the conclusion follows from Corollary 3.5. □

Problem 3.7. *Suppose an exchangeable random partition Π_∞ has block frequencies $(\tilde{P}_i)$ such that $0 < \tilde{P}_i < 1$ and $\tilde{P}_1$ is independent of the sequence $(\tilde{P}_i/(1-\tilde{P}_1), i \geq 2)$. Is Π_∞ necessarily some (α, θ) partition?*

Exercises

3.2.1. (Deletion of Classes.) Given a random partition Π_∞ of $\mathbb{N}$ with infinitely many classes, for each $k = 0, 1, \cdots$ let $\Pi_\infty(k)$ be the partition of $\mathbb{N}$ derived from Π_∞ by deletion of the first k classes. That is, first let $\Pi'_\infty(k)$ be the restriction of Π_∞ to $H_k := \mathbb{N} - G_1 - \cdots - G_k$ where $G_1, \cdots G_k$ are the first k classes of Π_∞ in order of appearance, then derive $\Pi_\infty(k)$ on $\mathbb{N}$ from $\Pi'_\infty(k)$ by renumbering the points of H_k in increasing order. The following are equivalent:
(i) for each k, $\Pi_\infty(k)$ is independent of the frequencies $(\tilde{P}_1, \cdots, \tilde{P}_k)$ of the first k classes of Π_∞;
(ii) Π_∞ is an (α, θ)-partition for some $0 \leq \alpha < 1$ and $\theta > -\alpha$, in which case $\Pi_\infty(k)$ is an $(\alpha, \theta + k\alpha)$-partition.

3.2.2. (Urn scheme for a $(\frac{1}{2}, 0)$ partition) Let an urn initially contain two balls of different colors. Draw 1 is a simple draw from the urn with replacement. Thereafter, balls are drawn from the urn, with replacement of the ball drawn, and addition of two more balls as follows. If the ball drawn is of a color never drawn before, it is replaced together with two additional balls of two distinct new colors, different to the colors of balls already in the urn. Whereas if the ball drawn is of a color that has been drawn before, it is replaced together with two balls of its own color. Let Π_n be the partition of $[n]$ generated by the colors of the first n draws from the urn. Then $\Pi_\infty := (\Pi_n)$ is a $(\frac{1}{2}, 0)$ partition.

3.2.3. (Number of blocks) Let $\mathbb{P}_{\alpha,\theta}$ govern $\Pi_\infty = (\Pi_n)$ as an (α, θ) partition, for some (α, θ) subject to the constraints (3.5). Let K_n be the number of blocks of Π_n:

$$K_n := |\Pi_n| = \sum_{j=1}^{n} |\Pi_n|_j = \sum_{i=1}^{n} X_i$$

where $|\Pi_n|_j$ is the number of blocks of Π_n of size j, and X_i is the indicator of the event that i is the least element of some block of Π_∞ (customer i sits at an unoccupied table). Under $\mathbb{P}_{\alpha,\theta}$ the sequence $(K_n)_{n\geq 1}$ is a Markov chain, starting at $K_1 = 1$, with increments in $\{0, 1\}$, and inhomogeneous transition probabilities

$$\mathbb{P}_{\alpha,\theta}(K_{n+1} = k+1 \,|\, K_1, \ldots, K_n = k) = \frac{k\alpha + \theta}{n + \theta} \tag{3.9}$$

$$\mathbb{P}_{\alpha,\theta}(K_{n+1} = k \,|\, K_1, \ldots, K_n = k) = \frac{n - k\alpha}{n + \theta}. \tag{3.10}$$

The distribution of K_n is given by

$$\mathbb{P}_{\alpha,\theta}(K_n = k) = \frac{(\theta+\alpha)_{k-1\uparrow\alpha}}{(\theta+1)_{n-1\uparrow}} S_\alpha(n,k) \tag{3.11}$$

where

$$S_\alpha(n,k) := B_{n,k}((1-\alpha)_{\bullet-1\uparrow}) = S_{n,k}^{-1,-\alpha} \tag{3.12}$$

is a generalized Stirling number of the first kind, as in (1.20). The expected value of K_n is

$$\mathbb{E}_{\alpha,\theta}(K_n) = \sum_{i=1}^{n} \frac{(\theta+\alpha)_{i-1\uparrow}}{(\theta+1)_{i-1\uparrow}} = \begin{cases} \displaystyle\sum_{i=1}^{n} \frac{\theta}{\theta+i-1} & \text{if } \alpha = 0 \\ \displaystyle\frac{(\theta+\alpha)_{n\uparrow}}{\alpha(\theta+1)_{n-1\uparrow}} - \frac{\theta}{\alpha} & \text{if } \alpha \neq 0. \end{cases} \tag{3.13}$$

3.2.4. (Serban Nacu [318]) **(Independent indicators of new blocks)** Compare with Exercise 2.1.4 and Exercise 4.3.4 . Let X_i be the indicator of the event that i is the least element of some block of an exchangeable random partition Π_n of $[n]$. Show that the $X_i, 1 \le i \le n$ are independent if and only if Π_n is a $(0,\theta)$ partition of $[n]$ for some $\theta \in [0,\infty]$, with the obvious definition by continuity in the two endpoint cases.

3.2.5. (Equilibrium of a coagulation/fragmentation chain)[301, 110, 361, 169] Let $\mathcal{P}_1^\downarrow$ be the space of real partitions of 1. Define a Markov kernel Q on $\mathcal{P}_1^\downarrow$ as follows. For $p = (p_i) \in \mathcal{P}_1^\downarrow$, let I and J be independent and identically distributed according to p. If $I = J$ then replace p_I by two parts $p_I U$ and $p_I(1-U)$ where U is uniform$(0,1)$ independent of I, J, and rerank, but if $I \neq J$ then replace the two parts p_I and p_J by a single part $p_I + p_J$, and rerank.

- (a) Show that PD$(0,1)$ is a Q-invariant measure.
- (b) [110] (hard). Show PD$(0,1)$ is the unique Q-invariant measure.
- (c) Modify the transition rule so that PD$(0,\theta)$ is an invariant measure.
- (d)(Open problem) Show that PD$(0,\theta)$ is the unique invariant measure for the modified rule.
- (d)(Open problem) Define some kind of coagulation/fragmentation kernel for which PD(α,θ) is an invariant measure.

3.2.6. The probabilities $q_{\alpha,\theta}(n,k) := P_{\alpha,\theta}(K_n = k)$ can be computed recursively from the forwards equations

$$q_{\alpha,\theta}(n+1,k) = \frac{n-k\alpha}{n+\theta} q_{\alpha,\theta}(n,k) + \frac{\theta+(k-1)\alpha}{n+\theta} q_{\alpha,\theta}(n,k-1),\ 1 \le k \le n. \tag{3.14}$$

and the boundary cases

$$q_{\alpha,\theta}(n,1) = \frac{(1-\alpha)_{n-1\uparrow}}{(\theta+1)_{n-1\uparrow}};\ q_{\alpha,\theta}(n,n) = \frac{(\theta+\alpha)_{n-1\uparrow\alpha}}{(\theta+1)_{n-1\uparrow\alpha}} \tag{3.15}$$

For instance, the distribution of K_3 is as shown in the following table:

k	1	2	3
$\mathbb{P}_{\alpha,\theta}(K_3 = k)$	$\frac{(1-\alpha)(2-\alpha)}{(\theta+1)(\theta+2)}$	$\frac{3(1-\alpha)(\theta+\alpha)}{(\theta+1)(\theta+2)}$	$\frac{(\theta+\alpha)(\theta+2\alpha)}{(\theta+1)(\theta+2)}$

3.2.7. Take $\theta = 0$ and use (3.11) to obtain a recursion for the $S_\alpha(n,k)$:

$$S_\alpha(n,1) = (1-\alpha)_{n-1\uparrow};\ S_\alpha(n,n) = 1 \tag{3.16}$$

$$S_\alpha(n+1,k) = (n-k\alpha)S_\alpha(n,k) + S_\alpha(n,k-1). \tag{3.17}$$

Toscano [415] used this recursion as his primary definition of these numbers, and obtained from it the formula

$$S_\alpha(n,k) = \frac{1}{k!}\Delta^k_{\alpha,x}(x)_{n\uparrow}\Bigg|_{x=0} \tag{3.18}$$

where $\Delta^k_{\alpha,x}$ is the kth iterate of the operator $\Delta_{\alpha,x}$, defined by $\Delta_{0,x} = \frac{d}{dx}$ (Jordan[222]) and for $\alpha \neq 0$,

$$(\Delta_{\alpha,x}f)(x) = \frac{f(x) - f(x-\alpha)}{\alpha}.$$

Check Toscano's formula

$$S_\alpha(n,k) = \frac{1}{\alpha^k k!}\sum_{j=1}^{k}(-1)^j\binom{k}{j}(-k\alpha)_{n\uparrow} \quad (\alpha \neq 0). \tag{3.19}$$

3.2.8. [132] Deduce the formula (3.13) for $\mathbb{E}_{\alpha,\theta}(K_n)$ by integration from the general formula (2.27) and the beta$(1-\alpha,\theta+\alpha)$ distribution of the frequency $\tilde{P}_1$ of the first block.

3.2.9. For real $p > 0$ let $[k]_p := \Gamma(k+p)/\Gamma(k)$ so that $[k]_p = (k)_{p\uparrow}$ for $p = 1, 2, \ldots$. For $0 < \alpha < 1$, and all real $p > 0$,

$$\mathbb{E}_{\alpha,0}[K_n]_p = \frac{\Gamma(p)[p\alpha]_n}{\Gamma(n)\alpha}. \tag{3.20}$$

3.2.10. Let $\Pi_\infty := (\Pi_n)$ be an exchangeable random partition of $\mathbb{N}$, with ranked frequencies denoted simply (P_j) instead of $(P_j^\downarrow)$. Let p be the EPPF of Π_∞, and let q be derived from p by (2.8).

- There is the formula

$$q(n_1,\ldots,n_k) = \mathbb{E}\left[\prod_{i=1}^{k}\left(\sum_{j=1}^{\infty}P_j^{n_i}\right)\right] \tag{3.21}$$

without further qualification if $\sum_j P_j = 1$ a.s., and with the qualification if $\mathbb{P}(\sum_j P_j < 1) > 0$ that $n_i \geq 2$ for all i.

- For each fixed $a > 0$, the distribution of (Π_n) on $\mathcal{P}_{\mathbb{N}}$, and that of (P_j) on $\mathcal{P}_1^{\downarrow}$, is uniquely determined by the values of $p(n_1, \ldots, n_k)$ for $n_i \geq a$ for all a. Similarly for q instead of p.
- (Π_n) is an (α, θ) partition, or equivalently (P_j) has $\mathrm{PD}(\alpha, \theta)$ distribution, iff p satisfies the recursion

$$p(n_1 + 1, \ldots, n_k) = \frac{n_1 - \alpha}{n + \theta} p(n_1, \ldots, n_k). \tag{3.22}$$

  Note that p is subject also to the constraints of an EPPF, that is symmetry, the addition rule, and $P(1) = 1$. These constraints and (3.22) imply $p = p_{(\alpha,\theta)}$ as in (3.6).
- (Π_n) is an (α, θ) partition, or equivalently (P_j) has $\mathrm{PD}(\alpha, \theta)$ distribution, iff q satisfies the recursion

$$q(n_1 + 1, \ldots, n_k) = \frac{n_1 - \alpha}{n + \theta} q(n_1, \ldots, n_k) + \sum_{s=2}^{n} q(n_1 + n_s, \ldots, n_k) \tag{3.23}$$

  where the number of arguments of $q(n_1 + 1, \ldots, n_k)$ and $q(n_1, \ldots, n_k)$ is k, and the number of arguments of $q(n_1 + n_s, \ldots, n_k)$ is $k - 1$, with n_s the missing argument. Note that q is subject also to the a priori constraints of symmetry, and $q(1, \ldots, 1) \equiv 1$. These constraints and (3.23) imply that $q = q_{(\alpha,\theta)}$ is given by formula (2.8) for $p = p_{(\alpha,\theta)}$ as in (3.6). There does not appear to be any simpler formula for $q_{(\alpha,\theta)}$.

In the case $\theta = 0$, the recursion (3.23) for $q = q_{(\alpha,0)}$ was derived by Talagrand [412, Proposition 1.2.2], using relations of Ghirlanda-Guerra [166] in the context of Derrida's random energy model [105] in the theory of spin glasses. The appearance of $\mathrm{PD}(\alpha, 0)$ in that setting is explained in Exercise 4.2.1 . Once the parallel between (3.22) and (3.23) has been observed for $\theta = 0$, the result for general θ is easily guessed, and can be verified algebraically using (2.8). The identities (2.8) and (3.22) have a transparent probabilistic meaning, the latter in terms of the Chinese Restaurant Process. Can (3.23) too be understood without calculation in some setting? Does (3.23) or $\mathrm{PD}(\alpha, \theta)$ have an interpretation in terms of spin glass theory for $\theta \neq 0$?

3.3. Asymptotics

The asymptotic properties of (α, θ) partitions of $[n]$ for large n depend on whether α is negative, 0, or positive. Recall the notations $K_n := |\Pi_n|$ for the number of blocks of Π_n, and $|\Pi_n|_j$ for the number of blocks of Π_n of size j. So

$$K_n := |\Pi_n| = \sum_{j=1}^{n} |\Pi_n|_j$$

Case $(\alpha < 0)$. Then $\theta = -m\alpha$ for some positive integer m, and $K_n = m$ for all sufficiently large n almost surely.

Case ($\alpha = 0$). Immediately from the prediction rule, for a $(0,\theta)$ partition, the X_i are independent Bernoulli$(\theta/(\theta+i-1)$ variables. Hence [263]

$$\lim_{n\to\infty} \frac{K_n}{\log n} = \theta, \quad \text{a.s. } \mathbb{P}_{0,\theta} \text{ for every } \theta > 0. \tag{3.24}$$

Moreover, it follows easily from Lindeberg's theorem that the $\mathbb{P}_{0,\theta}$ distribution of $(K_n - \theta \log n)/\sqrt{\theta \log n}$ converges to standard normal as $n \to \infty$. By consideration of the Ewens sampling formula (2.20), for each fixed k

$$\{(|\Pi_n|_j, j \geq 1); \mathbb{P}_{0,\theta}\} \xrightarrow{d} (Z_{\theta,j}, j \geq 1) \tag{3.25}$$

meaning that under $\mathbb{P}_{0,\theta}$ which governs Π_∞ as a $(0,\theta)$ partition, the finite dimensional distributions of the counts $(|\Pi_n|_j, j \geq 1)$ converge without normalization to those of $(Z_{\theta,j}, j \geq 1)$, where the $Z_{\theta,j}$ are independent Poisson variables with parameters θ/j. See [27] for various generalizations and refinements of these results.

Case ($0 < \alpha < 1$). Now K_n is a sum of dependent indicators X_i. It is easily seen from (3.13) and Stirling's formula that

$$\mathbb{E}_{\alpha,\theta} K_n \sim \frac{\Gamma(\theta+1)}{\alpha\Gamma(\theta+\alpha)} n^\alpha$$

which indicates the right normalization for a limit law.

Theorem 3.8. *For $0 < \alpha < 1$, $\theta > -\alpha$, under $\mathbb{P}_{\alpha,\theta}$ as $n \to \infty$,*

$$K_n/n^\alpha \to S_\alpha \text{ almost surely} \tag{3.26}$$

and in pth mean for every $p > 0$, for a strictly positive random variable S_α, with continuous density

$$\frac{d}{ds}\mathbb{P}_{\alpha,\theta}(S_\alpha \in ds) = g_{\alpha,\theta}(s) := \frac{\Gamma(\theta+1)}{\Gamma(\frac{\theta}{\alpha}+1)} s^{\frac{\theta}{\alpha}} g_\alpha(s) \qquad (s > 0) \tag{3.27}$$

where $g_\alpha = g_{\alpha,0}$ is the Mittag-Leffler density (0.43) *of the $\mathbb{P}_{\alpha,0}$ distribution of S_α, whose pth moment is $\Gamma(p+1)/\Gamma(p\alpha+1)$.*

Proof. Fix $\alpha \in (0,1)$. Let $\mathcal{F}_n$ be the field of events generated by Π_n. The formula (3.6) for the EPPF of Π_n under $\mathbb{P}_{\alpha,\theta}$ gives the likelihood ratio

$$M_{\alpha,\theta,n} := \left.\frac{d\mathbb{P}_{\alpha,\theta}}{d\mathbb{P}_{\alpha,0}}\right|_{\mathcal{F}_n} = \frac{f_{\alpha,\theta}(K_n)}{f_{1,\theta}(n)} \tag{3.28}$$

where for $\theta > -\alpha$

$$f_{\alpha,\theta}(k) := \frac{(\theta+\alpha)_{k-1\uparrow\alpha}}{(\alpha)_{k-1\uparrow\alpha}} = \frac{\Gamma(\frac{\theta}{\alpha}+k)}{\Gamma(\frac{\theta}{\alpha}+1)\Gamma(k)} \sim \frac{k^{\theta/\alpha}}{\Gamma(\frac{\theta}{\alpha}+1)} \text{ as } k \to \infty. \tag{3.29}$$

Thus, for each $\theta > -\alpha$,

$$(M_{\alpha,\theta,n}, \mathcal{F}_n; n = 1, 2, \ldots) \text{ is a positive } \mathbb{P}_{\alpha,0}\text{-martingale.}$$

By the martingale convergence theorem $M_{\alpha,\theta,n}$ has a limit $M_{\alpha,\theta}$ almost surely $(\mathbb{P}_{\alpha,0})$. Theorem 3.2 shows that Π_∞ has infinitely many blocks with strictly positive frequencies, and hence $K_n \to \infty$ almost surely $(\mathbb{P}_{\alpha,0})$ so (3.29) gives

$$M_{\alpha,\theta,n} \sim \frac{\Gamma(\theta+1)}{\Gamma(\frac{\theta}{\alpha}+1)} \left(\frac{K_n}{n^\alpha}\right)^{\theta/\alpha} \quad \text{almost surely} (\mathbb{P}_{\alpha,0}) \tag{3.30}$$

Moreover the ratio of the two sides in (3.30) is bounded away from 0 and ∞. Using (3.20), it follows that for each $\theta > -\alpha$, the martingale $M_{\alpha,\theta,n}$ is bounded in $L^p(\mathbb{P}_{\alpha,0})$, hence convergent in $L^p(\mathbb{P}_{\alpha,0})$ to $M_{\alpha,\theta}$ for every $p > 1$. Hence

$$\mathbb{E}_{\alpha,0} M_{\alpha,\theta} = 1. \tag{3.31}$$

But also by (3.30),

$$\frac{\Gamma(\theta+1)}{\Gamma(\frac{\theta}{\alpha}+1)} \left(\frac{K_n}{n^\alpha}\right)^{\theta/\alpha} \to M_{\alpha,\theta} \;=\; \frac{\Gamma(\theta+1)}{\Gamma(\theta/\alpha+1)} S_\alpha^{\theta/\alpha} \tag{3.32}$$

$\mathbb{P}_{\alpha,0}$ almost surely and in L^p, where $S_\alpha := M_{\alpha,\alpha}/\Gamma(\alpha+1)$. Now (3.31) and (3.32) yield the moments of the $\mathbb{P}_{\alpha,0}$ distribution of S. Since these are the moments (0.42) of the Mittag-Leffler distribution, the conclusions of the theorem in case $\theta = 0$ are evident. The corresponding results for $\theta > 0$ follow immediately from the results for $\theta = 0$, due to the following corollary of the above argument. □

Corollary 3.9. *Let $\mathbb{P}_{\alpha,\theta}$ denote the distribution on $\mathcal{P}_\mathbb{N}$ of an (α,θ)-partition $\Pi_\infty := (\Pi_n)$. For each $0 < \alpha < 1$, $\theta > -\alpha$, the laws $\mathbb{P}_{\alpha,\theta}$ and $\mathbb{P}_{\alpha,0}$ are mutually absolutely continuous, with density*

$$\frac{d\mathbb{P}_{\alpha,\theta}}{d\mathbb{P}_{\alpha,0}} = \frac{\Gamma(\theta+1)}{\Gamma(\frac{\theta}{\alpha}+1)} S_\alpha^{\frac{\theta}{\alpha}} \tag{3.33}$$

where S_α is the almost sure limit of $|\Pi_n|/n^\alpha$ under $\mathbb{P}_{\alpha,\theta}$ for every $\theta > -\alpha$.

Proof. This is read from the previous argument, by martingale theory. □

In view of Corollary 3.9, the limit random variable

$$S_\alpha := \lim_n |\Pi_n|/n^\alpha \tag{3.34}$$

plays a key role in describing asymptotic properties of an (α,θ) partition Π_∞.

Definition 3.10. Say that Π_∞, an exchangeable partition of $\mathbb{N}$ *has α-diversity* S_α if the limit (3.34) exists and is strictly positive and finite almost surely.

This limit random variable S_α can be characterized in a number of different ways, by virtue of the following lemma. According to Theorem 3.8 and Corollary 3.9, the conditions of the Lemma apply to an (α, θ) partition Π_∞, for each $\alpha \in (0,1)$, and each $\theta > -\alpha$.

Write $A_i \sim B_i$ if $A_i/B_i \to 1$ almost surely as $i \to \infty$.

Lemma 3.11. *Fix $\alpha \in (0,1)$, An exchangeable random partition Π_∞ has α-diversity S_α, defined as an almost sure limit (3.34), which is strictly positive and finite, if and only if*

$$P_i^\downarrow \sim Z i^{-1/\alpha} \text{ as } i \to \infty \tag{3.35}$$

for some random variable Z with $0 < Z < \infty$. In that case S_α and Z determine each other by

$$Z^{-\alpha} = \Gamma(1-\alpha) S_\alpha$$

and the following conditions also hold:

$$(1 - \sum_{i=1}^k \tilde{P}_i) \sim \alpha \Gamma(1-\alpha)^{1/\alpha} Z k^{1-1/\alpha} \text{ as } k \to \infty \tag{3.36}$$

where $\tilde{P}_i$ is the frequency of the ith block of Π_∞ in order of appearance;

$$|\Pi_n|_j \sim p_\alpha(j) S_\alpha n^\alpha \text{ for each } j = 1, 2, \ldots \tag{3.37}$$

where $|\Pi_n|_j$ is the number of blocks of Π_n of size j, and $(p_\alpha(j), j = 1, 2, \ldots)$ is the discrete probability distribution defined by

$$p_\alpha(j) = (-1)^{j-1}\binom{\alpha}{j} = \frac{\alpha(1-\alpha)_{j-1\uparrow}}{j!} \tag{3.38}$$

and

$$|\Pi_n|_j / |\Pi_n| \to p_\alpha(j) \text{ for every } j = 1, 2, \ldots \text{ a.s. as } n \to \infty. \tag{3.39}$$

Sketch of proof. By Kingman's representation, it suffices to establish the Lemma for Π_∞ with deterministic frequencies $P_i^\downarrow$. Most of the claims in this case can be read from the works of Karlin [232] and Rouault [392], results in the theory of regular variation [66], and large deviation estimates for sums of bounded independent random variables obtained by Poissonization [158]. □

The discrete probability distribution (3.38) arises in other ways related to the positive stable law of index α. See the exercises below, and [351, 355] for further references.

The ranked frequencies

Theorem 3.12. Case ($\alpha = 0$) [153]. *A random sequence $(P_i^\downarrow)$ has* PD$(0, \theta)$ *distribution iff for Γ_θ a gamma(θ) variable independent of $(P_i^\downarrow)$, the sequence $(\Gamma_\theta P_i^\downarrow)$ is the ranked sequence of points of a Poisson process on $(0, \infty)$ with intensity $\theta x^{-1} e^{-x} dx$.*

Proof. This follows from previous discussion.

Theorem 3.13. Case ($0 < \alpha < 1$).
(i) [341] *A random sequence* $(P_i^\downarrow)$ *with* $\sum_i P_i^\downarrow = 1$ *has* $\text{PD}(\alpha, 0)$ *distribution iff the limit*

$$S_\alpha := \lim_{i \to \infty} i\Gamma(1-\alpha)(P_i^\downarrow)^\alpha \tag{3.40}$$

exists almost surely, and the sequence $(S_\alpha^{-1/\alpha} P_i^\downarrow)$ *is the ranked sequence of points of a Poisson process on* $(0, \infty)$ *with intensity* $\alpha\Gamma(1-\alpha)^{-1}x^{-\alpha-1}dx$.
(ii) [351] *For* $\theta > -\alpha$, *and* $(P_i^\downarrow)$ *the* $\text{PD}(\alpha, \theta)$ *distributed sequence of ranked frequencies of an* (α, θ)*-partition* Π_∞, *the limit* S_α *defined by* (3.40) *exists and equals almost surely the* α*-diversity of* Π_∞, *that is*

$$S_\alpha = lim_{n \to \infty} |\Pi_n| / n^\alpha. \tag{3.41}$$

(iii) [341] *For* $\theta > -\alpha$, *the* $\text{PD}(\alpha, \theta)$ *distribution is absolutely continuous with respect to* $\text{PD}(\alpha, 0)$, *with density*

$$\frac{d\,\text{PD}(\alpha, \theta)}{d\,\text{PD}(\alpha, 0)} = \frac{\Gamma(\theta+1)}{\Gamma(\frac{\theta}{\alpha}+1)} S_\alpha^{\frac{\theta}{\alpha}}$$

for S_α *as in* (3.40).

Proof. Part (i) follows from results of [341] which are reviewed in Section 4.1. If a sequence of ranked frequencies admits the limit (3.40) almost surely in $(0, \infty)$, then it can be evaluated as in (3.41) using the associated random partition Π_∞. This was shown by Karlin [232] for Π_∞ with deterministic frequencies, and the general result follows by conditioning on the frequencies. This gives (ii), and (iii) is just a translation of Corollary 3.9 via Kingman's correspondence, using (ii). □

In particular, parts (i) and (ii) imply that if $S := S_\alpha$ is the α-diversity of an $(\alpha, 0)$ partition Π_∞, then $S^{-1/\alpha}$ has the stable(α) law whose Lévy density is $\alpha\Gamma(1-\alpha)^{-1}x^{-\alpha-1}dx$. This can also be deduced from Theorem 3.8, since we know from (0.43) that a random variable S has Mittag-Leffler(α) law iff $S^{-1/\alpha}$ has this stable(α) law. It must also be possible to establish the Poisson character of the random set of points $\{S^{-1/\alpha} P_i^\downarrow\} = \{S^{-1/\alpha} \tilde{P}_i\}$ by some direct computation based on the prediction rule for an $(\alpha, 0)$ partition, but I do not know how to do this.

Exercises

3.3.1. [351] **(Poisson subordination)** Fix $\alpha \in (0, 1)$, and let Z be the closure of the range of a stable subordinator of index α. Let N be a homogeneous Poisson point process on $\mathbb{R}_{>0}$ and let X_i be the number of points of N in the ith interval component of the complement of Z that contains at least one point of N. Then the X_i are independent and identically distributed with distribution $(p_\alpha(j), j = 1, 2, \ldots)$ as in (3.38). Generalize to a drift-free subordinator that is not stable.

3.3.2. If $\mathbb{P}_\alpha$ governs independent $X_1, X_2, \ldots$ with distribution (3.38), as in the previous exercise, then

$$\mathbb{E}_\alpha(z^{X_i}) = 1 - (1 - z)^\alpha. \tag{3.42}$$

Let $S_k := X_1 + \cdots + X_k$. Then

$$\mathbb{P}_\alpha(S_k = n) = [z^n](1 - (1 - z)^\alpha)^k \tag{3.43}$$

so the generalized Stirling number $S_\alpha(n, k)$ in (3.11), (3.12), (3.18), (3.19), acquires another probabilistic meaning as

$$S_\alpha(n, k) = \frac{n!}{k!}\alpha^{-k}\mathbb{P}_\alpha(S_k = n) \tag{3.44}$$

and the distribution of K_n for an (α, θ) partition is represented by the formula

$$\mathbb{P}_{\alpha,\theta}(K_n = k) = \frac{(\frac{\theta}{\alpha} + 1)_{k-1\uparrow}}{\alpha(\theta + 1)_{n-1\uparrow}} \frac{n!}{k!}\mathbb{P}_\alpha(S_k = n). \tag{3.45}$$

3.3.3. (A local limit theorem) [355] In the setting of Theorem 3.8, establish the local limit theorem

$$\mathbb{P}_{\alpha,\theta}(K_n = k) \sim g_{\alpha,\theta}(s)n^{-\alpha} \text{ as } n \to \infty \text{ with } k \sim sn^\alpha. \tag{3.46}$$

Deduce from (3.11) and (3.46) an asymptotic formula for $S_\alpha(n, k)$ as $n \to \infty$ with $k \sim sn^\alpha$.

3.3.4. For $0 < \alpha < 1$, as $n \to \infty$, for each $p > 0$

$$\mathbb{E}_{\alpha,\theta}(K_n^p) \sim n^{\alpha p} \frac{\Gamma(\frac{\theta}{\alpha} + p + 1)\Gamma(\theta + 1)}{\Gamma(\theta + p\alpha + 1)\Gamma(\frac{\theta}{\alpha} + 1)}. \tag{3.47}$$

Notes and comments

Lemma 3.11 is from unpublished work done jointly with Ben Hansen. There is much interest in power law behaviour, such as described by Lemma 3.11, in the literature of physical processes of fragmentation and coagulation. See [306] and papers cited there.

3.4. A branching process construction of the two-parameter model

This section offers an interpretation of the (α, θ) model for $0 \leq \alpha \leq 1, \theta > -\alpha$, in terms of a branching process in continuous time, which generalizes the model of Tavaré [414] in case $\theta = 0$. This brings out some interesting features of (α, θ) partitions which are hidden from other points of view.

Fix $0 \le \alpha \le 1$. Consider a population of individuals of two types, *novel* and *clone*. Each individual is assigned a color, and has infinite lifetime. Starting from a single novel individual at time $t = 0$, of some first color, suppose that each individual produces offspring throughout its infinite lifetime as follows:

- Novel individuals produce novel offspring according to a Poisson process with rate α, and independently produce clone offspring according to a Poisson process with rate $1 - \alpha$.
- Clones produce clone offspring according to a Poisson process with rate 1.

Each novel individual to appear is assigned a new color, distinct from the colors of all individuals in the current population. Each clone has the same color as its parent. Let

$$N_t := \text{ number of all individuals at time } t$$

$$N_t^* := \text{ number of novel individuals at time } t.$$

Thus $N_0^* = N_0 = 1$, and $1 \le N_t^* \le N_t$ for all $t \ge 0$. The process $(N_t^*, t \ge 0)$ is a *Yule process* with rate α, that is a pure birth process with transition rate $i\alpha$ from state i to state $i + 1$. Similarly, $(N_t, t \ge 0)$ is a Yule process with rate 1. Think of the individuals as colored balls occupying boxes labelled by $\mathbb{N} := \{1, 2, \ldots\}$. So the nth individual to be born into the population is placed in box n. The colors of individuals then induce a random partition of $\mathbb{N}$. Each novel individual appears in the first of an infinite subset of boxes containing individuals of the same color.

Proposition 3.14. *The random partition of Π of $\mathbb{N}$, generated by the colors of successive individuals born into the population described above, is an $(\alpha, 0)$ partition. The number of blocks in the induced random partition of $[n]$ is the value of N_t^* at every time t such that $N_t = n$. For each $t > 0$, the conditional distribution of N_t^* given $N_t = n$ is identical to the distribution of K_n, the number of blocks of the partition of $[n]$, for an $(\alpha, 0)$ partition.*

Proof. Let Π_n be the partition of $[n]$ induced by Π. It follows easily from the description of the various birth rates that $(\Pi_n, n = 1, 2, \ldots)$ is a Markov chain with transition probabilities described by the (α, θ) urn scheme, independent of the process $(N_t, t \ge 0)$. □

According to a standard result for the Yule process

$$e^{-t} N_t \overset{a.s.}{\to} W$$

$$e^{-\alpha t} N_t^* \overset{a.s.}{\to} W^*,$$

where W and W^* are both exponentially distributed with mean 1. Combined with Proposition 3.14 this implies

Corollary 3.15. $W^* = SW^\alpha$ *where* $S := \lim_{n \to \infty} K_n / n^\alpha$ *is independent of* W.

A formula for the moments of S follows immediately, confirming the result of Theorem 3.8 that S has Mittag-Leffler distribution with parameter α.

To present a continuous time variation of the residual allocation model, let $N_t^{(k)}$ = be the number of individuals of the kth color to appear that are present in the population at time t. From the previous analysis, as $t \to \infty$

$$\left(e^{-t}N_t, \frac{N_t^{(1)}}{N_t}, \frac{N_t^{(2)}}{N_t}, \frac{N_t^{(3)}}{N_t}, \ldots\right) \stackrel{a.s.}{\to} (W,\ X_1,\ \overline{X}_1 X_2,\ \overline{X}_1 \overline{X}_2 X_3, \ldots) \quad (3.48)$$

where W, $X_1, X_2, \ldots$ are independent, W has exp(1) distribution, and X_i (denoted W_i in (3.8)) has beta$(1-\alpha, i\alpha)$ distribution. Equivalently,

$$e^{-t}(N_t, N_t^{(1)}, N_t^{(2)}, \ldots) \stackrel{a.s.}{\to} (W, WX_1, W\overline{X}_1 X_2, \ldots). \quad (3.49)$$

In particular, the limit law of $e^{-t}(N_t^{(1)}, N_t - N_t^1)$ is that of WX_1 and $W\overline{X}_1$, which are independent gamma$(1-\alpha)$ and gamma(α) respectively. The subsequent terms have more complicated joint distributions.

Case $0 \leq \alpha \leq 1$, $\theta > -\alpha$. Define a population process with two types of individuals, exactly as in the case $\theta = 0$ treated as above, but with the following modification of the rules for the offspring process of the first novel individual only. This first individual produces novel offspring at rate $\alpha + \theta$ (instead of α as before) and clone offspring at rate $1 - \alpha$ (exactly as before). Both clone and novel offspring of the first individual reproduce just as before. And the rules for coloring are just as before. It is easily checked that the transition rules when the partition is extended from n individuals to $n+1$ individuals are exactly those of the (α, θ) prediction rule. So the random partition Π of $\mathbb{N}$ induced by this population process is an (α, θ) partition.

Case $0 \leq \alpha \leq 1$, $\theta \geq 0$. This can be described more simply by a slight modification of the rules for the above scheme. The modified scheme is then a generalization of the process described by Tavaré [414] in case $\alpha = 0$, $\theta > 0$. Instead of letting the first individual produce novel offspring at rate $\alpha + \theta$, let the first individual produce novel offspring at rate α, and let an independent Poisson migration process at rate θ bring further novel individuals into the population. Otherwise the process runs as before. Now the first novel individual follows the same rules as all other novel individuals.

If the distinction between novel and clone individuals is ignored, we just have a Yule process with immigration, where all individuals produce offspring at rate 1, and there is immigration at rate θ. If we keep track of the type of individuals, since each immigrant is novel by definition, it is clear that the partition generated by all the colors is a refinement of the partition whose classes are the progeny of the first individual, the progeny of the first immigrant, the progeny of the second immigrant, and so on. Each of these classes is created by a Yule process with rate 1, whose individuals are partitioned by coloring exactly as before in case $\theta = 0$. This structure reveals the following result:

Proposition 3.16. *Let $0 \leq \alpha \leq 1, \theta \geq 0$. Let a stick of length 1 be broken into lengths $P_n = \overline{X}_1 \dots \overline{X}_{n-1} X_n$ according to the GEM $(0, \theta)$ distribution, as in (3.8). Then, let each of these stick be broken further, independently of each other, according to the GEM $(\alpha, 0)$, to create a countable array of sticks of lengths*

$$P_1 X_{11}, P_1 \overline{X}_{11} X_{12}, P_1 \overline{X}_{11} \overline{X}_{12} X_{13}, \dots$$

$$P_2 X_{21}, P_2 \overline{X}_{21} X_{22}, P_2 \overline{X}_{21} \overline{X}_{22} X_{23}, \dots$$

where $X_1, X_2, \dots, X_{11}, X_{12}, \dots, X_{21}, X_{22}, \dots$ are independent, with $X_j \sim \text{beta}(1, \theta)$ for all j, and $X_{ij} \sim \text{beta}(1 - \alpha, j\alpha)$ for all i and j. Let $Q_1, Q_2, \dots$ be a size-biased random permutation of the lengths in this array. Then the Q_n are distributed according to GEM(α, θ), that is:

$$Q_n = \overline{Y}_1 \overline{Y}_2 \dots \overline{Y}_{n-1} Y_n$$

where the Y_j are independent beta$(1 - \alpha, \theta + j\alpha)$.

By arguing as in Hoppe [202], Proposition 3.16 can be restated as follows:

Proposition 3.17. *Let $0 < \alpha <, \theta \geq 0$. Let $\{A_i\}$ be a $(0, \theta)$ random partition of $[n]$. Given $\{A_i\}$, with say k blocks, let $\{A_{ij}\}$, $j = 1, \dots, k$ be independent $(\alpha, 0)$ random partitions of A_i. Then $\{A_{ij}\}$ is an (α, θ) random partition of $[n]$.*

In view of Theorem 3.2, either of these propositions follow easily from the other. A direct calculation shows that the result for finite partitions reduces to the following variant of formula (1.16) for the generating function of numbers of cycles in a random permutation of $[n]$:

$$\sum_{j=1}^{n} \theta^j \sum_{\{C_i, 1 \leq i \leq j\}} \Pi_{i=1}^{j} (|C_i| - 1)! \alpha^{|C_i| - 1} = \theta(\theta + \alpha) \dots (\theta + (n-1)\alpha) \quad (3.50)$$

where the second sum is over all partitions $\{C_i, 1 \leq i \leq j\}$ of $[n]$ into j parts, and $|C_i|$ is the number of elements of C_i. See also [368, (67)], [371, Proposition 22] for further discussion, and (5.26) for a more refined result.

Notes and comments

This section is based on an unpublished supplement to the technical report [346], written in November 1992. See also Feng and Hoppe [152] for a similar approach, with reference to an earlier model of Karlin. See Dong, Goldschmidt and Martin [113] for some recent developments.

4

Poisson constructions of random partitions

This chapter is a review of various constructions of random partitions from Poisson point processes of random lengths, based on the work of Kingman and subsequent authors [249, 341, 371, 351, 362]. The lengths can be interpreted as the jumps of a subordinator, or as the lengths of excursions of some Markov process. The treatment is organized by sections as follows:

4.1. Size-biased sampling This section presents results of Perman, Pitman and Yor [341], characterizing the distribution of the size-biased permutation of relative lengths derived from the Poisson point process of jumps $(T_t - T_{t-}; 0 \leq t \leq 1)$ of a subordinator $(T_t, t \geq 0)$.

4.2. Poisson representation of the two-parameter model In the particular case when (T_t) is a stable subordinator of index $\alpha \in (0,1)$, this provides a natural approach to the two-parameter family of (α, θ) random partitions whose frequencies in size-biased order have the characteristic property that they may be represented as

$$(W_1, \overline{W}_1 W_2, \overline{W}_1 \overline{W}_2 W_3, \ldots) \tag{4.1}$$

where the W_i are independent random variables with values in $(0,1)$, and $\overline{W}_i := 1 - W_i$.

4.3. Representation of infinite Gibbs partitions This section shows how exchangeable random partitions of $\mathbb{N}$, whose partition probability functions are of a particular product form, may be represented in terms of partitions derived from suitable subordinators.

4.4. Lengths of stable excursions This section treats various features of the two-parameter Poisson-Dirichlet distributions PD$(\alpha, 0)$ and PD(α, α) which are induced by the ranked lengths of excursions of a Bessel process and Bessel bridge of dimension $2 - 2\alpha$.

4.5. Brownian excursions The special case $\alpha = \frac{1}{2}$, corresponds to excursions of a Brownian motion or Brownian bridge. Various integrals, which are intractable for general α, can be expressed for $\alpha = \frac{1}{2}$ in terms of classical special functions. This provides a detailed description of random partitions of $[0,1]$ generated by Brownian excursions, with conditioning on the local time at 0. As shown in [17, 18], and discussed further in Chapter 9 and Chapter 10, these random partitions arise naturally in studying the asymptotic distribution of partitions derived in various ways from random forests, random mappings, and the additive coalescent.

4.1. Size-biased sampling

Following McCloskey [302], Kingman [249], Engen [132], Perman, Pitman and Yor [339, 341, 371], consider the ranked random discrete distribution $(P_i^\downarrow) := (J_i^\downarrow/T)$ derived from an inhomogeneous Poisson point process of random lengths

$$J_1^\downarrow \geq J_2^\downarrow \geq \cdots \geq 0$$

by normalizing by the total length $T := \sum_{i=1}^\infty J_i^\downarrow$. For each interval I it is assumed that $N_I := \sum_i 1(J_i^\downarrow \in I)$ is a Poisson variable with mean $\Lambda(I)$, for some Lévy measure Λ on $(0,\infty)$, and that the counts $N_{I_1}, \cdots, N_{I_k}$ are independent for every finite collection of disjoint intervals $I_1, \cdots, I_k$. The following assumption will also be made so that various conditional probabilities can be defined easily:
Regularity assumption. *The Lévy measure Λ has a density $\rho(x)$ and the distribution of T is absolutely continuous with density*

$$f(t) := \mathbb{P}(T \in dt)/dt$$

which is strictly positive and continuous on $(0,\infty)$ with $\int_0^\infty f(t)dt = 1$.

For $\lambda \geq 0$ let

$$\Psi(\lambda) := \int_0^\infty (1 - e^{-\lambda x})\rho(x)dx. \tag{4.2}$$

Since $\mathbb{P}(T = 0) = \exp[-\Psi(\infty-)]$ the regularity assumption implies

$$\Psi(\infty-) = \int_0^\infty \rho(x)dx = \infty. \tag{4.3}$$

It is well known that f is uniquely determined by ρ, either via its Laplace transform

$$\mathbb{E}(e^{-\lambda T}) = \int_0^\infty e^{-\lambda x} f(x)dx = \exp[-\Psi(\lambda)] \tag{4.4}$$

or as the unique solution of the integral equation

$$f(t) = \int_0^t \rho(v) f(t-v)\frac{v}{t}dv \tag{4.5}$$

which is obtained by differentiation of (4.4) with respect to λ. Let $(\tilde{P}_j)$ be a size-biased permutation of the normalized lengths $(P_i^\downarrow) := (J_i^\downarrow/T)$ and let $(\tilde{J}_j) = (T\tilde{P}_j)$ be the corresponding size-biased permutation of $(J_i^\downarrow)$. Then [302, 341]

$$\mathbb{P}(\tilde{J}_1 \in dv, T \in dt) = \rho(v)dv f(t-v)dt\frac{v}{t}, \tag{4.6}$$

which can be understood informally as follows. The left side of (4.6) is the probability that among the Poisson lengths there is some length in dv near v, and the sum of the rest of the lengths falls in an interval of length dt near $t-v$, and finally that the interval of length about v is the one picked by length-biased sampling. Formally, (4.6) is justified by the description of a Poisson process in terms of its family of Palm measures [341]. Note that (4.5) is recovered from (4.6) be integrating out v. See also Exercise 4.1.1 for another interpretation of (4.5).

Immediately from (4.6), there is the following formula for the density of the structural distribution of $\tilde{P}_1 := \tilde{J}_1/T$ given $T=t$: for $0<p<1$ with $\bar{p} := 1-p$

$$\frac{\mathbb{P}(\tilde{P}_1 \in dp \mid T=t)}{dp} = \tilde{f}(p \mid t) := pt\,\rho(pt)\,\frac{f(\bar{p}t)}{f(t)}. \tag{4.7}$$

For $j=0,1,2,\cdots$ let

$$\tilde{T}_j := T - \sum_{k=1}^{j} \tilde{J}_k = \sum_{k=j+1}^{\infty} \tilde{J}_k \tag{4.8}$$

which is the total length remaining after removal of the first j Poisson lengths $\tilde{J}_1,\ldots,\tilde{J}_j$ chosen by length-biased sampling. Note that $\tilde{T}_0 := T$. Then a calculation similar to (4.6) yields:

Lemma 4.1. [341, Theorem 2.1] *The sequence $(\tilde{T}_0, \tilde{T}_1, \tilde{T}_2, \ldots)$ is a Markov chain with stationary transition probabilities*

$$\mathbb{P}(\tilde{T}_{j+1} \in dt_1 \mid \tilde{T}_j = t) = \frac{\rho_*(t-t_1)}{t}\frac{f(t_1)}{f(t)}dt_1 \tag{4.9}$$

where $\rho_*(x) := x\rho(x)$.

Now

$$\tilde{P}_j = W_j \prod_{i=1}^{j-1}(1-W_i) \tag{4.10}$$

where $1-W_i = \tilde{T}_i/\tilde{T}_{i-1}$ with $\tilde{T}_0 = T$ and the joint law of the $\tilde{T}_i$ given $T=t$ is described by Lemma 4.1. This description of the conditional law of $(\tilde{P}_j)$ given $T=t$ determines corresponding conditional distributions for the associated ranked sequence $(P_i^{\downarrow})$ and for an associated exchangeable random partition Π_∞ of positive integers. Each of these families of conditional distributions is weakly continuous in t. For the size-biased frequencies $(\tilde{P}_j)$ this is clear by inspection of formula (4.9). For the others it follows by continuity of the associations. A formula for the joint density of $(P_1^{\downarrow},\cdots,P_n^{\downarrow})$ given $T=t$ was obtained by Perman [339] in terms of the joint density of T and $J_1^{\downarrow}$. This joint density can be described in terms of ρ and f as the solution of an integral equation [339], or as a series of repeated integrals [371]. The distribution of an exchangeable partition (Π_n) of positive integers, whose ranked frequencies are distributed like $(P_i^{\downarrow})$ given $T=t$, is now determined by either of the formulae (2.14) and (3.4). But a simpler way to compute partition probabilities to use the following formula:

Lemma 4.2. [362] *For each particular partition* $\{A_1, \ldots, A_k\}$ *of* $[n]$, *and arbitrary* $x_i > 0$ *and* $t > 0$,

$$\mathbb{P}(\Pi_n = \{A_1, \ldots, A_k\}, \tilde{J}_i \in dx_i, 1 \leq i \leq k, T \in dt)$$

$$= \left(\prod_{i=1}^{k} \rho(x_i) dx_i\right) f(t - \Sigma_{i=1}^{k} x_i) dt \prod_{i=1}^{k} \left(\frac{x_i}{t}\right)^{|A_i|}. \tag{4.11}$$

Proof. This is derived in much the same way as (4.6) and (4.9). □

See [362] and exercises below for various applications of this formula.

Exercises

4.1.1. (Size-biasing) [417] For a non-negative random variable X with $\mathbb{E}(X) < \infty$ let X^* denote a random variable with the *size-biased* distribution of X, that is

$$\mathbb{P}(X^* \in dx) = x\mathbb{P}(X \in dx)/\mathbb{E}(X).$$

The distribution of X^* arises naturally both as an asymptotic distribution in renewal theory, and in the theory of infinitely divisible laws. For $\lambda \geq 0$ let $\varphi_X(\lambda) := \mathbb{E}[e^{-\lambda X}]$. Then the Laplace transform of X^* is

$$\mathbb{E}[e^{-\lambda X^*}] = -\varphi'_X(\lambda)/\mathbb{E}(X)$$

where φ'_X is the derivative of φ_X. According to the Lévy-Khintchine representation, the distribution of X is infinitely divisible iff

$$-\frac{\varphi'_X(\lambda)}{\varphi_X(\lambda)} = c + \int_0^\infty x e^{-\lambda x} \Lambda(dx)$$

for some $c \geq 0$ and some Lévy measure Λ, that is iff

$$\mathbb{E}[e^{-\lambda X^*}] = \varphi_X(\lambda)\varphi_Y(\lambda)$$

where Y is a random variable with

$$\mathbb{P}(Y \in dy) = (c\delta_0(dy) + y\Lambda(dy))/\mathbb{E}(X). \tag{4.12}$$

Hence, for a non-negative random variable X with $\mathbb{E}(X) < \infty$, the equation $X^* \stackrel{d}{=} X + Y$ is satisfied for some Y independent of X if and only if the law of X is infinitely divisible, in which case the distribution of Y is given by (4.12) for c and Λ the Lévy characteristics of X. See also [27, 375] for further developments.

4.1.2. [362] Derive from (4.11) a formula for the infinite exchangeable partition probability function (EPPF) Section 2.2 associated with random ranked frequencies $(\tilde{J}_i/T, i \geq 1)$ given $T = t$. Deduce that the infinite EPPF associated with random ranked frequencies $(\tilde{J}_i/T, i \geq 1)$ is given by the $(k+1)$-fold integral

$$p(n_1, \ldots, n_k) = \int_0^\infty \cdots \int_0^\infty \frac{f(v)dv \prod_{i=1}^k \rho(x_i) x_i^{n_i} dx_i}{(v + \Sigma_{i=1}^k x_i)^n}, \tag{4.13}$$

where $n = \Sigma_i n_i$. Use $b^{-n} = \Gamma(n)^{-1} \int_0^\infty \lambda^{n-1} e^{-\lambda b} d\lambda$ to recast this as

$$p(n_1, \ldots, n_k) = \frac{(-1)^{n-k}}{(n-1)!} \int_0^\infty \lambda^{n-1} d\lambda \, e^{-\Psi(\lambda)} \prod_{i=1}^k \Psi^{(n_i)}(\lambda) \tag{4.14}$$

where $\Psi(\lambda)$ is the Laplace exponent (4.2) and $\Psi^{(j)}(\lambda)$ is the jth derivative of $\Psi(\lambda)$. Check from (4.14) that the gamma(θ) Lévy density (4.15) corresponds to a $(0, \theta)$ partition, and the stable(α) Lévy density (0.39) corresponds to an $(\alpha, 0)$ partition.

4.1.3. Show that formula (4.14) and an obvious generalization of (4.13) give the infinite EPPF associated with random ranked frequencies $(\tilde{J}_i/T, i \geq 1)$ without any regularity assumption on the Lévy measure besides $\Psi(\lambda) < \infty$ and $\Psi(\infty-) = \infty$.

4.1.4. Check that formula (4.14) satisfies the addition rule (2.9) for an EPPF.

4.1.5. (Problem) What conditions on a function Ψ are necessary for (4.14) to define an EPPF? Must Ψ be the Laplace exponent of some infinitely divisible distribution on $(0, \infty)$?

Notes and comments

This section is based on [341] and [362]. See James [212] for a closely related approach to partition probabilities associated with sampling from random measures derived from Poisson processes, and applications to Bayesian inference.

4.2. Poisson representation of the two-parameter model

Consider now the two-parameter family of (α, θ) random partitions, defined as in Section 3.2 for $0 < \alpha < 1$ and $\theta > -\alpha$ by size-biased frequencies of the product form (4.1) for independent W_i with beta$(1-\alpha, \theta + i\alpha)$ distributions.

Case$(\alpha = 0)$ Consider the particular choice of Lévy density

$$\rho(x) = \theta \, x^{-1} \, e^{-bx} \tag{4.15}$$

where $\theta > 0$ and $b > 0$. This makes $T \stackrel{d}{=} \Gamma_\theta / b$ where $(\Gamma_s, s \geq 0)$ is a standard gamma process. The formulae of the preceding section confirm that in this case the W_i in (4.10) are independent beta$(1, \theta)$ variables, so $(\tilde{P}_j)$ has the GEM$(0, \theta)$ distribution of frequencies of blocks of a $(0, \theta)$-partition of positive integers, as described in Theorem 3.2. The corresponding distribution of ranked frequencies $(P_i^\downarrow)$ is Poisson-Dirichlet$(0, \theta)$, by Definition 3.3 . See [27] for further study of this distribution, which has numerous applications. Two special properties, each known to be characteristic [341] of this Lévy density (4.15), are that

$$T \text{ is independent of the sequence of ranked frequencies } (P_i^\downarrow), \tag{4.16}$$

$$T \text{ is independent of the size-biased frequency } \tilde{P}_1. \tag{4.17}$$

Case$(0 < \alpha < 1)$ Consider next the stable(α) Lévy density $\rho(x) = s\rho_\alpha(x)$ for some $s > 0$ where

$$\rho_\alpha(x) := \frac{\alpha}{\Gamma(1-\alpha)} \frac{1}{x^{\alpha+1}} \quad (x > 0) \tag{4.18}$$

In contrast to (4.16), a feature of this choice of $\rho(x)$ is that

$$T \text{ is a measurable function of the ranked frequencies } (P_i^\downarrow), \tag{4.19}$$

specifically

$$T = \frac{(s/\Gamma(1-\alpha))^{1/\alpha}}{\lim_{i \to \infty} i^{1/\alpha} P_i^\downarrow}. \tag{4.20}$$

This formula holds, by the law of large numbers for the underlying Poisson process, whenever $\rho(x) \sim s\rho_\alpha(x)$ as $x \downarrow 0$. A special feature of $\rho(x) = s\rho_\alpha(x)$, shown in [341] to be characteristic of this case, is that

$$\tilde{T}_1 := T(1 - \tilde{P}_1) \text{ is independent of the size-biased frequency } \tilde{P}_1. \tag{4.21}$$

To see this, and derive the distributions of $\tilde{T}_1$ and $\tilde{P}_1$, observe from (4.6) that for general $\rho(x)$ the joint density of $\tilde{T}_1 := T - \tilde{J}_1$ and $U_1 := \tilde{T}_1/T = 1 - \tilde{P}_1$ is

$$f_{\tilde{T}_1, U_1}(t_1, u_1) = u_1^{-1} \rho_* \left(\frac{\overline{u}_1}{u_1} t_1 \right) f(t_1) \tag{4.22}$$

where $\overline{u}_1 := 1 - u_1$ and $\rho_*(x) := x\rho(x)$. For $\rho(x) = s\rho_\alpha(x)$, by scaling there is no loss of generality in supposing $s = 1$, when (4.22) simplifies to

$$f_{\tilde{T}_1, U_1}(t_1, u_1) = K_\alpha u_1^{\alpha-1} \overline{u}_1^{-\alpha} t_1^{-\alpha} f_\alpha(t_1) \tag{4.23}$$

for some constant K_α, where $f_\alpha(t)$ is the stable(α) density of T. This confirms the independence result (4.21), and shows that

$$U_1 := 1 - \tilde{P}_1 \text{ has beta}(\alpha, 1-\alpha) \text{ distribution}, \tag{4.24}$$

$$\tilde{T}_1 \text{ has density at } t \text{ proportional to } t^{-\alpha} f_\alpha(t). \tag{4.25}$$

In view of the independence of U_1 and $\tilde{T}_1$, formula (4.25), and the homogeneous Markov property of $T, \tilde{T}_1, \tilde{T}_2, \ldots$ provided by Lemma 4.1, the random variable $U_1 := \tilde{T}_1/T$ is independent of $(\tilde{T}_1, \tilde{T}_2, \ldots)$, and the joint density of $\tilde{T}_2$ and $U_2 := \tilde{T}_2/\tilde{T}_1$ at (t_1, u_1) relative to that of $\tilde{T}_1$ and $U_1 := \tilde{T}_1/T$ at (t_1, u_1), is some constant times $(t_1/u_1)^{-\alpha}$. Hence U_1, U_2 and $\tilde{T}_2$ are independent, the distribution of U_2 is beta$(2\alpha, 1-\alpha)$, and

$$\tilde{T}_2 \text{ has density at } t \text{ proportional to } t^{-2\alpha} f_\alpha(t). \tag{4.26}$$

Continuing like this, it is clear that the U_i are independent, the distribution of U_i is beta$(i\alpha, 1-\alpha)$, and

$$\tilde{T}_i \text{ has density at } t \text{ proportional to } t^{-i\alpha} f_\alpha(t). \tag{4.27}$$

A little more generally, if $\mathbb{P}_\alpha$ denotes the probability measure governing the original scheme with T distributed according to the stable(α) density, and $\mathbb{P}_{\alpha,\theta}$ denotes the probability measure which is absolutely continuous with respect to $\mathbb{P}_\alpha$, with density $C_{\alpha,\theta} T^{-\theta}$ for some $\theta > -\alpha$, then since $T = \tilde{T}_1/U_1$ we find from (4.23) that the joint density of $(\tilde{T}_1, U_1)$ at (t_1, u_1) under $\mathbb{P}_{\alpha,\theta}$ is

$$K_\alpha u_1^{\alpha-1} \overline{u}_1^{-\alpha} t_1^{-\alpha} f(t_1) C_{\alpha,\theta} \left(\frac{t_1}{u_1}\right)^{-\theta} = K_\alpha C_{\alpha,\theta}\, u_1^{\theta+\alpha-1} \overline{u}_1^{-\alpha} t_1^{-\theta-\alpha} f(t_1).$$

That is to say, under $\mathbb{P}_{\alpha,\theta}$ the random variables U_1 and $\tilde{T}_1$ are independent, U_1 with beta$(\alpha+\theta, 1-\alpha)$ distribution, and $\tilde{T}_1$ distributed like T under $\mathbb{P}_{\alpha,\theta+\alpha}$. Combine this calculation with the homogeneous Markov property of $(T = \tilde{T}_0, \tilde{T}_1, \tilde{T}_2, \ldots)$ under $\mathbb{P}_{\alpha,\theta}$ for each θ, and set $W_i = 1 - \tilde{T}_i/\tilde{T}_{i-1}$ to obtain the following conclusion:

Theorem 4.3. *Let $\mathbb{P}_\alpha$ govern a Poisson process of lengths with intensity $s\rho_\alpha(x)$ for some $s > 0$ and $0 < \alpha < 1$, so the sum T of these lengths has a stable law of index α, and let $\tilde{P}_j = \tilde{J}_j/T$ where $(\tilde{J}_j)$ is the sequence of lengths in a length-biased random order. Let $\mathbb{P}_{\alpha,\theta}$ denote the probability measure which is absolutely continuous with respect to $\mathbb{P}_\alpha$, with density proportional to $T^{-\theta}$ for some $\theta > -\alpha$. Then under $\mathbb{P}_{\alpha,\theta}$ the sequence $(\tilde{P}_j)$ admits the representation*

$$\tilde{P}_j = W_j \prod_{i=1}^{j-1} (1 - W_i) \tag{4.28}$$

where the W_i are independent and W_i has beta$(1-\alpha, \theta + i\alpha)$ distribution.

Compare with Theorem 3.2 to see that if Π_∞ is the exchangeable random partition obtained by sampling from random frequencies defined by normalizing the Poisson lengths with sum T, then $\mathbb{P}_{\alpha,\theta}$ governs Π_∞ as an (α, θ) partition. To complete the proof of Theorem 3.13, it has to be seen that the α-diversity $\lim_n |\Pi_n|/n^\alpha$ of such a partition equals $T^{-\alpha}$ almost surely, but this can be seen as in [171].

Exercises

4.2.1. (Low temperature asymptotics for Derrida's random energy model) This exercise is taken almost verbatim from Neveu [322]. See also Talagrand [412, §1.2] for a similar treatment. Let $X, X_1, X_2, \ldots$ be independent standard Gaussian variables. For $n = 1, 2, \ldots$ define a_n by $\mathbb{P}(X > a_n) = 1/n$. Let $\ell_n := \sqrt{2 \log n}$ so that $a_n/\ell_n \to 1$ as $n \to \infty$. By the theory of extreme values [279],

$$\lim_{n\to\infty} \mathbb{P}(\max_{1\le i\le n} X_i \le a_n + x/\ell_n) = \exp(-e^{-x}) \qquad (x \in \mathbb{R}). \tag{4.29}$$

More precisely, if the sequence $(Y_{n,k}, 1 \le k \le n)$ with $Y_{n,1} \ge Y_{n,2} \ge \cdots \ge Y_{n,n}$ is the ranked rearrangement of $(\ell_n(X_i - a_n), 1 \le i \le n)$, then as $n \to \infty$ there is the convergence of finite dimensional distributions

$$(Y_{n,k}, 1 \le k \le n) \xrightarrow{d} (-\log(\tau_k), k \ge 1) \tag{4.30}$$

where $\tau_k := \sum_{i=1}^k \varepsilon_i$ for independent standard exponential variables ε_i. So the right side of (4.30) is the sequence of points of a Poisson process on $\mathbb{R}$ with intensity $e^{-x}dx$, in descending order. Consider next the sums of Boltzman exponentials

$$Z_n^{(\alpha)} := \sum_{i=1}^{n} \exp(\alpha^{-1}\ell_n X_i) \tag{4.31}$$

where $\alpha > 0$ is a temperature parameter, and note that $\mathbb{E}(Z_n^{(\alpha)}) = n^{1+\alpha^{-2}}$. In the low temperature case $\alpha < 1$ the leading terms in (4.31) are those corresponding to the greatest X_i, and

$$Z_n^{(\alpha)} \exp(-\alpha^{-1}\ell_n a_n) \xrightarrow{d} \sum_{k=1}^{\infty} \tau_k^{-1/\alpha} \stackrel{d}{=} T_{\Gamma(1-\alpha)}^{(\alpha)} \tag{4.32}$$

for $(T_s^{(\alpha)}, s \ge 0)$ the stable(α) subordinator with $\mathbb{E}[\exp(-\lambda T_s^{(\alpha)})] = \exp(-s\lambda^\alpha)$. Hence for $\alpha \in (0,1)$ the asymptotic distribution of the ranked Gibbs weights $\exp(\alpha^{-1}\ell_n X_i)/Z_n^{(\alpha)}, 1 \le i \le n$ is PD(α, 0).

See [371, p. 861] for other references to the appearance of PD(α, 0) by random normalization of arrays of independent random variables whose sum is asymptotically stable with index α. See [322, 412] regarding the asymptotics of the Gibbs weights in the high temperature case $\alpha > 1$, which is quite different. See [322, 72, 50, 54, 357] for further developments of the low temperature case, related to continuous state branching processes, Ruelle's probability cascades, and the Bolthausen-Sznitman coalescent. See [412] for a recent review of the rigorous mathematical theory of spin glasses. The paper of Derrida [106] draws parallels between the spin-glass theory and the Poisson-Dirichlet asymptotics for random permutations and random mappings discussed in this course.

4.3. Representation of infinite Gibbs partitions

Consider the setup of Corollary 3.9, with $\mathbb{P}_{\alpha,\theta}$ the distribution of an (α,θ) partition Π_∞ of positive integers, for $\alpha \in (0,1)$, and $\theta > -\alpha$. That is to say, $\mathbb{P}_{\alpha,\theta}$ makes the frequencies of classes of Π_∞, in order of their least elements, distributed as in (4.28). According to Corollary 3.9, the density of $\mathbb{P}_{\alpha,\theta}$ with respect to $\mathbb{P}_{\alpha,0}$ is a function of the α-diversity $S_\alpha(\Pi_\infty)$. Consequently, for each fixed $\alpha \in (0,1)$, the one-parameter family of laws $\{\mathbb{P}_{\alpha,\theta}, \theta > -\alpha\}$ shares a common conditional distribution for Π_∞ given $S_\alpha(\Pi_\infty)$. To be more precise:

Lemma 4.4. *For each $\alpha \in (0,1)$ there exists a unique family of probability laws $\{\mathbb{P}_{\alpha|s}, 0 < s < \infty\}$ on partitions of positive integers such that*

$$\mathbb{P}_{\alpha|s}(\cdot) = \mathbb{P}_{\alpha,\theta}(\cdot \mid S_\alpha = s) \text{ for all } \theta > -\alpha \tag{4.33}$$

and $s \to \mathbb{P}_{\alpha|s}$ is weakly continuous in s, where $S_\alpha := \lim_n K_n/n^\alpha$ with K_n the number of blocks of the partition of $[n]$. Under $\mathbb{P}_{\alpha|s}$ the frequencies of classes of Π_∞ are distributed like

$$\text{the jumps of } (T_u, 0 \le u \le s) \text{ given } T_s = 1,$$

or, equivalently, like

$$\text{the jumps of } (T_u/T_1, 0 \le u \le 1) \text{ given } T_1 = s^{-1/\alpha},$$

where $(T_u, u \ge 0)$ is the stable(α) subordinator with

$$\mathbb{E}[e^{-\lambda T_u}] = \exp(-u\lambda^\alpha).$$

Moreover, $\mathbb{P}_{\alpha|s}(S_\alpha = s) = 1$ for all s, and $\mathbb{P}_{\alpha|s}(\cdot)$ is the weak limit of $\mathbb{P}_{\alpha,\theta}(\cdot \mid K_n = k_n)$ as $n \to \infty$ for any sequence k_n with $k_n/n^\alpha \to s$.

Proof. This is read from Theorem 4.3, with the last sentence a particular case of Theorem 2.5. □

By combining Theorem 3.8 and Lemma 4.4 , for each $\alpha \in (0,1)$ there is the disintegration

$$\mathbb{P}_{\alpha,\theta}(\cdot) = \int_0^\infty \mathbb{P}_{\alpha|s}(\cdot) g_{\alpha,\theta}(s) ds \tag{4.34}$$

where $g_{\alpha,\theta}$ as in (3.27) is the probability density of S_α under $\mathbb{P}_{\alpha,\theta}$. In particular, the EPPF $p_{\alpha,\theta}$ of an (α,θ) partition, displayed in (3.6), is disintegrated as

$$p_{\alpha,\theta}(\cdot) = \int_0^\infty p_{\alpha|s}(\cdot) g_{\alpha,\theta}(s) ds \tag{4.35}$$

where $p_{\alpha|s}$ is the EPPF of Π_∞ governed by $\mathbb{P}_{\alpha|s}$. The form of this EPPF is made explicit by the following theorem:

Theorem 4.5. [362] *For each $\alpha \in (0,1)$ and each $s > 0$, the EPPF $p_{\alpha|s}$ of Π_∞ governed by $\mathbb{P}_{\alpha|s}$ is of the Gibbs form*

$$p_{\alpha|s}(n_1, \ldots, n_k) = v_k^{n,\alpha|s} \prod_{i=1}^{k} (1-\alpha)_{n_i - 1\uparrow} \tag{4.36}$$

where $n = \sum_i n_i$ and

$$v_k^{n,\alpha|s} = \alpha^k s^{n/\alpha} G_\alpha(n - k\alpha, s^{-1/\alpha}), \tag{4.37}$$

where

$$G_\alpha(q, t) := \frac{1}{\Gamma(q) f_\alpha(t)} \int_0^t f_\alpha(t - v) v^{q-1} dv \tag{4.38}$$

with f_α the the stable(α) density of the $\mathbb{P}_{\alpha,0}$ distribution of $(S_\alpha)^{-1/\alpha}$, as in (0.37).

Proof. That the EPPF $p_{\alpha|s}$ must be of the Gibbs form

$$v_{n,k} \prod_{i=1}^{k} (1-\alpha)_{n_i - 1\uparrow} \tag{4.39}$$

for some weights $v_{n,k}$ depending on α and s can be seen without calculation as follows. Let $\Pi_\infty := (\Pi_n)$ be an (α, θ) partition, with $K_n := |\Pi_n|$. From the prediction rule for construction of (Π_n), described in Section 3.2, the random partition Π_n of $[n]$ and the inhomogeneous Markov chain $(K_n, K_{n+1}, \ldots)$ are conditionally independent given K_n. So for each $n < m$ and each choice of $k_m \in [m]$, the conditional law of Π_n given $K_m = k_m$ is some mixture over k of the conditional laws of Π_n given $K_n = k$, for $1 \leq k \leq n$. That is to say, by (3.6), Π_n given $K_m = k_m$ is an exchangeable partition of $[n]$ with EPPF of the Gibbs form (4.39) for all compositions $(n_1, \ldots, n_k)$ of n, for some $v_{n,k}$ depending on m and k_m. According to Lemma 4.4, the $\mathbb{P}_{\alpha|s}$ distribution of Π_n is some weak limit of such laws, and any such limit is evidently of the same form. The precise formula (4.37) is deduced from Lemma 4.2. □

Theorem 4.6. *An exchangeable random partition of positive integers Π_∞ with infinite number of blocks has an EPPF of the Gibbs form*

$$p(n_1, \cdots, n_k) = c_{n,k} \prod_{i=1}^{k} w_{n_i}, \quad \text{with } n = \textstyle\sum_{i=1}^k n_i \tag{4.40}$$

for some positive weights $w_1 = 1, w_2, w_3, \ldots$ and some $c_{n,k}$ if and only if $w_j = (1-\alpha)_{j\uparrow}$ for all j for some $0 \leq \alpha < 1$. If $\alpha = 0$ then the distribution of Π_∞ is $\int_0^\infty \mathbb{P}_{0,\theta}\gamma(d\theta)$ for some probability distribution γ on $(0, \infty)$, whereas if $0 < \alpha < 1$ then the distribution of Π_∞ is $\int_0^\infty \mathbb{P}_{\alpha|s}\gamma(ds)$ for some such γ.

Sketch of proof. That the weights w_j chosen with $w_1 = 1$ are necessarily of the form $w_j = (1-\alpha)_{j\uparrow}$ for some $\alpha \in [0, 1)$ can be seen by elementary consideration of the addition rules of an EPPF, much as in Kerov [240] and Zabell [441]. For given α, the Gibbs prescription specifies the law of Π_n given $K_n, K_{n+1}, \ldots$. The extreme laws for this specification are then identified using the general theory of such problems [108], combined with the facts that $K_n/\log n \to \theta$ under $\mathbb{P}_{0,\theta}$ and $K_n/n^\alpha \to s$ under $\mathbb{P}_{\alpha|s}$ for $0 < \alpha < 1$. □

Exercises

4.3.1. [362] Check that the function defined by (4.36) satisfies the addition rule (2.9) for an EPPF.

4.3.2. (Problem) In Theorem 4.6 it is supposed that the exchangeable partition $\Pi_\infty = (\Pi_n)$ is such that Π_n is a $\text{Gibbs}_{[n]}(v_\bullet^{(n)}, w_\bullet)$ for some weight sequence $w_\bullet$ which does not depend on n, and $v_\bullet^{(n)}$ which does. What if Π_n is a $\text{Gibbs}_{[n]}(v_\bullet^{(n)}, w_\bullet^{(n)})$ partition, allowing both weight sequences to depend on n. Is a larger family of distributions of Π_∞ obtained? I guess no.

4.3.3. (Problem) Formulae (4.59) and (4.67) in Section 4.5 show that in the case $\alpha = \frac{1}{2}$ the integral $G_\alpha(q, t)$ in (4.38) can be simply expressed in terms of an entire function of a complex variable, the *Hermite function*, which has been extensively studied. It is natural to ask whether $G_\alpha(q, t)$ might be similarly represented in terms of some entire function with a parameter α, which reduces to the Hermite function for $\alpha = \frac{1}{2}$.

4.3.4. (Problem: Number of components a Markov chain) (Compare Exercise 2.1.4 and Exercise 3.2.4 Suppose Π_∞ is an exchangeable partition of $\mathbb{N}$ such that $(|\Pi_n|, n = 1, 2, \ldots)$ is a Markov chain, with possibly inhomogeneous transition probabilities, which increases to infinity. Is Π_∞ necessarily of the form described by Theorem 4.6?

4.4. Lengths of stable excursions

Following Lamperti [267], Wendel [431], Kingman [249], Knight [256], Perman, Pitman and Yor [339, 341, 366], consider the following construction of random partitions based on a random closed subset Z of $[0, \infty)$ such that Z has Lebesgue measure 0 almost surely. Without loss of generality, Z may be regarded as the closure of the zero set of some random process. For fixed or random $T > 0$ let

$$V_1^\downarrow(T) \geq V_2^\downarrow(T) \geq \cdots \geq 0 \tag{4.41}$$

be the ranked lengths of component intervals of the set $[0, T] \backslash Z$. For $i \geq 1$ let

$$P_i^\downarrow(T) := V_i^\downarrow(T)/T. \tag{4.42}$$

Then $(P_i^{\downarrow}(T), i \geq 1)$ is a sequence of ranked relative lengths with sum 1, which may be regarded as the frequencies of an exchangeable random partition $\Pi_\infty(T)$ of positive integers. Such a random partition could be generated by the random equivalence relation $i \sim j$ if TU_i and TU_j fall in the same component interval of $[0,T]\backslash Z$, where the U_i are independent and identically distributed uniform $[0,1]$ variables, independent of Z and T.

This construction is of particular interest when Z is the closure of the zero set of some strong Markov process X started at zero. If X is recurrent, and Z is not discrete, then it is well known that Z is the closure of the range of a subordinator $(T_\ell, \ell \geq 0)$ which is the inverse of a continuous local time process of X at zero, say $(L_t, t \geq 0)$.

The stable case If X is self-similar, e.g. a Brownian motion or Bessel process, or a stable Lévy process, then $(T_\ell, \ell \geq 0)$ is necessarily a stable(α) subordinator for some $0 < \alpha < 1$, so

$$\mathbb{E}[\exp(-\lambda T_\ell)] = \exp(-\ell K \lambda^\alpha) \tag{4.43}$$

for some $K > 0$.

Theorem 4.7. [366, 341, 371] *If Z is the closure of the range of a stable(α) subordinator then*

$$(P_i^{\downarrow}(T))_{i\geq 1} \text{ has } PD(\alpha, 0) \text{ distribution} \tag{4.44}$$

1. *for each fixed time $T > 0$, and*
2. *for each inverse local time $T = T_\ell$ for some $\ell > 0$.*

Proof. Recall from Definition Definition 3.3 that PD(α, θ) is the distribution of the sequence derived by ranking $(W_j \prod_{i=1}^{j-1}(1 - W_i))_{j\geq 1}$ for W_j independent with $W_j \overset{d}{=} \beta_{1-\alpha,\alpha+j\theta}$. Case 2 of the theorem is read from Theorem 4.3. Case 1 is left as an exercise. □

It is shown in [370] that (4.44) also holds for a number of other random times T besides inverse local times, for instance T the first time t that $V_n^{\downarrow}(t) = v$, for arbitrary fixed $v > 0$ and $n = 1, 2, \ldots$. As observed in [370], it follows from (4.20) that for any fixed or random T, a random partition $\Pi_\infty(T)$ with frequencies $(P_i^{\downarrow}(T), i \geq 1)$ has α-diversity

$$S_\alpha(\Pi_\infty(T)) = \Gamma(1-\alpha) \lim_{i\to\infty} i(P_i^{\downarrow}(T))^\alpha = \frac{KL_T}{T^\alpha} \text{ almost surely.} \tag{4.45}$$

In particular, if T is a constant time, the α-diversity of $\Pi_\infty(T)$ is just a constant multiple of the local time L_T, whereas if T is an inverse local time, the α-diversity is a constant multiple of $T^{-\alpha}$. So the two cases of (4.44) described in Theorem 4.7 are consistent with the consequence of (0.41) that L_T/T^α has the same distribution for any fixed time T as for $T = T_\ell$ for any $\ell > 0$, when $L_T = \ell$. Let

$$G_T = \sup(Z \cap [0,T]); \qquad D_T = \inf(Z \cap (T, \infty)). \tag{4.46}$$

A fundamental difference between fixed times and inverse local times is that if T is fixed then $G_T < T < D_T$ almost surely whereas if $T = T_\ell$ then $G_T = T = D_T$ almost surely. In the former case, the *meander interval* (G_T, T) is one of the component intervals whose lengths are ranked to form the sequence $(V_i^\downarrow(T), i \geq 1)$, so $T - G_T = V_{N_T}^\downarrow(T)$ for some random index N_T. To prove (4.44) for fixed times T it seems essential to understand the joint distribution of the length of the meander interval and the ranked lengths of the remaining intervals. This is specified by the following corollary of all known proofs of (4.44) for fixed T:

Corollary 4.8. [366, 341, 371] *For each fixed time T, the length of the meander interval $T - G_T$ is a size-biased pick from the whole collection of ranked interval lengths:*

$$\mathbb{P}(N_T = n \mid P_i^\downarrow(T), i \geq 1) = P_n^\downarrow(T), \tag{4.47}$$

$$(P_i^\downarrow(G_T))_{i\geq 1} \textit{ has PD}(\alpha, \alpha) \textit{ distribution,} \tag{4.48}$$

and

$$\textit{the meander length } T - G_T \textit{ is independent of } (P_i^\downarrow(G_T))_{i\geq 1}. \tag{4.49}$$

As a check, recall the well known fact that the distribution of G_T/T is beta$(\alpha, 1-\alpha)$ for each fixed T. So the distribution of $(T-G_T)/T$ is beta$(1-\alpha, \alpha)$. According to Corollary 4.8, this is the structural distribution of a size-biased pick from the ranked relative lengths $(P_i^\downarrow(T))_{i\geq 1}$. This agrees with (4.44) for fixed T, because the structural distribution of PD$(\alpha, 0)$ is beta$(1-\alpha, \alpha)$. Granted (4.44) and (4.47), the assertions (4.48) and (4.49) follow from Exercise 3.2.1 because $(P_i^\downarrow(G_T), i \geq 1)$ is derived from $(P_i^\downarrow(T), i \geq 1)$ by deleting a term picked by size-biased sampling, then renormalizing the rest of the terms.

Recall the definition of standardized bridges and excursions derived from a self-similar Markov process X. Then $(P_i^\downarrow(G_T), i \geq 1)$ is the sequence of ranked lengths of excursions of the bridge of length 1 derived by rescaling the path of X on $[0, G_T]$. Then (4.44) and (4.48) yield:

Corollary 4.9. *For each $\alpha \in (0, 1)$,*

- *the sequence of ranked lengths of excursions of a standard Bessel process of dimension $2-2\alpha$, up to time 1 and including the meander length $1-G_1$, has PD$(\alpha, 0)$ distribution;*
- *the sequence of ranked lengths of excursions of a standard Bessel bridge of dimension $2-2\alpha$ has PD(α, α) distribution.*

Exercises

4.4.1. (Proof of Case 1 of (4.44) and Corollary 4.8) See [366], [341] and [351] for three different approaches. Basically, what is required is a good understanding of, on the one hand, the joint law of $(P_i^\downarrow(T))_{i\geq 1}$ and $(T-G_T)/T$, for fixed T, which is provided by the last exit decomposition at time G_T given by excursion theory, and on the other hand, the joint law of $(P_i^{\downarrow}(T_\ell))_{i\geq 1}$ and a size-biased pick $\tilde{P}_1(T_\ell)$, which is provided by Lemma 4.1. If both are computed they are found to be identical, and the conclusion follows, along with Corollary 4.8. The approach of [351] is to use sampling with points of a Poisson process to reduce the result to an elementary combinatorial analog, involving sums of independent variables X_i as in Exercise 3.3.1 .

Notes and comments

This section is based on [366, 341, 371]. See [371] for a thorough treatment of the two-parameter family of Poisson-Dirichlet distributions. The study of ranked lengths of excursion intervals has a long history which is reviewed there. See [370, 368, 369] for various generalizations to do with ranked lengths, and [374] regarding ranked values of other functionals of self-similar Markovian excursions, such as their heights or areas.

4.5. Brownian excursions

Suppose in this section that $\mathbb{P}$ governs $B = (B_t, 0 \leq t \leq 1)$ as standard Brownian motion started at 0. Let $\Pi_\infty := (\Pi_n)$ be the *Brownian excursion partition.* That is the random partition of positive integers defined by the random equivalence relation $i \sim j$ iff U_i and U_j fall in the same excursion interval of B away from 0, where the U_i are independent and identically distributed uniform $[0,1]$ independent of B. According to the result of [366] and [347] recalled at the end of the last section,

$$\Pi_\infty \text{ is a } (\tfrac{1}{2}, 0) \text{ partition.} \tag{4.50}$$

That is, the sequence of partitions $(\Pi_n, n = 1, 2, \ldots)$ develops according to a variation of Pólya's urn scheme for random sampling with double replacement, described in Exercise 3.2.2 . Equivalently, the sequence $(\tilde{P}_j)$ of lengths of excursions of B on $[0,1]$, in the size-biased order of discovery of excursions by the sampling process, has the GEM$(\frac{1}{2}, 0)$ distribution defined by the products of independent variables (4.28) for W_i with beta$(\frac{1}{2}, \frac{1}{2}i)$ distribution. And the sequence $(P_i^{\downarrow})$ of ranked lengths of excursions of B on $[0,1]$ has PD$(\frac{1}{2}, 0)$ distribution. Note that each of the sequences $(\tilde{P}_j)$ and $(P_i^{\downarrow})$ has a term equal to the length $1 - G_1$ of the final meander interval. The common distribution of $\tilde{P}_1$ and $1 - G_1$ is beta$(\frac{1}{2}, \frac{1}{2})$, commonly known as the *arcsine law.*

Conditioned on $B_1 = 0$, the process B becomes a standard Brownian bridge, and we find instead that

$$(\Pi_\infty \,|\, B_1 = 0) \text{ is a } (\tfrac{1}{2}, \tfrac{1}{2}) \text{ partition.} \tag{4.51}$$

Let L_1 be the local time of B at 0 up to time 1, with the usual normalization of Brownian local time as occupation density relative to Lebesgue measure. Note from (4.45) that the $\frac{1}{2}$-diversity of Π_∞ is $\sqrt{2}L_1$. So the number K_n of components of Π_n grows almost surely like $\sqrt{2n}L_1$ as $n \to \infty$, both for the unconditioned Brownian excursions under $\mathbb{P}$, and for the excursions of Brownian bridge under $\mathbb{P}(\cdot \,|\, B_1 = 0)$. According to well known results of Lévy, unconditionally, L_1 has the same law as $|B_1|$, that is

$$\mathbb{P}(L_1 \in d\lambda) = \mathbb{P}(|B_1| \in d\lambda) = 2\varphi(\lambda)d\lambda \qquad (\lambda > 0)$$

where $\varphi(z) := (1/\sqrt{2\pi})\exp(-\frac{1}{2}z^2)$ for $z \in \mathbb{R}$ is the standard Gaussian density of B_1, whereas the conditional law of L_1 given $B_1 = 0$ is the *Rayleigh distribution* which is derived by size-biasing the unconditional law of L_1.

$$\mathbb{P}(L_1 \in d\lambda \,|\, B_1 = 0) = \sqrt{\pi/2}\,\lambda\,\varphi(\lambda)d\lambda \qquad (\lambda > 0). \tag{4.52}$$

Conditioning on the local time For $\lambda \geq 0$ let $\Pi_\infty(\lambda)$ denote a random partition of positive integers with

$$\Pi_\infty(\lambda) \stackrel{d}{=} (\Pi_\infty \,|\, L_1 = \lambda) \stackrel{d}{=} (\Pi_\infty \,|\, L_1 = \lambda, B_1 = 0) \tag{4.53}$$

where $\stackrel{d}{=}$ denotes equality in distribution, the second equality in distribution is due to Lemma 4.4, and the law of $\Pi_\infty(\lambda)$ is easily seen to be weakly continuous as a function of λ. The following lemma presents $\Pi_\infty(\lambda)$ in terms of a family of conditioned Brownian bridges, which turns out to be of interest in a number of contexts.

Lemma 4.10. [358] *Let $L_1(B^{\mathrm{br}})$ denote the local time at 0 up to time 1 for a standard Brownian bridge B^{br}. There exists for each $\lambda \geq 0$ a unique law on $C[0,1]$ of a process $B^{\mathrm{br}}_\lambda := (B^{\mathrm{br}}_\lambda(u), 0 \leq u \leq 1)$ such that*

$$B^{\mathrm{br}}_\lambda \stackrel{d}{=} (B^{\mathrm{br}} \,|\, L_1(B^{\mathrm{br}}) = \lambda) \tag{4.54}$$

and the law of B^{br}_λ is a weakly continuous function of λ. The law of B^{br}_λ is uniquely determined by the following two properties:

1. *The complement of the zero set of B^{br}_λ is an exchangeable random partition of $[0,1]$ into open intervals such that the corresponding partition $\Pi_\infty(\lambda)$ of positive integers, whose frequencies are the ranked lengths of these intervals, has the distribution denoted $\mathbb{P}_{\frac{1}{2}|\sqrt{2}\lambda}$ in Lemma 4.4, so $\Pi_\infty(\lambda)$ has $\frac{1}{2}$-diversity $\sqrt{2}\lambda$.*
2. *Conditionally given the interval partition generated by its zero set, the excursion of B^{br}_λ over each interval of length t is distributed as a Brownian excursion of length t, independently for the different intervals, with the signs of excursions chosen by a further independent process of fair coin tossing.*

In particular, B^{br}_0 is a standard signed Brownian excursion, whose sign is chosen by a fair coin-toss independent of $|B^{\mathrm{br}}_0|$.

Let $\tilde{P}_j(\lambda)$ denote the frequency of the jth class of $\Pi_\infty(\lambda)$. So $(\tilde{P}_j(\lambda), j = 1, 2 \ldots)$ is distributed like the lengths of excursions of B over $[0,1]$ given $L_1 = \lambda$, as discovered by a process of length-biased sampling. In view of Lévy's formula for the stable($\frac{1}{2}$) density, the general formula (4.7) reduces for $\alpha = \frac{1}{2}$ to the following more explicit formula for the structural density of $\Pi_\infty(\lambda)$:

$$\mathbb{P}(\tilde{P}_1(\lambda) \in dp) = \frac{\lambda}{\sqrt{2\pi}} p^{-\frac{1}{2}}(1-p)^{-\frac{3}{2}} \exp\left(-\frac{\lambda^2}{2}\frac{p}{(1-p)}\right) dp \qquad (0 < p < 1) \tag{4.55}$$

or equivalently

$$\mathbb{P}(\tilde{P}_1(\lambda) \leq y) = 2\Phi\left(\lambda\sqrt{\frac{y}{1-y}}\right) - 1 \qquad (0 \leq y < 1) \tag{4.56}$$

where $\Phi(z) := \mathbb{P}(B_1 \leq z)$ is the standard Gaussian distribution function. Now (4.56) amounts to

$$\tilde{P}_1(\lambda) \stackrel{d}{=} \frac{B_1^2}{\lambda^2 + B_1^2} \stackrel{d}{=} \frac{1}{T_\lambda + 1} \tag{4.57}$$

where $(T_\lambda, \lambda \geq 0)$ is the stable($\frac{1}{2}$) subordinator with $\mathbb{E}(e^{-\xi T_\lambda}) = \exp(-\lambda\sqrt{2\xi})$. Furthermore [18, Corollary 5], by a similar analysis using Lemma 4.1 , the whole sequence $(\tilde{P}_j(\lambda), j \geq 1)$ may be represented as

$$\tilde{P}_j(\lambda) = R_{j-1}(\lambda) - R_j(\lambda) \text{ with } R_j(\lambda) = \frac{\lambda^2}{\lambda^2 + S_j} \tag{4.58}$$

for $S_j = \sum_{i=1}^{j} X_i$ with X_i independent and identically distributed copies of B_1^2. Then $\Pi_\infty(\lambda)$ can be constructed from $(\tilde{P}_j(\lambda), j \geq 1)$ as in part (ii) of Theorem 3.1.

Moments of the structural distribution of $\Pi_\infty(\lambda)$ From (4.57), for $q > -\frac{1}{2}$ the qth moment of the structural distribution of $\tilde{P}_1(\lambda)$ is

$$\mathbb{E}[(\tilde{P}_1(\lambda))^q] = \mathbb{E}\left[\left(\frac{B_1^2}{\lambda^2 + B_1^2}\right)^q\right] = \mathbb{E}(|B_1|^{2q})\, h_{-2q}(\lambda) \tag{4.59}$$

where

$$\mathbb{E}(|B_1|^{2q}) = 2^q \frac{\Gamma(q+\frac{1}{2})}{\Gamma(\frac{1}{2})} = 2^{-q}\frac{\Gamma(2q+1)}{\Gamma(q+1)} \qquad (q > -\tfrac{1}{2}) \tag{4.60}$$

by the gamma($\frac{1}{2}$) distribution of $2B_1^2$ and the duplication formula for the gamma function, and h_{-2q} is the *Hermite function of index* $-2q$, that is $h_0(\lambda) = 1$ and for $q \neq 0$

$$h_{-2q}(\lambda) := \frac{1}{2\Gamma(2q)} \sum_{j=0}^{\infty} \Gamma(q + j/2) 2^{q+j/2} \frac{(-\lambda)^j}{j!}. \tag{4.61}$$

The second equality in (4.59) is the integral representation of the Hermite function provided by Lebedev [280, Problem 10.8.1], and (4.61) is read from [280,

(10.4.3)]. Note that h_n for $n = 0, 1, 2, \ldots$, which can be evaluated from (4.61) by continuity as $-2q$ approaches n, is the usual sequence of Hermite polynomials orthogonal with respect to the standard Gaussian density $\varphi(x)$. The function $h_{-1}(x)$ for real x is *Mill's ratio* [221, 33.7]:

$$h_{-1}(x) = \frac{\mathbb{P}(B_1 > x)}{\varphi(x)} = e^{\frac{1}{2}x^2} \int_x^\infty e^{-\frac{1}{2}z^2}\, dz. \tag{4.62}$$

For all complex ν and z, the Hermite function satisfies the recursion

$$h_{\nu+1}(z) = z h_\nu(z) - \nu h_{\nu-1}(z), \tag{4.63}$$

which combined with (4.62) and $h_0(x) = 1$ yields

$$h_{-2}(x) = 1 - x h_{-1}(x) \tag{4.64}$$

$$2! h_{-3}(x) = -x + (1 + x^2) h_{-1}(x) \tag{4.65}$$

$$3! h_{-4}(x) = 2 + x^2 - (3x + x^3) h_{-1}(x) \tag{4.66}$$

and so on, as discussed further in [362]. Theorem 4.5 now yields:

Corollary 4.11. *The distribution of $\Pi_\infty(\lambda)$, a Brownian excursion partition conditioned on $L_1 = \lambda$, is determined by the EPPF*

$$p_{\frac{1}{2}|\sqrt{2}\lambda}(n_1, \ldots, n_k) = 2^{n-k} \lambda^{k-1} h_{k+1-2n}(\lambda) \prod_{i=1}^k (\tfrac{1}{2})_{n_i - 1\uparrow}. \tag{4.67}$$

To illustrate, according to (2.25) and (4.59), or (4.67) for $n = 2$, given $L_1 = \lambda$, two independent uniform $[0, 1]$ variables fall in the same excursion interval of a Brownian motion or Brownian bridge conditioned on $L_1 = \lambda$ with probability

$$p_{\frac{1}{2}|\sqrt{2}\lambda}(2) = \mathbb{E}[\tilde{P}_1(\lambda)] = h_{-2}(\lambda) = 1 - \lambda h_{-1}(\lambda) \tag{4.68}$$

and in different components with probability

$$p_{\frac{1}{2}|\sqrt{2}\lambda}(1, 1) = \lambda h_{-1}(\lambda) = \frac{\lambda \mathbb{P}(B_1 > \lambda)}{\varphi(\lambda)}. \tag{4.69}$$

Note that this function increases from 0 to 1 as λ increases from 0 to ∞, as should be intuitively expected: given L_1 is close to 0 the Brownian excursion partition most likely contains one large interval of length close to 1, whereas given L_1 is large the Brownian excursion partition most likely has maximum interval length close to 0. Note that (4.69) implies $\mathbb{P}(B_1 > \lambda) < \varphi(\lambda)/\lambda$ for $\lambda > 0$, which is a standard estimate for the Gaussian tail probability.

Exercises

4.5.1. (Proof of Lemma 4.10)

4.5.2. (The Brownian pseudo-bridge) [59] Let $\tau_\ell := \inf\{t : L_t(B) > \ell\}$ where $L_t(B), t \geq 0$ is the local time process at 0 of an unconditioned Brownian motion B. Let $B_*[0, \tau_\ell]$ be the standardized path defined as in (0.9) . Then for each $\ell > 0$

$$\mathbb{P}(B_*[0,\tau_\ell] \in \cdot) = \int_0^\infty \mathbb{P}(B_\lambda^{\mathrm{br}} \in \cdot)\sqrt{\frac{2}{\pi}} e^{-\frac{1}{2}\lambda^2} d\lambda. \tag{4.70}$$

Deduce that the common distribution of the *Brownian pseudo-bridge* $B_*[0, \tau_\ell]$ for all $\ell > 0$ is mutually absolutely continuous with respect to that of B^{br}, and describe the density. To check (4.70), use the basic *switching identity*

$$(B_*[0,\tau_\ell] \text{ given } \tau_\ell = t) \stackrel{d}{=} (B^{\mathrm{br},t} \text{ given } L_t(B^{\mathrm{br},t}) = \ell) \tag{4.71}$$

where $B^{\mathrm{br},t}$ is a Brownian bridge of length t, and both conditional distributions in (4.71) are everywhere determined by weak continuity t and ℓ.

4.5.3. Show that the function $h_\nu(\lambda)$ defined by (4.59) for $\nu < 1$ and $\lambda > 0$ satisfies the recursion (4.63) for the Hermite function.

4.5.4. Check that $\Pi_\infty(\lambda)$ can be constructed using the following seating plan in the Chinese Restaurant, which obeys the rules of probability by virtue of the recursion (4.63) for the Hermite function: given that at stage n there are $n_j \geq 1$ customers at table j for $1 \leq j \leq k$ with $\sum_{j=1}^k n_j = n$, the $(n+1)$th customer sits at table j with probability $(2n_j - 1)h_{k-2n-1}(\lambda)/h_{k-2n+1}(\lambda)$, and at a new table with probability $\lambda h_{k-2n}(\lambda)/h_{k-2n+1}(\lambda)$.

4.5.5. [351, Corollary 3], [362]. Let $1 \leq k \leq n$. For the partition of n uniform random sample points generated the excursions of an unconditioned Brownian motion

$$\mathbb{P}(|\Pi_n| = k) = \binom{2n-k-1}{n-1} 2^{k+1-2n} \tag{4.72}$$

whereas for the Brownian bridge

$$\mathbb{P}(|\Pi_n| = k \,|\, B_1 = 0) = \frac{k(n-1)!}{(\frac{1}{2})_{n-1\uparrow}} \mathbb{P}(|\Pi_n| = k) \tag{4.73}$$

and for $\Pi_n(\lambda)$ distributed like either of these partitions given $L_1 = \lambda$

$$\mathbb{P}(|\Pi_n(\lambda)| = k) = \frac{(2n-k-1)!\,\lambda^{k-1} h_{k+1-2n}(\lambda)}{(n-k)!(k-1)!2^{n-k}}. \tag{4.74}$$

Some insight into formula (4.72) is provided by Chapter 5.

4.5.6. (Problem: existence of α-fragmentations) It is shown in Chapter 9 that for $\alpha = \frac{1}{2}$ there is a natural construction of $(\Pi_\infty(\lambda), \lambda \geq 0)$ as a partition valued *fragmentation process*, meaning that $\Pi_\infty(\lambda)$ is constructed for each λ on the same probability space, in such a way that $\Pi_\infty(\lambda)$ is a coarser partition than $\Pi_\infty(\mu)$ whenever $\lambda < \mu$. The question of whether a similar construction is possible for index α instead of index $\frac{1}{2}$ remains open. A natural guess is that such a construction might be made with one of the self-similar fragmentation processes of Bertoin [46], but Miermont and Schweinsberg [312] have shown that a construction of this form is possible only for $\alpha = \frac{1}{2}$.

Notes and comments

See [91] for more about conditioning B^{br} on its local time at 0, and [10], [427] regarding the more difficult problem of conditioning a Brownian path fragment on its entire local time process.

5

Coagulation and fragmentation processes

The work of Kingman and others shows how the theory of exchangeable random partitions and associated random discrete distributions provides a natural mathematical framework for the analysis of coagulation processes (also called coalescents) and fragmentation processes.

This chapter provides an introduction to this framework. Later chapters provide more detailed analysis of two particular coalescent processes, the *multiplicative* and *additive* coalescents, which are of special combinatorial significance.

5.1. Coalescents Kingman's coalescent is a Markov process with state space the set $\mathcal{P}_{\mathbb{N}}$ of all partitions of $\mathbb{N} := \{1, 2, \ldots\}$. Markovian coalescent processes are considered with more general transition mechanisms and state space either $\mathcal{P}_{\mathbb{N}}$ or the set $\mathcal{P}_{[n]}$ of partitions of $[n] := \{1, \ldots, n\}$. In a coalescent process with collision rate $K(x, y)$, each pair of blocks of size x and y may merge to form a combined block of size $x + y$. The case $K(x, y) \equiv 1$ gives Kingman's coalescent. Taking $K(x, y) = xy$ gives the *multiplicative coalescent*, studied further in Section 6.4, and $K(x, y) = x+y$ gives the *additive coalescent* studied further in Section 10.2. A special feature of Kingman's coalescent is that the $\mathcal{P}_{[n]}$-valued processes can be defined consistently as n varies. Successive generalizations of this property lead to the study of *coalescents with multiple collisions* [357] including the *Bolthausen-Sznitman coalescent* [72], and *coalescents with simultaneous multiple collisions* [396]. The transition mechanisms of these coalescents involve a coagulation operator on $\mathcal{P}_{\mathbb{N}}$, denoted p-COAG, which is associated with an arbitrary probability distribution p on $\mathcal{P}_{\mathbb{N}}$.

5.2. Fragmentations This section initiates discussion of a fundamental duality between processes of coagulation and fragmentation. A fragmentation operator on $\mathcal{P}_{\mathbb{N}}$, denoted p-FRAG is associated with an arbitrary probability distribution p on $\mathcal{P}_{\mathbb{N}}$.

5.3. Representations of infinite partitions This section reviews some basic facts and terminology related to different representations of infinite partitions which are useful in the study of processes of coagulation and fragmentation.

5.4. Coagulation and subordination The basic coagulation operator p-COAG has a natural interpretation in terms of subordination (composition) of increasing processes with exchangeable increments. In particular, taking p to be the Poisson-Dirichlet distribution with parameters $(\alpha, 0)$, and substituting $\alpha = e^{-t}$, yields the semigroup of the Bolthausen-Sznitman coalescent [72].

5.5. Coagulation – fragmentation duality The semigroup property of the Bolthausen-Sznitman coalescent is related to a more general duality formula involving coagulation and fragmentation kernels associated with the two-parameter Poisson-Dirichlet family. Another instance of this duality involves the asymptotic features of the combinatorial structure of random mappings.

5.1. Coalescents

The first paragraph introduces *Kingman's coalescent.* The following paragraphs recall how this process has been generalized to construct various other partition-valued coalescent processes.

Kingman's coalescent Motivated by applications in the theory of genetic diversity, concerning the evolution over time of the distribution of different genetic types in a large population, Kingman [253] discovered the remarkable $\mathcal{P}_{\mathbb{N}}$-valued process described by the following theorem. As shown by Kingman, this process arises naturally as a limit process governing lines of descent, viewed backwards in time, from numerous natural models of population genetics, with the first n integers labelling the first n individuals sampled from a large population.

Theorem 5.1. Kingman [253] *There exists a uniquely distributed $\mathcal{P}_{\mathbb{N}}$-valued process $(\Pi_\infty(t), t \geq 0)$, called* Kingman's coalescent, *with the following properties:*

- $\Pi_\infty(0)$ *is the partition of* $\mathbb{N}$ *into singletons;*
- *for each n the restriction $(\Pi_n(t), t \geq 0)$ of $(\Pi_\infty(t), t \geq 0)$ to $[n]$ is a Markov chain with càdlàg paths with following transition rates: from state $\Pi = \{A_1, \ldots, A_k\} \in \mathcal{P}_{[n]}$, the only possible transitions are to one of the $\binom{k}{2}$ partitions $\Pi_{i,j}$ obtained by merging blocks A_i and A_j to form $A_i \cup A_j$, and leaving all other blocks unchanged, for some $1 \leq i < j \leq k$, with*

$$\Pi \to \Pi_{i,j} \text{ at rate } 1 \tag{5.1}$$

Proof. This follows easily from the consistency of the descriptions for different values of n, and Kolmogorov's extension theorem. Or see Exercise 5.1.1 . □

If $(\Pi_\infty(t), t \geq 0)$ is Kingman's coalescent, then it is easily shown that $\Pi_\infty(t)$ is an exchangeable random partition of $\mathbb{N}$ for each $t \geq 0$, with proper frequencies for each $t > 0$, which may be described as follows. Let $D_t := |\Pi_\infty(t)|$. The process $(D_t, t > 0)$ is the homogeneous Markovian death process with state space $\mathbb{N}$ whose only transitions are $n \to n-1$ at rate $\binom{n}{2}$, with initial condition $D(0+) = \infty$. This process can be constructed in an obvious way from its sequence of holding times ε_n in state $n \geq 2$, where $\sum_{n=2}^{\infty} \varepsilon_n < \infty$ a.s.. The frequencies of $\Pi_\infty(t)$, in exchangeable random order, can then be constructed as the lengths of intervals obtained by cutting $[0,1]$ at $D_t - 1$ independent points $U_1, \ldots, U_{D_t-1}$, where $U_1, U_2, \ldots$ is a sequence of independent uniform $[0,1]$ variables independent of $(D_t, t \geq 0)$. So the conditional distribution of $\Pi_\infty(t)$ given $D_t = m$ does not depend on t: for each *particular* partition of $[n]$ into k subsets of sizes $(n_1, \ldots, n_k)$, the probability that the restriction of $\Pi_\infty(t)$ to $[n]$ coincides with that partition is given by the formula

$$p(n_1, \ldots, n_k) = (m)_{k\downarrow} \frac{\prod_{i=1}^{k} (\kappa)_{n_i\uparrow}}{(m\kappa)_{n\uparrow}} \tag{5.2}$$

for $\kappa = 1$, as in (2.17) . Unconditional partition probabilities for $\Pi_\infty(t)$ can then be obtained by mixing with respect to the distribution of D_t, which is known explicitly but complicated. If $\mathcal{I}_t$ denotes the interval partition of $[0,1]$ into D_t intervals, so defined, then $\Pi_\infty(t)$ can be constructed as the partition generated by $\sim_t$, where $i \sim_t j$ iff V_i and V_j fall in the same component interval of $\mathcal{I}_t$, where $V_1, V_2, \ldots$ is a further independent sequence of independent uniform $[0,1]$ variables.

Coalescents defined by a collision kernel One natural generalization of Kingman's coalescent, considered already by Marcus [296] and Lushnikov [291], is to allow the rate of collisions between two blocks to depend on the number of elements in the blocks. More generally, it is convenient to allow the collision rate to depend on some other measure of the blocks besides their size.

Let μ be a *mass distribution* on $[n]$, say $\mu(A) = \sum_{i \in A} \mu_i$ for some μ_i, and assume $\mu_i > 0$ to avoid annoyances. Let $K(x,y)$ be a non-negative symmetric measurable function of $x, y \in [0,1]$, called a *collision rate kernel.* Let $\mathcal{P}_{[n]}$ denote the set of partitions of the set $[n]$, and $\mathcal{P}_n$ the set of partitions of the integer n.

Definition 5.2. Call a $\mathcal{P}_{[n]}$-valued process $(\Pi(t), t \geq 0)$ a *K-coalescent with mass distribution* μ if $(\Pi(t), t \geq 0)$ is a continuous time Markov chain, with right-continuous step function paths, with the following time-homogeneous transition rates: from state $\Pi = \{A_1, \ldots, A_k\} \in \mathcal{P}_{[n]}$ the only possible transitions are to one of the $\binom{k}{2}$ partitions $\Pi_{i,j}$ obtained by merging blocks A_i and A_j to form $A_i \cup A_j$, and leaving all other blocks unchanged, for some $1 \leq i < j \leq k$, with

$$\Pi \to \Pi^{i,j} \text{ at rate } K(\mu(A_i), \mu(A_j)). \tag{5.3}$$

To illustrate, three cases of special interest are

- *Kingman's coalescent* with $K(x,y) \equiv 1$;

- the *multiplicative coalescent* with $K(x,y) = xy$, discussed further in Section 6.4;
- the *additive coalescent* with $K(x,y) = x + y$, discussed further in Section 10.2.

Let $\mathcal{P}^{\downarrow}_{\text{finite}}$ be the set of *finite real partitions*, meaning decreasing sequences of non-negative reals $(x_1, x_2, \ldots)$ with only a finite number of non-zero terms. Then $(x_1, x_2, \ldots)$ may be called a finite real partition of $\sum_i x_i$. Let $X(t)$ be the sequence of ranked μ-masses of blocks of $\Pi(t)$. If $(\Pi(t), t \geq 0)$ is a $\mathcal{P}_{[n]}$-valued K-coalescent with mass distribution μ then the corresponding *ranked coalescent* $(X(t), t \geq 0)$, with values in $\mathcal{P}^{\downarrow}_{\text{finite}}$, is a Markov chain with the following transition mechanism, which depends neither on n nor on the choice of the mass distribution μ.

Definition 5.3. For K a collision kernel as in Definition 5.2, a *$\mathcal{P}^{\downarrow}_{\text{finite}}$-valued K-coalescent* is a continuous time Markov chain with state-space $\mathcal{P}^{\downarrow}_{\text{finite}}$, with right-continuous step function paths governed by the following time-homogeneous transition rates: from state $x = (x_1, x_2, \ldots) \in \mathcal{P}^{\downarrow}_{\text{finite}}$ the only possible transitions are to one of the states $x^{\oplus(i,j)} := (x_1^{\oplus(i,j)}, x_2^{\oplus(i,j)}, \ldots)$ derived from x by picking indices $i < j$ with $x_i > 0$ and $x_j > 0$, and replacing the two terms x_i and x_j by a single term $x_i + x_j$, and re-ranking, with

$$x \to x^{\oplus(i,j)} \text{ at rate } K(x_i, x_j). \tag{5.4}$$

Note that for each n the set $\mathcal{P}_n$ of partitions of the integer n may be regarded as a subset of $\mathcal{P}^{\downarrow}_{\text{finite}}$. Then for μ the counting measure on $[n]$, the ranked coalescent $(X(t), t \geq 0)$ provides a natural representation of the process of integer partitions induced by a $\mathcal{P}_{[n]}$-valued K-coalescent $(\Pi(t), t \geq 0)$ with mass defined by counting measure. This model for a coalescing process of set partitions and associated integer partitions was first studied by Marcus [296]. Lushnikov [291] showed that for kernels K of the special form $K(x,y) = xk(y) + yk(x)$ for some function k, and initial state $\Pi(0)$ that is the partition of $[n]$ into singletons, $\Pi(t)$ has a Gibbs distribution with weights $w_j(t)$ determined as the solutions of a system of differential equations.

The advantage of working with a $\mathcal{P}_{[n]}$-valued coalescent $(\Pi(t), t \geq 0)$ is that this process keeps track of the *merger history* of collisions of blocks over time, with a nice consistent labeling system for all time. In passing to the ranked coalescent $(X(t), t \geq 0)$ there is some loss of information in the merger history, which makes these processes difficult to handle analytically. But ranked coalescents allow coalescent processes derived from sets of different sizes to be readily compared by suitable scaling. Note the special feature of Kingman's coalescent that the $\mathcal{P}_{[n]}$-valued processes can be defined consistently as n varies to induce a $\mathcal{P}_{\mathbb{N}}$-valued coalescent whose ranked frequencies are then governed by the corresponding ranked-mass coalescent with state-space $\mathcal{P}^{\downarrow}_{\text{finite}}$, provided the time parameter t is restricted to $t > 0$. As shown in the next paragraph, this construction can be generalized to a larger class of coalescent processes, but this class does not appear to include K-coalescents for any non-constant K.

Given an arbitrary collision rate kernel $K(x, y)$, subject only to joint symmetry and measurability in x and y, the above discussion shows how to define a Markov process of jump-hold type with state space $\mathcal{P}^{\downarrow}_{\text{finite}}$, the set of finite real partitions. Since $\mathcal{P}^{\downarrow}_{\text{finite}}$ is dense in various spaces of *finite or infinite partitions* meaning decreasing sequences (x_i), equipped with suitable metrics, it is natural to expect that under suitable regularity conditions on K it should be possible to define a nice Markov process with such a state space, by continuous extension of the coalescent transition mechanism defined on $\mathcal{P}^{\downarrow}_{\text{finite}}$. For a general kernel K subject to a Lipschitz condition, such a result was obtained in [141], but only for restricted spaces of partitions equipped with particular metrics. Particular results for the multiplicative and additive coalescents will be discussed in later chapters, as indicated above. See [356, 141] for further background on Markovian coalescent processes, and [11] for a broader review of stochastic and deterministic models for coalescent evolutions.

Consistent Markovian coalescents Consider now a $\mathcal{P}_{\mathbb{N}}$-valued coalescent process, which has the property of Kingman's coalescent that for each n its restriction to $[n]$, say $(\Pi_n(t), t \geq 0)$, is a Markov chain with stationary transition probabilities. To start with a simple case, suppose that each of these processes evolves according to the following dynamics:

- when $\Pi_n(t)$ has b blocks each k-tuple of blocks of $\Pi_n(t)$ is merging to form a single block at rate $\lambda_{b,k}$

Clearly, the law of such a process is determined by the rates $\{\lambda_{b,k} : 2 \leq k \leq b\}$. Not all collections of rates $\{\lambda_{b,k} : 2 \leq k \leq b\}$ are possible however. For instance, you cannot have both $\lambda_{3,3} = 1$ and $\lambda_{2,2} = 0$.

Theorem 5.4. [357] *A $\mathcal{P}_{\mathbb{N}}$-valued coalescent process, whose restriction $(\Pi_n(t), t \geq 0)$ to $[n]$ is a Markov chain with stationary transition transition rates $\lambda_{b,k}$ as described above, exists for all starting partitions of $\mathbb{N}$ iff the rates satisfy the* consistency condition

$$\lambda_{b,k} = \lambda_{b+1,k} + \lambda_{b+1,k+1} \qquad (2 \leq k \leq b),$$

in which case

$$\lambda_{b,k} = \int_0^1 x^{k-2}(1-x)^{b-k}\Lambda(dx) \tag{5.5}$$

for some non-negative and finite measure Λ on $[0,1]$. Then the sequence of coalescents $(\Pi_n(t), t \geq 0)$, with $\Pi_n(0)$ the partition of $[n]$ into singletons, defines an exchangeable $\mathcal{P}_{\mathbb{N}}$-valued coalescent $(\Pi_\infty(t), t \geq 0)$, called a Λ-coalescent. The frequencies of $\Pi_\infty(t)$ sum to 1 for all $t > 0$ iff

$$\int_0^1 x^{-1}\Lambda(dx) = \infty,$$

in which case these frequencies define a ranked Markovian coalescent with state-space the set of real partitions of 1.

Sketch of proof. The necessity of the consistency condition is evident by consideration of the following rates:

$$\{1\},\ldots,\{b\} \to \{1,\ldots,k\},\{k+1\},\ldots,\{b\} \text{ at rate } \lambda_{b,k}$$

$$\{1\},\ldots,\{b+1\} \to \{1,\ldots,k,b+1\},\{k+1\},\ldots,\{b\} \text{ at rate } \lambda_{b+1,k+1}$$

$$\{1\},\ldots,\{b+1\} \to \{1,\ldots,k\},\{k+1\},\ldots,\{b+1\} \text{ at rate } \lambda_{b+1,k}$$

Sufficiency is clear by the elementary criterion for a function of a Markov chain to be Markov [389, Section IIId]. The integral representation (5.5) follows from the consistency condition by de Finetti's theorem. □

Examples:

- If $\Lambda = \delta_0$, then $\lambda_{b,k} = 1(k = 2)$ (Kingman's coalescent)
- If $\Lambda(dx) = dx$, then $\lambda_{b,k} = \frac{(k-2)!(b-k)!}{(b-1)!}$ (the Bolthausen-Sznitman coalescent [72])

The Bolthausen-Sznitman coalescent has the following property: when started with initial state that is all singletons, the distribution of ranked frequencies of $\Pi_\infty(t)$ is that of the ranked jumps of a stable(α) subordinator $(T_s^{(\alpha)}, 0 \le s \le 1)$ normalized by $T_1^{(\alpha)}$, for $\alpha = e^{-t}$. See [72, 357, 54, 50, 176, 29] for various proofs of this fact, and more about this remarkable process.

The general probabilistic meaning of the measure Λ is clarified as follows:

Corollary 5.5. [357] *Let $\tau_{i,j}$ be the least t such that i and j are in the same block of $\Pi_\infty(t)$. Given $\tau_{i,j} > 0$ and $|\Pi_\infty(\tau_{i,j}-)| = \infty$ a random variable $X_{i,j}$ with distribution $\Lambda(\cdot)/\Lambda[0,1]$ is obtained as the almost sure relative frequency of blocks of $\Pi_\infty(\tau_{i,j}-)$ which merge at time $\tau_{i,j}$ to form the block containing both i and j.*

See [357] for the proof, and a more careful statement which applies also to the case when $|\Pi_\infty(\tau_{i,j}-)| < \infty$. In particular, the corollary applies to the Bolthausen-Sznitman coalescent, and shows that in this process, each time a collision occurs, the relative frequency of the blocks which merge to form the new block is a uniform$(0,1)$ variable. Note that in this case, and whenever the number of blocks stays infinite (see Exercise 5.1.3), the set of collision times is a.s. dense in $\mathbb{R}_{\ge 0}$.

Coagulation operators

Definition 5.6. For a partition π of $[n]$, where $n \in \mathbb{N} \cup \{\infty\}$ and $[\infty] := \mathbb{N}$, write $\pi = \{A_1, A_2, \ldots\}$ to indicate that the blocks of π in increasing order of their least elements are $A_1, A_2, \ldots$, with the convention $A_i = \emptyset$ for $i > |\pi|$. For a partition $\pi = \{A_1, A_2, \ldots\}$ of $\mathbb{N}$ and a partition $\Pi := \{B_1, B_2, \ldots\}$ of $[n]$ with $n \ge |\pi|$ let the *Π-coagulation of π* be the partition of $\mathbb{N}$ whose blocks are the non-empty sets of the form $\cup_{j \in B_i} A_j$ for some $i = 1, 2, \ldots$. For each probability distribution p on $\mathcal{P}_{\mathbb{N}}$, define a Markov kernel p-COAG on $\mathcal{P}_{\mathbb{N}}$, the *p-coagulation kernel*, as follows: for $\pi \in \mathcal{P}_{\mathbb{N}}$ let $p\text{-COAG}(\pi, \cdot)$ be the distribution of the Π-coagulation of π for Π with distribution p.

Think of Π as describing a coagulation of singleton subsets into the blocks $B_1, B_2, \ldots$. Then the Π-coagulation of π describes a corresponding coagulation of blocks of π.

Let Π_∞^π be a $\mathcal{P}_{\mathbb{N}}$-valued coalescent process with $\Pi_\infty^\pi(0) = \pi$ for some π with $|\pi| = n \in \mathbb{N} \cup \{\infty\}$. Then it is easily seen that

$$\Pi_\infty^\pi(t) = \text{the } \Pi_n(t)\text{-coagulation of } \pi \text{ for } t \geq 0 \tag{5.6}$$

for some uniquely defined $\mathcal{P}_{[n]}$-valued coalescent process Π_n with initial state 1^n, the partition of $[n]$ into singletons.

Theorem 5.7. [357, Theorem 6] *A coalescent process Π_∞^π starting at π with $|\pi| = n$ for some $1 \leq n \leq \infty$ is a Λ-coalescent if and only if Π_n defined by* (5.6) *is distributed as the restriction to $[n]$ of a Λ-coalescent. The semigroup of the Λ-coalescent on $\mathcal{P}_{\mathbb{N}}$ is thus given by*

$$\mathbb{P}^{\Lambda,\pi}(\Pi_\infty(t) \in \cdot) = p_t^\Lambda\text{-COAG}(\pi, \cdot) \tag{5.7}$$

where $p_t^\Lambda(\cdot) := \mathbb{P}^{\Lambda,1^\infty}(\Pi_\infty(t) \in \cdot)$ is the distribution of an exchangeable random partition of $\mathbb{N}$ with the EPPF $p_t^\Lambda(n_1, \ldots, n_k)$ which is uniquely determined by Kolmogorov equations for the finite state chains Π_n for $n = 2, 3, \ldots$.

Recall from Section 2.2 that the EPPF (exchangeable partition probability function) is a symmetric function of $(n_1, \ldots, n_k)$ giving the probability of each *particular* partition of an n element set into k subsets of sizes $(n_1, \ldots, n_k)$. Unfortunately, it seems possible to describe the EPPF $p_t^\Lambda(n_1, \ldots, n_k)$ explicitly only in very special cases, most notably the Kingman and Bolthausen-Sznitman coalescents. See [357] for further discussion.

Note from Definition 5.6 that no matter what the distribution p on $\mathcal{P}_{\mathbb{N}}$, each of the kernels $K = p\text{-COAG}$ acts *locally* on $\mathcal{P}_{\mathbb{N}}$, meaning that if Π^π denotes a random partition of $\mathbb{N}$ with distribution $K(\pi, \cdot)$, and R_n denotes the operation of restriction of a partition of $\mathbb{N}$ to $[n]$, then for each n the distribution of $\mathsf{R}_n\Pi^\pi$ depends on π only through $\mathsf{R}_n\pi$. It follows that any $\mathcal{P}_{\mathbb{N}}$-valued Markov process Π_∞, each of whose transition kernels is of the form $p\text{-COAG}$ for some p, is such that the $\mathcal{P}_n$-valued process $\mathsf{R}_n\Pi_\infty$ is a Markov chain. Such a coalescent process Π_∞ with càdlàg paths can therefore be constructed more generally from a consistent family of Markov chains with more complex transition rules, allowing not just multiple collisions in which several blocks merge to form one block, but simultaneous multiple collisions, in which several new blocks might be formed, each from the merger of two or more smaller blocks. The description of all possible semigroups of such coagulation operators is then provided by the larger class of *coalescents with simultaneous collisions* described in the next paragraph. Note that there is a *composition rule* for coagulation kernels associated with exchangeable distributions p_i on $\mathcal{P}_{\mathbb{N}}$ which induces a semigroup operation on these distributions: $(p_1\text{-COAG})(p_2\text{-COAG}) = p_3\text{-COAG}$ where p_3 is determined explicitly by [357, Lemma 34].

Coalescents with simultaneous multiple collisions Following Schweinsberg [396], consider a coalescent process in which each restriction to a finite set has the only transitions of the following kind: if the current partition has b blocks, there may be a $(b; k_1, \ldots, k_r; s)$-*collision*, which takes b blocks down to $r+s$ blocks in such a way that

- s of the new blocks are identical to s of the original blocks,
- r other new blocks contain $k_1, \ldots, k_r \geq 2$ of the original blocks.

where $b = k_1 + \cdots + k_r + s$, and the order of $k_1, \ldots, k_r$ is irrelevant.

Definition 5.8. [396] A *coalescent with simultaneous multiple collisions* is a $\mathcal{P}_{\mathbb{N}}$-valued process $\Pi_\infty = (\Pi_\infty(t), t \geq 0)$ such that for all n, when $\Pi_n(t)$ has b blocks, each $(b; k_1, \ldots, k_r; s)$-collision is occurring at some fixed rate $\lambda_{b;k_1,\ldots,k_r;s}$.

As shown in [396], all possible collections of collision rates $\lambda_{b;k_1,\ldots,k_r;s}$ can be characterized by an integral representation over $\mathcal{P}^{\downarrow}_{[0,1]}$ which is a generalization of (5.5), and there are corresponding generalizations of Corollary 5.5.

Limits of ancestral processes These coalescents with simultaneous multiple collisions can also be characterized as the weak limits obtained from Cannings' model [84] for the evolution of the genetic makeup of a population of fixed size. In Cannings' model,

- there are N individuals in each generation;
- there are infinitely many generations, forwards and backwards in time;
- in each generation, the family sizes have the same distribution as some exchangeable sequence $(\nu_{1,N}, \ldots, \nu_{N,N})$;
- the family sizes for different generations are independent;

Now sample n individuals at random from the 0th generation. Construct a $\mathcal{P}_{[n]}$-valued Markov chain $(\Pi_{n,N}(a))_{a=0}^{\infty}$ such that i and j are in the same block of $\Pi_{n,N}(a)$ if and only if the ith and jth individuals in the sample have the same ancestor in the ath generation backwards in time. Let

$$c_N := \mathbb{P}(\text{two random individuals have the same ancestor in the previous generation}).$$

Then there is the following generalization of Kingman's derivation of his coalescent process from classical genetics models:

Theorem 5.9. Möhle and Sagitov [393], [313] *Suppose that*

$$\lim_{N\to\infty} \frac{E[(\nu_{1,N})_{k_1\downarrow} \cdots (\nu_{r,N})_{k_r\downarrow}]}{N^{k_1+\cdots+k_r-r} c_N} \tag{5.8}$$

exists for all $r \geq 1$, $k_1, \ldots, k_r \geq 2$, *and that*

$$\lim_{N\to\infty} c_N = 0.$$

Then as $N \to \infty$

$$(\Pi_{n,N}(\lfloor t/c_N \rfloor), t \geq 0) \xrightarrow{d} (\Pi_{n,\infty}(t), t \geq 0)$$

where the limit process is the collection of consistent Markovian restrictions to $[n]$ *of some exchangeable* $\mathcal{P}_{\mathbb{N}}$*-valued Markovian coalescent process.*

In particular, if

$$\lim_{N\to\infty} \frac{E[(\nu_{1,N})_3]}{NE[(\nu_{1,N})_2]} = 0$$

the limit is Kingman's coalescent, if

$$\lim_{N\to\infty} N^{-2} c_N^{-1} E[(\nu_{1,N})_2(\nu_{2,N})_2] = 0$$

the limit is a Λ-coalescent as in Theorem 5.4 for some Λ, and without such assumptions the coalescent is one of Schweinsberg's coalescents with simultaneous multiple collisions, whose transition rates can be identified in terms of the limits (5.8). See [396] for details, and [398, 67, 126] for futher work on how coalescents with multiple and simultaneous multiple collisions arise as limits in various discrete or continuous population models.

Exercises

5.1.1. (Construction of Kingman's coalescent from a Poisson process) [396]. Show how to construct $\Pi_\infty(t)$ explicitly for each $t > 0$ as a function of points of a Poisson random measure on a suitable space.

5.1.2. (The Ewens sampling formula derived from Kingman's coalescent) [253] This construction has well known interpretations in terms of standard models of population genetics, due to Wright and Fisher, Moran, and Cannings [84]. See [413, 429], and [325] for more recent developments. Fix n. Let the integers represent individuals located along a horizontal line. Time is vertical. Given a realization of the restriction of $[n]$ of Kingman's coalescent, say $(\Pi_n(t), t \geq 0)$, starting with the partition into singletons, start by drawing *lines of descent* vertically up from each individual, until the first coalescent event at time T_1 say. If say $\{i\}$ merges with $\{j\}$ to form $\{i, j\}$, identify the tips of these two lines of descent at time T_1, and continue upwards with $n - 1$ lines, until the next coalescence, when the tips of two lines are identified, continue upwards with $n - 2$ lines, and so on, until the time T_{n-1} when the last two lines meet. This defines a random tree $\mathcal{T}$ with edge-lengths, whose set of leaves is $[n]$, with $n - 2$ internal nodes at levels T_i for $1 \leq i < n - 1$, each of degree 3, and an exceptional node at level T_{n-1} of degree 2. Now let N_θ denote a Poisson process of marks, called *mutations* along all branches of this treee at rate $\theta/2$ per unit length. Define a random partition Π_n^θ of $[n]$ to be the partition generated by the random equivalence relation $i \sim^\theta j$ iff there is no mutation on the unique path in $\mathcal{T}$ that joins i to j. Then Π_n^θ is an exchangeable random partition of $[n]$, whose distribution is given by the EPPF (2.19) corresponding to the Ewens sampling formula (2.20). Note that the random trees defined by this construction are consistent in an obvious sense as n varies. The completion of their union (relative to the metric defined by distance measures along branches) is a relatively simple kind of continuum random tree, discussed in [7, §4.1] and [138].

5.1.3. (Coming down from infinity) [357, 395] Let $(\Pi_\infty(t), t \geq 0)$ be a Λ-coalescent starting with all singletons. If $|\Pi_\infty(t)| = \infty$ for all $t > 0$, say the process *stays infinite*, whereas if $|\Pi_\infty(t)| < \infty$ for all $t > 0$, say it *comes down from infinity.* Assume Λ has no atom at 1. Then either the Λ-coalescent either comes down from infinity almost surely or it stays infinite almost surely. Let γ_b be the rate at which the number of blocks is decreasing:

$$\gamma_b = \sum_{k=2}^{b} (k-1) \binom{b}{k} \lambda_{b,k}.$$

Then the Λ-coalescent comes down from infinity if and only if

$$\sum_{b=2}^{\infty} \gamma_b^{-1} < \infty.$$

For instance, if Λ is the beta(a, b) distribution the Λ-coalescent comes down from infinity if and only if $a < 1$. In particular, for $a = b = 1$, the Bolthausen-Sznitman coalescent stays infinite. Whereas, for $\Lambda = \delta_0$, Kingman's coalescent comes down from infinity.

Notes and comments

This section is based on [141, 357] and [396]. Thanks to Jason Schweinsberg for his help in summarizing some results of [396].

5.2. Fragmentations

The operation of time reversal introduces a fundamental duality between a process of coalescence and a process of fragmentation. If one is a Markov process, then so will be the other. However, the dual by time-reversal of a time-homogeneous Markov process will usually not be time-homogeneous. Also, a key assumption about fragmentation processes is the branching property according to which different fragments evolve independently. The problem with time-reversing a coalescent process is that in general there is no such branching property, and this is a much more serious obstacle to its study than the Markovian inhomogeneity.

This complicates considerably the discussion of duality relations between processes of fragmentation and coagulation. The idea of looking for some kind of duality by time reversal (or inversion, or some other transformation that reverses the direction of time) has proved very fruitful, though difficult to formalize as part of any general theory. It was argued in [357] that the following definition of a *fragmentation operator* associated with a probability distribution p on $\mathcal{P}_{\mathbb{N}}$ is dual in a number of intuitive ways to the Definition 5.6 of the coagulation operator associated with p.

Definition 5.10. For each probability measure p on $\mathcal{P}_{\mathbb{N}}$, define a Markov kernel p-FRAG on $\mathcal{P}_{\mathbb{N}}$, the *p-fragmentation kernel* as follows. Let $p\text{-FRAG}(\pi, \cdot)$ be the distribution of a random refinement of π whose restriction to the mth block of π is the restriction of $\Pi^{(m)}$ to that block, where the $(\Pi^{(m)}, m = 1, 2, \ldots)$ are independent random partitions of $\mathbb{N}$ with distribution p.

Note the key fact that this definition is *local*, in the sense discussed after Theorem 5.7. So if such kernels are used to make a Markov process, then its restrictions to $[n]$ will be Markovian for every n.

Suppose that p and q are distributions on $\mathcal{P}_{\mathbb{N}}$ of exchangeable random partitions, and that $(P_i^{\downarrow})$ and $(Q_j^{\downarrow})$ denote the ranked frequencies of random partitions governed by p and q respectively. Define $r(\cdot) = \int p(d\pi) q\text{-FRAG}(\pi, \cdot)$, and call it the distribution of a *q-fragmentation of a p-partition*. Then the ranked frequencies associated with r are the ranked rearrangement of the collection of frequencies

$$\{P_i^{\downarrow} Q_{i,j}^{\downarrow}, i, j \geq 1\}$$

where the $(Q_{i,j}^{\downarrow}, j \geq 1)$ are a collection of independent copies of $(Q_j^{\downarrow}, j \geq 1)$ which are also independent of $(P_i^{\downarrow})$. Thus these fragmentation operators have a natural action on the space $\mathcal{P}_{[0,1]}^{\downarrow}$ of proper or improper ranked frequencies. Some examples of identities involving such fragmentation kernels and the two-parameter family of random partitions introduced in the next lecture were pointed out in [371], [368]. These identities were generalized and related by time-reversal to corresponding identities for coagulation kernels in [357]. The general problem of characterizing semigroups of both coagulation and fragmentation operators of the form introduced in Definitions 5.6 and 5.10 was posed in [357, §3.3]. The solution in the case of coagulation operators is provided by Schweinsberg's coalescents with simultaneous multiple collisions, discussed in the previous section, while the solution for fragmentation operators is provided by the theory of *homogeneous fragmentation processes* due to Bertoin [45]. In some loose sense, this theory is dual to the theory of [357, 396] for coagulation semigroups. The method of analysis, using consistent rates to define Lévy measures on partitions, and the Poisson constructions for both kinds of processes are very similar. The fragmentation theory introduces the concept of *erosion* of mass in the fragmentation process, which is roughly dual to the way in which there can be *creation* of mass in the coalescent theory, as in [357, Proposition 26]. Bertoin [46]. characterized a more general class of *self-similar fragmentation processes*, where objects split at a rate proportional to a power α of their mass. The case $\alpha = 0$ corresponds to homogeneous fragmentation fragmentation processes. When the index α is non-negative, Bertoin [47] gave strong limit theorems for the behavior at large times of self-similar fragmentation processes. Some further references are Filipov [154], Bertoin [43], Haas [183, 184], and Jeon [218].

It is not clear what if any class of coalescent processes should be regarded as the dual of self-similar fragmentation processes. Berestycki [36] considers exchangeable coalescent-fragmentations in equilibrium, providing a synthesis of the works by Schweinsberg and Bertoin. See also Haas [182] where an equilibrium is obtained for a fragmentation process with immigration. Some related studies are [123, 110, 361].

See [357], [141], [35] and papers of Bertoin cited above, for treatment of various technical problems related to regularity of different partition-valued Markov processes, transformations between different representations, and so on. There is also a connection between homogeneous fragmentations and branching random walks. If $X(t)$ is the process of ranked frequencies, and

$$Z^{(t)}(dx) = \sum_{i=1}^{\infty} \delta_{-\log X_i(t)}(dx)$$

then discrete time skeletons of Z are branching random walks. This idea and its applications are developed in Bertoin and Rouault [56]. For connections between fragmentation processes and continuum random trees, see Section 10.4 and [185, 310, 311].

Exercises

5.2.1. (Problem: Fragmentation processes associated with the Ewens family) The construction of Exercise 5.1.2 can easily be arranged to construct a family of random partitions $(\Pi_n^\theta, \theta \geq 0)$ which is *refining* or *fragmenting* as θ increases from 0 to ∞, such that $\Pi_n^0 = \{[n]\}$ and Π_n^θ for $\theta > 0$ has the Ewens EPPF (2.19) with parameter θ. Call a process with these properties a *Ewens fragmentation process.* Compare with the Brownian fragmentation process derived similarly from the tree in a Brownian excursion in Section 10.4. Show that when regarded for fixed n, the Ewens fragmentation process just defined is not Markovian. What if the state is regarded as an infinite partition? Does this Ewens fragmentation process admit any nice autonomous description? Is there any natural construction of a Markovian Ewens fragmentation process? Clearly, there exists some such process, by copying the inhomogeneous transition rates of a non-Markovian one. But it does not seem easy to compute these rates for any construction. Another construction, probably also non-Markovian and different to the above, can be made as follows, by application of Theorem 3.2. Let N^θ be a Poisson process on $[0,1]$ with intensity $\theta x^{-1}dx$, arranged to be intensifying as θ increases. Let $\mathcal{I}^\theta$ be the interval partition generated by cutting $[0,1]$ at the points of N^θ, and construct Π_n^θ simultaneously for all θ, by uniform random sampling from $[0,1]$ partitioned by $\mathcal{I}^\theta$. Note that if $(\Pi_n^\theta, \theta \geq 0)$ is a Ewens fragmentation process, then for each fixed $\theta > 0$, Π_n^θ given $|\Pi_n^\theta| = k$ has the microcanonical Gibbs distribution on $\mathcal{P}_{[n]}$ with weights $w_j = (j-1)!$. If one could construct the fragmentation so this was true also for the fixed random time θ replaced by a random times $\Theta_1 > \Theta_2 > \cdots$ so that $|\Pi_n^{\Theta_k}| = k$, then one would have constructed a Gibbs $((\bullet - 1)!)$ fragmentation on $[n]$. But it is not obvious that this is possible.

Notes and comments

Mekjian and others [90, 281, 307, 308] have considered Ewens partitions with parameter θ as a model for fragmentation phenomena, with the intuitive notion that increasing θ corresponds to further fragmentation, but it does not seem obvious how to construct a nice Markovian fragmentation process corresponding to this idea. Bertoin and Goldschmidt [49] provide an example of a fragmentation chain for which the one-dimensional distributions are $PD(0, \theta + k)$, $k = 0, 1, \ldots$. This paper also provides other examples of fragmentation/coalescence duality.

5.3. Representations of infinite partitions

As a preliminary to further study of coagulation and fragmentation operators on various spaces of partitions, this section reviews some terminology and basic facts related to different representations of infinite partitions. Recall that $\mathcal{P}_1^\downarrow$ denotes the space of *real partitions of* 1, that is decreasing non-negative sequences with sum 1. Probability distributions on $\mathcal{P}_1^\downarrow$ can be used to describe the distribution of sizes of components of many different kinds of partitions of an infinite set, subject to the constraint that the sum of sizes of components is always equal to 1. Sometimes, especially in the discussion of weak limits, and in the theory of processes of coagulation and fragmentation, it is convenient to work in the larger space $\mathcal{P}_{[0,1]}^\downarrow := \cup_{0 \le x \le 1} \mathcal{P}_x^\downarrow$ where $\mathcal{P}_x^\downarrow$ is the space of real partitions of x. For $x > 0$ there is an obvious bijection between $\mathcal{P}_x^\downarrow$ and $\mathcal{P}_1^\downarrow$ by scaling, and $\mathcal{P}_0^\downarrow := \{(0, 0, \ldots)\}$. To avoid largely trivial complications, the following discussion will be restricted to probability measures on $\mathcal{P}_1^\downarrow$. With a little care everything can be adapted to probability measures on $\mathcal{P}_{[0,1]}^\downarrow$ or even $\cup_{x \ge 0} \mathcal{P}_x^\downarrow$. Much more care is required to handle probability distributions over decreasing sequences (x_i) with $\sum_i x_i = \infty$, for which many of the following constructions don't make sense. See [9, 13] regarding what can be done in that case.

Let P be a probability distribution on $\mathcal{P}_1^\downarrow$. Associated with P are the distributions of various stochastic processes determined by P, each constructed from some random element (P_i) of $\mathcal{P}_1^\downarrow$ with distribution P and some further randomization, as indicated in the following definitions. See [347, 350, 357] for background and further discussion of the relations between these various processes. Note well the slightly unusual but efficient convention of notation, that P is *not* the random sequence (P_i), rather the *distribution on* $\mathcal{P}_1^\downarrow$ of this sequence. Each of the following processes encodes the distribution P in a more manageable way, and facilitates the description of some basic operations on probability distributions on $\mathcal{P}_1^\downarrow$.

Definition 5.11. An *exchangeable random discrete distribution on* $[0,1]$ *governed by* P is a random measure τ_P on $[0,1]$ which puts mass P_j at $U_j \in [0,1]$, for some sequence of independent and identically distributed uniform $(0,1)$ variables (U_j) independent of (P_j):

$$\tau_P(\cdot) := \sum_{j=1}^{\infty} P_j 1(U_j \in \cdot) \tag{5.9}$$

Definition 5.12. *A process with exchangeable increments governed by* P is the process $(\tau_P(u), 0 \le u \le 1)$ obtained as the cumulative distribution function of an exchangeable random discrete distribution τ_P as above, that is

$$\tau_P(u) := \tau_P([0,u]) := \sum_{j=1}^{\infty} P_j 1(U_j \le u). \tag{5.10}$$

Evidently, the distribution P of (P_i) on $\mathcal{P}_1^{\downarrow}$, and that of $(\tau_P(u), 0 \le u \le 1)$ on $D[0,1]$, determine each other uniquely. According to Kallenberg's theory of processes with exchangeable increments [226, 228], every pure jump process $(\tau(u), 0 \le u \le 1)$ with increasing paths in $D[0,1]$, which has exchangeable increments with $\tau(0) = 0, \tau(1) = 1$, has the same distribution as τ_P derived as above from (P_i) the sequence of ranked jumps of $(\tau(u), 0 \le u \le 1)$.

For $(x_1, x_2, \ldots)$ a sequence of non-negative real numbers with sum $x < \infty$, let $\text{RANK}(x_1, x_2, \ldots) \in \mathcal{P}_x^{\downarrow}$ be the decreasing rearrangement of terms of the sequence.

Definition 5.13. An *interval partition* of $[0,1]$ is a sequence of random disjoint open intervals (I_j) with $\sum_i \lambda_{I_j} = 1$, where λ_I is the length of I. A P*-partition of* $[0,1]$, often denoted (I_j^P) below, is an interval partition (I_j) of $[0,1]$ such that

$$\text{RANK}(\lambda_{I_1}, \lambda_{I_2}, \ldots)$$

has distribution P.

Note that an interval partition will always be regarded as *ordered.* So $I_j = (G_j, D_j)$ say is simply identified as a pair of random variables specifying its endpoints, and (I_j) is just a sequence of such pairs. To completely specify the distribution of a P-partition of $[0,1]$, it is necessary to specify the order of intervals in some way. If the order is not mentioned, it can be assumed by default that the intervals are laid out from left to right in decreasing order of size. But other orderings, such as the exchangeable ordering discussed later in Definition 5.16, are much more useful.

Definition 5.14. A P*-partition of* $\mathbb{N}$ is an exchangeable random partition Π_∞ of $\mathbb{N}$ whose ranked frequencies (P_i) have distribution P.

The distribution of Π_∞ is determined by the distribution of its ranked frequencies via (2.14). The EPPF p of Π_∞ and the distribution of ranked frequencies P determine each other uniquely, by Kingman's correspondence Theorem

2.2. According to that correspondence, the blocks of a P-partition of $\mathbb{N}$ can be made by *Kingman's paintbox construction* as the random sets $\{i : U_i \in I_j^P\}$ for any P-partition (I_j^P) of $[0,1]$, and (U_i) a sequence of independent uniform$(0,1)$ variables, assumed independent of (I_j^P).

Definition 5.15. A *size-biased presentation of P* is a size-biased ordering of a P-distributed sequence.

Such a sequence is obtained from an exchangeable P-partition of $\mathbb{N}$ as the frequencies of classes in order of their first elements. In terms of Kingman's paintbox construction, this is the sequence of lengths of intervals I_j^P in the order they are discovered by the process of random sampling with the U_i. Think of each I_j^P being painted a different color, then observing the sequence of colors found by the U_i.

Interval partitions of $[0,1]$ may be constructed as in Chapter 4 as the collection of excursion intervals of a suitable stochastic process parameterized by $[0,1]$. Many such interval partitions have the following property:

Definition 5.16. Let (I_j) be an interval partition of $[0,1]$, with sequence of ranked lengths

$$(\lambda_{I_{(n)}}, n \geq 1) := \text{RANK}(\lambda_{I_j}, j \geq 1)$$

whose non-zero terms are distinct almost surely. Call (I_j) *exchangeable* if for each $n = 2, 3, \ldots$ such that $\mathbb{P}(\lambda_{I_{(n)}} > 0) > 0$, conditionally given $\lambda_{I_{(n)}} > 0$, the ordering in $[0,1]$ of the longest n intervals $I_{(j)}, 1 \leq j \leq n$ is equally likely to be any one of the $n!$ possible orders, independently of the lengths of these n longest intervals. Call (I_j) *infinite* if $P(\lambda_{I_{(n)}} > 0) = 1$ for all n.

The extension of this definition to the case with ties is pedantic but obvious. As shown by Kallenberg [227], for an infinite exchangeable interval partition (I_j), for each $u \in [0,1]$ the fraction of the longest n intervals that lie to the left of u has an almost sure limit L_u as $n \to \infty$. The process $(L_u, 0 \leq u \leq 1)$ is a continuous increasing process, called the *normalized local time process of* (I_j).

Note that the distribution of an *exchangeable* P-partition (I_j) of $[0,1]$ is uniquely determined by P provided the indexing is ranked, meaning $I_j = I_{(j)}$ for all j, with strictly decreasing lengths, as will be assumed for simplicity from now on. An exchangeable P partition of $[0,1]$ can be constructed as

$$I_j^P := (\tau_P(U_j-), \tau_P(U_j)) = (\tau_P([0,U_j)), \tau_P([0,U_j]))$$

with notation as in (5.9) and (5.10), for τ_P regarded as a process with exchangeable increments in the first expression, and as a random measure in the second. Inversely, assuming that (I_j) is an infinite exchangeable P-partition of $[0,1]$, the increasing process with exchangeable increments $(\tau_P(u), 0 \leq u \leq 1)$ is the right-continuous inverse of the normalized local time process of (I_j).

The following lemma is a well known consequence of Itô's description of excursions of a Markov process and the general construction of bridges of a nice recurrent Markov process, as in [155].

Lemma 5.17. *Let X^{br} be the bridge of length 1 from 0 to 0 derived from any nice recurrent strong Markov process X with recurrent state 0 which is regular for itself. Then the interval partition (I_j) defined by excursions of X^{br} away from 0 is an infinite exchangeable interval partition of $[0,1]$, whose normalized local time process is $L_u = L_u^0/L_1^0, 0 \le u \le 1$ for any of the usual Markovian definitions of a bridge local time process $L_u^0 := L_u^0(X^{\text{br}})$.*

In particular, this lemma applies to a self-similar recurrent Markov process $X = B$ as in Section 4.4. Then, according to Corollary 4.9 the distribution of ranked lengths of excursion intervals (I_j) is PD(α,α) for some $\alpha \in (0,1)$, with $\alpha = (2-\delta)/2$ if B is BES(δ) for $\delta \in (0,2)$, and $\alpha = \frac{1}{2}$ if B is standard Brownian motion.

The two-parameter family Quickly reviewed here are some of the above representations in the case of the two-parameter family. The distribution P on $\mathcal{P}_1^{\downarrow}$ is PD(α,θ) iff the corresponding partition of $\mathbb{N}$ is an (α,θ) partition, as defined via the Chinese Restaurant Process in Section 3.2. In particular, the distribution P is PD$(0,\theta)$ iff the corresponding process with exchangeable increments τ_P is a Dirichlet process with parameter θ, obtained by normalization and scaling of a gamma subordinator $(\Gamma_s, 0 \le s \le \theta)$. The distribution of P is PD$(\alpha,0)$ iff the corresponding process with exchangeable increments τ_P is a normalized stable subordinator of index α.

5.4. Coagulation and subordination

Recall from Section 5.1 and Section 5.2 that for each probability measure Q on $\mathcal{P}_1^{\downarrow}$, two Markov transition kernels Q-COAG and Q-FRAG can be defined on $\mathcal{P}_1^{\downarrow}$ as follows. For $p \in \mathcal{P}_1^{\downarrow}$,

- $(Q\text{-COAG})(p,\cdot)$ is the distribution on $\mathcal{P}_1^{\downarrow}$ of

$$\text{RANK}\left(\sum_i p_i 1(U_i \in I_j^Q), j \ge 1\right) \tag{5.11}$$

  where (I_j^Q) is a Q-partition of $[0,1]$, and the U_i are i.i.d. uniform on $(0,1)$ independent of (I_j^Q).
- $(Q\text{-FRAG})(p,\cdot)$ is the distribution of

$$\text{RANK}(p_i Q_{i,j}, i, j \ge 1) \tag{5.12}$$

  where $(Q_{i,j})_{j\ge 1}$ has distribution Q for each i, and these sequences are independent as i varies.

For a probability measure P on $\mathcal{P}_1^{\downarrow}$, let $R := P(Q\text{-COAG})$. That is the probability measure on $\mathcal{P}_1^{\downarrow}$ defined by

$$R(\cdot) := \int_{\mathcal{P}_1^{\downarrow}} P(dr) Q\text{-COAG}(r,\cdot), \tag{5.13}$$

which may be called P *coagulated by* Q. Notice that (5.13) is equivalent to

$$R\text{-COAG} = (P\text{-COAG})(Q\text{-COAG})$$

in the usual sense of composition of Markov kernels. The same remark applies to fragmentation kernels too.

The space $\mathcal{P}_1^{\downarrow}$ does not have a group structure, but these operations on probability measures P and Q on $\mathcal{P}_1^{\downarrow}$ to obtain P coagulated by Q, or P fragmented by Q, are similar to the more familiar operation of convolution of measures on a group. For instance there is a representation of both coagulation and fragmentation semigroups, analogous to a Lévy-Khintchine representation, as has been shown by Schweinsberg [396] and Bertoin [45]. The following Lemma reduces the notion of composition of coagulation operators to the operation of composition of increasing processes, commonly known as *subordination*, which was considered by Kallenberg [228] for increasing processes with exchangeable increments of the kind involved here. This is seen by associating each probability distribution P on $\mathcal{P}_1^{\downarrow}$ with its corresponding exchangeable random discrete distribution on $[0, 1]$, defined as in (5.9) by putting mass P_j at $U_j \in [0, 1]$, for some sequence of independent and identically distributed uniform $(0, 1)$ variables (U_j) independent of (P_j) with distribution P. Now R in (5.13) is the distribution of the ranked rearrangement of the terms

$$\tau_P(I_j^Q) := \sum_{i=1}^{\infty} P_i 1(U_i \in I_j^Q), \quad j = 1, 2, \ldots, \tag{5.14}$$

where (I_j^Q) is Q partition of $(0, 1)$, assumed independent of (P_i) and (U_i). The terms in (5.14) are the masses assigned by τ_P to the intervals I_j^Q. Think of Q as determining the sizes of *bins* I_j^Q, and P as determining the sizes of masses to be sprinkled into the bins by uniform random allocation. Then P coagulated by Q is the distribution of ranked P-masses collected in the bins. Note that the locations of the intervals I_j^Q in $[0, 1]$ have no effect in this construction, due to the assumed independence of (I_j^Q), τ_P, and (U_i).

For $0 \leq u \leq 1$ let $(\tau_P(u), 0 \leq u \leq 1)$ as in (5.10) be the process with exchangeable increments whose ranked jumps have distribution P, that is the cumulative distribution function of the random discrete distribtion τ_P. Recall that the distribution P on $\mathcal{P}_1^{\downarrow}$ determines that of $(\tau_P(u), 0 \leq u \leq 1)$ on $D[0, 1]$, and vice versa.

Lemma 5.18. *For each pair of probability distributions P and Q on $\mathcal{P}_1^{\downarrow}$, the distribution $R = P(Q\text{-COAG})$ is the unique probability distribution on $\mathcal{P}_1^{\downarrow}$ such that*

$$(\tau_R(u), 0 \leq u \leq 1) \stackrel{d}{=} (\tau_P(\tau_Q(u)), 0 \leq u \leq 1),$$

where it is assumed that $(\tau_P(u), 0 \leq u \leq 1)$ and $(\tau_Q(u), 0 \leq u \leq 1)$ are independent.

Proof. It may be assumed that the Q-partition (I_j^Q) in (5.14) is exchangeable. The union of intervals I_j^Q is then almost surely identical to the union of open intervals $(\tau_Q(u-), \tau_Q(u))$ as u ranges over $[0,1]$. The process $(\tau_P(\tau_Q(u)), 0 \leq u \leq 1)$ is easily seen to be increasing with exchangeable increments, and its collection of jumps is almost surely identical to the collection of strictly positive increments of τ_P over the non-empty intervals $(\tau_Q(u-), \tau_Q(u))$. So the conclusion follows from (5.14). □

Lemma 5.18 generalizes the connection pointed out in [54] between the coagulation operators of the Bolthausen-Sznitman coalescent [72] and the operation of subordination of stable increasing processes. To see this connection, let $Q_\alpha :=$ PD$(\alpha, 0)$. According to Theorem 4.3, Q_α is the distribution of ranked jumps of $(T_\alpha(s)/T_\alpha(S), 0 \leq s \leq S)$, for arbitrary fixed $S > 0$, where $(T_\alpha(s), s \geq 0)$ is a stable(α) subordinator. For two independent stable subordinators $(T_\alpha(t), t \geq 0)$ and $(T_\beta(s), s \geq 0)$ there is the well known *subordination identity*

$$(T_\alpha(T_\beta(s)), s \geq 0) \stackrel{d}{=} (T_{\alpha\beta}(s), s \geq 0). \tag{5.15}$$

By application of Lemma 5.18, $Q_\alpha(Q_\beta\text{-COAG})$ is the distribution of ranked jumps of the process

$$\left(\frac{T_\alpha(T_\beta(u)/T_\beta(1))}{T_\alpha(1)}, \ 0 \leq u \leq 1\right) \stackrel{d}{=} \left(\frac{T_{\alpha\beta}(u)}{T_{\alpha\beta}(1)}, \ 0 \leq u \leq 1\right) \tag{5.16}$$

where the identity in distribution is justified by (5.15) and the scaling property $(T_\alpha(ct), t \geq 0) \stackrel{d}{=} (c^{1/\alpha}T_\alpha(t), t \geq 0)$. That is to say, $Q_\alpha(Q_\beta\text{-COAG}) = Q_{\alpha\beta}$, which is equivalent to the identity of coagulation kernels

$$(Q_\alpha\text{-COAG})(Q_\beta\text{-COAG}) = Q_{\alpha\beta}\text{-COAG} \qquad (0 < \alpha, \beta < 1). \tag{5.17}$$

If we set $P_t := Q_{\exp(-t)}\text{-COAG}$ then (5.17) reads

$$P_s P_t = P_{s+t} \qquad (s, t > 0). \tag{5.18}$$

Thus we deduce:

Theorem 5.19. Bolthausen-Sznitman [72] *The family of Markov kernels*

$$(PD(e^{-t}, 0)\text{-COAG}, t > 0), \text{ is a semigroup of Markov kernels on } \mathcal{P}_1^{\downarrow}. \tag{5.19}$$

See [72, 357, 50, 54] for further analysis of the corresponding coalescent process, which may be constructed for all $t \geq 0$ with state space $\mathcal{P}_{\mathbb{N}}$, or just for $t > 0$ with state space $\mathcal{P}_1^{\downarrow}$. Bertoin and Le Gall [51] show that the correspondence between operations of coagulation and subordination, described by Lemma 5.18, can be extended in a natural way to associate Schweinsberg's coalescent processes with multiple collisions (Definition 5.8) with a class of stochastic flows of bridges with exchangeable increments. See also [52] for further developments.

Coagulation by PD(0,1) The PD$(0,1)$ distribution on $\mathcal{P}_1^{\downarrow}$ has some unique properties implied by the fact that it is the asymptotic distribution of ranked relative lengths of cycles of a uniform random permutation of $[n]$, as discussed in Section 3.1. This gives the uniform stick-breaking representation of the size-biased presentation of PD$(0,1)$, which is the simplest possible description of the size-biased presentation of any probability measure on $\mathcal{P}_1^{\downarrow}$ which concentrates on infinite partitions. Moreover, the entire structure of random partitions generated by uniform permutations of $[n]$, for every n, is embedded in a $(0,1)$ partition of $\mathbb{N}$ whose ranked frequencies have PD$(0,1)$ distribution. The following Theorem 5.21, presents a remarkable consequence of this intimate link between PD$(0,1)$ and random permutations. First, a definition:

Definition 5.20. [17, 24] Let $(I_j)_{j=1,2,\ldots}$ be a random interval partition of $[0,1]$. The *D-partition derived from* (I_j) is the interval partition (I_j^D) defined as follows. Let $U_1, U_2, \ldots$ denote a sequence of independent uniform $(0,1)$ variables, independent of (I_j). Let $I_j^D := (D_{V_{j-1}}, D_{V_j})$ where $V_0 = D_{V_0} = 0$ and V_j is defined inductively along with the D_{V_j} for $j \geq 1$ as follows: given that D_{V_i} and V_i have been defined for $0 \leq i < j$, let

$$V_j := D_{V_{j-1}} + U_j(1 - D_{V_{j-1}}),$$

so V_j is uniform on $[D_{V_{j-1}}, 1]$ given (V_i, D_{V_i}) for $0 \leq i < j$, and let D_{V_j} be the right end of the interval $I_{k(j)}$ which contains V_j.

This definition was first introduced in [17] for (I_j) the exchangeable interval partition defined by excursions of a standard Brownian bridge, when (I_j) is a PD$(\frac{1}{2}, \frac{1}{2})$ partition Corollary 4.9. It was shown in [17], in connection with the asymptotics of partitions of $[n]$ generated by trees and basins in the digraph of a uniform random mapping of $[n]$, that the D-partition derived from excursion intervals of Brownian bridge is a PD$(0, \frac{1}{2})$ partition, with intervals in length-biased order. See the next section for further discussion of this example. It was shown in [24] that many of the properties of the D-partition derived from intervals of Brownian bridge are in fact shared by the D-partition derived from any exchangeable interval partition of $[0,1]$. Following is a formulation of one such property:

Theorem 5.21. *For each probability distribution P on $\mathcal{P}_1^{\downarrow}$, the sequence of lengths of the D-partition derived from an exchangeable P-partition of $[0,1]$ is a size-biased presentation of P coagulated by PD$(0,1)$. In particular, the distribution of ranked lengths of this D-partition is $P(PD(0,1)$-COAG$)$.*

Proof. Let (P_i) be distributed according to P. By conditioning, it is enough to consider one of the following two cases: either

- $P_i > 0$ for all i almost surely, or
- $P_i > 0$ for $1 \leq i \leq n$ and $\sum_{i=1}^n P_i = 1$ almost surely.

But in the first case, the conclusion can be read from [24, Theorem 25], and in the second case, the conclusion can be read from [24, Lemma 26]. □

Corollary 5.22. *In the setting of Theorem 5.21 with* (I_j) *an infinite exchangeable interval partition of* $[0,1]$, *for each* $k = 1, 2, \ldots$ *the interval* (G_{V_k}, D_{V_k}) *that contains* V_k *is a length-biased pick from the restriction of* (I_j) *to* $[D_{V_{k-1}}, D_{V_k}]$. *In particular,* (G_{V_1}, D_{V_1}) *is a length-biased pick from the restriction of* (I_j) *to* $[0, D_{V_1}]$.

Exercises

5.4.1. (Proof of Theorem 5.21)

5.4.2. (Proof of Corollary 5.22)

5.4.3. (Problem) Is there a generalization of Theorem 5.21 to some other law Q instead of $\mathrm{PD}(0,1)$, for instance $\mathrm{PD}(0,\theta)$ or $\mathrm{PD}(\alpha,\theta)$?

5.5. Coagulation – fragmentation duality

Let us start with some general considerations. For two random variables $X : \Omega \to \Omega_X$ and $Y : \Omega \to \Omega_Y$, with values in arbitrary measurable spaces $(\Omega_X, \mathcal{F}_X)$ and $(\Omega_Y, \mathcal{F}_Y)$, and a Markov transition kernel $Q : (\Omega_X, \mathcal{F}_X) \to (\Omega_Y, \mathcal{F}_Y)$, either of the displays

$$X \overset{Q}{\longrightarrow} Y \qquad \text{or} \qquad Y \underset{Q}{\longleftarrow} X \tag{5.20}$$

means that $\mathbb{P}(Y \in \cdot \,|\, X) = Q(X, \cdot)$. Assuming this, and that $(\Omega_X, \mathcal{F}_X)$ is a nice measurable space, the general theory of regular conditional distributions [122] provides an essentially unique Markov transition kernel $\hat{Q} : (\Omega_Y, \mathcal{F}_Y) \to (\Omega_X, \mathcal{F}_X)$, such that

$$X \underset{\hat{Q}}{\longleftarrow} Y \tag{5.21}$$

The dual pair of relations (5.20) and (5.21) can then be indicated by a single diagram

$$X \; \underset{\hat{Q}}{\overset{Q}{\rightleftarrows}} \; Y \tag{5.22}$$

which should be read clockwise starting from X. Some things to keep in mind here:

- the notation is simplistic: $\hat{Q}$ depends on both Q and the law of X.
- If $\mathbb{P}(X \in dx) = f_X(x)\mu(dx)$ and $Q(x, dy) = q(x,y)\nu(dy)$ for some reference measures μ and ν, then $\mathbb{P}(Y \in dy) = f_Y(y)\nu(dy)$ with $f_Y(y) = \int_x q(x,y) f_X(x)\mu(dx)$, and there is *Bayes rule* $\hat{Q}(y, dx) = f_X(x)q(x,y)\,\mu(dx)/f_Y(y)$.

- Without densities with respect to some reference measures, the abstract theory gives you existence of $\hat{Q}$. But when it comes to approximating or computing it, or identifying it with some other kernel you might have seen before, you are on your own.

So when X and Y have values in some infinite dimensional space like $\mathcal{P}_1^\downarrow$, where there is no natural reference measure, given some law of X and some Markov kernel Q for the law of Y given X, the problem of identifying the law of Y and the inverse kernel $\hat{Q}$ is usually a non-trivial one. A useful general method for handling such situations is the following:

Finessing Bayes rule

- Find some manageable class of jointly measurable functions f such that a joint law of (X, Y) of the kind under consideration is determined by $\mathbb{E}f(X, Y)$ for all f in the class.
- Compute $\mathbb{E}f(X, Y)$ for all f in the class, using the law of X and
$$\mathbb{P}(Y \in \cdot \,|\, X) = Q(X, \cdot).$$
- Now anticipate or guess some parametric form for what the law of Y and $\hat{Q}$ might be. Recompute $\mathbb{E}f(X, Y)$, using your guess for the law of Y, and
$$\mathbb{P}(X \in \cdot \,|\, Y) = \hat{Q}(Y, \cdot).$$
- If you can find values of the parameters which get you the same value of $\mathbb{E}f(X, Y)$ as you got the other way, for all f in your class, the problem is solved.

This method is well illustrated by the following example. Suppose X has $\mathrm{PD}(\alpha_0, \theta_0)$ distribution on $\mathcal{P}_1^\downarrow$, and Q is the (α, θ)-coagulation kernel, meaning the kernel p-COAG derived from $p = \mathrm{PD}(\alpha, \theta)$. It is not at all obvious how to describe the distribution of Y on $\mathcal{P}_1^\downarrow$ in any direct way, let alone how to find the family of conditional laws $\hat{Q}$ of X given Y. The kernel $\hat{Q}$ must obviously be some kind of fragmentation kernel, meaning that $\hat{Q}(y, \cdot)$ should concentrate on real partitions obtained by shattering each part of y into fragments in some way, then re-ranking all the fragments. But the exact form of this kernel is unknown except in the special case, discovered using the above method, and discussed in more detail below, when the distribution of Y is $\mathrm{PD}(\alpha_1, \theta_1)$, and the inverse kernel is $\mathrm{PD}(\hat{\alpha}, \hat{\theta})$-fragmentation for some (α_1, θ_1) and $(\hat{\alpha}, \hat{\theta})$ determined by (α_0, θ_0) and (α, θ).

Example: partitions generated by random mappings The first indication of such a duality relation is provided by the combinatorial structure associated with random mappings. The digraph of a uniform random mapping from $[n]$ to $[n]$ partitions $[n]$ coarsely into *connected components* called *basins*, each of which is further partitioned into *tree components*. See Section 9.1 for further discussion. The partition of $[n]$ by tree components is thus obtained by a combinatorial fragmentation operation on the partition of $[n]$ by basins. Moreover, given the tree components, they are tied together into cycles by the action of a uniform permutation of their root vertices, which provides a combinatorial coagulation of trees. In the large n limit, the various partitions involved have weak limits in the sense of Section 2.4, all of which can be identified as members of the two-parameter family of random partitions Section 3.2 parameterized by (α, θ) as indicated below. See Chapter 9 for further discusssion. As $n \to \infty$,

- the weak limit of the partition of $[n]$ by tree components is a $(\frac{1}{2}, \frac{1}{2})$ partition.
- the weak limit of the partition of tree roots, by cycles of the random permutation of what is most likely a large number of cyclic points, is a $(0, 1)$ partition;
- the weak limit of the partition of $[n]$ by basins is a $(0, \frac{1}{2})$ partition;
- the weak limit of the partition of a large basin by its tree components is a $(\frac{1}{2}, 0)$ partition.

Moreover, the combinatorial fragmentation and coagulation operators, which relate the two random partitions generated by the mapping digraph, converge weakly to their asymptotic counterparts, which are the $(0,1)$-coagulation and $(\frac{1}{2}, 0)$-fragmentation operators on $\mathcal{P}_1^{\downarrow}$. To summarize: the joint weak limit of the tree and basin partitions generated by a random mapping is a $(\frac{1}{2}, \frac{1}{2})$-partition which is a refinement of $(0, \frac{1}{2})$-partition, according to the following prescription, where the symbol (α, θ) may represent

- *either* the law of an (α, θ) partition of $\mathbb{N}$,
- *or* the corresponding law $\mathrm{PD}(\alpha, \theta)$ on $\mathcal{P}_1^{\downarrow}$:

$$\left(\tfrac{1}{2}, \tfrac{1}{2}\right) \quad \underset{(\frac{1}{2},0)\text{-FRAG}}{\overset{(0,1)\text{-COAG}}{\rightleftarrows}} \quad \left(0, \tfrac{1}{2}\right) \tag{5.23}$$

To read around the diagram, and review its interpretation in terms of laws on $\mathcal{P}_1^{\downarrow}$. Starting on the left side, $\mathrm{PD}(\frac{1}{2}, \frac{1}{2})$ is the limit distribution of ranked relative sizes of tree components of the mapping. The $\mathrm{PD}(0,1)$-COAG operation describes how the trees are bundled into basins, according to the $\mathrm{PD}(0,1)$ asymptotic distribution of relative sizes of cycles of the random permutation of the roots. The result is the asymptotic basin partition, which is $\mathrm{PD}(0, \frac{1}{2})$. Finally, the asymptotic mechanism by which the basins are partitioned into trees is a $\mathrm{PD}(\frac{1}{2}, 0)$-fragmentation. The reader will find it instructive to read around the diagram, interpreting each step instead in terms of the joint law of a nested pair of random partitions of $[n]$, where the joint laws are consistent in an obvious sense as n varies. It should then become clear that the problem of discovering or proving a duality relation like (5.23) is essentially combinatorial in nature, as it reduces entirely to computations involving partitions of $[n]$.

Brownian bridge interpretation Of course, the interesting thing about a duality relation like (5.23) is that while it can be proved combinatorially, it has meaning in the continuum limit as a fact about kernels on real partitions of 1. Recall from Corollary 4.9 that PD($\frac{1}{2}, \frac{1}{2}$) and PD($\frac{1}{2}$,0) are the laws of the partitions of 1 generated by the lengths of excursions away from 0 of B^{br} and $B[0,1]$ respectively, where B is standard Brownian motion and B^{br} is $B[0,1]$ conditioned on $B(1) = 0$. Consequently, the diagram (5.23) can be interpreted in terms of first coagulating and then fragmenting an ensemble of Brownian excursions, using the representation of PD(0, 1)-coagulation provided by Theorem 5.21. See Chapter 1 and [24] for details.

Generalization The coagulation/fragmentation duality (5.23) generalizes as follows:

Theorem 5.23. [357] *For* $0 < \alpha < 1, 0 \le \beta < 1, -\beta < \theta/\alpha$

$$(\alpha,\theta) \quad \underset{(\alpha,-\alpha\beta)\text{-FRAG}}{\overset{(\beta,\theta/\alpha)\text{-COAG}}{\rightleftarrows}} \quad (\alpha\beta,\theta) \tag{5.24}$$

Sketch of proof. As indicated in the discussion below the special case (5.23), the key is that for any probability law Q on $\mathcal{P}_1^{\downarrow}$, the kernels Q-COAG and Q-FRAG, regarded as Markov kernels on $\mathcal{P}_{\mathbb{N}}$ rather than $\mathcal{P}_1^{\downarrow}$, act locally on the partitions of $[n]$ obtained by restriction of a partition of $\mathbb{N}$, in the sense discussed below Theorem 5.7 and Definition 5.10 . Moreover their action can be described in terms of EPPFs by explicit combinatorial formulae [357, §4]. Consequently, an identity such as (5.24), once guessed, can be proved by a computation with EPPFs, according to the general method of finessing Bayes rule, discussed above. See [357, §4] for details. □

Some special cases of (5.24) are worthy of note.

• ($\beta = 0, 0 < \alpha = \theta < 1$)

$$(\alpha,\alpha) \quad \underset{(\alpha,0)\text{-FRAG}}{\overset{(0,1)\text{-COAG}}{\rightleftarrows}} \quad (0,\alpha). \tag{5.25}$$

This has an interpretation, developed in [24], in terms of excursions of the bridge of a self-similar Markov process whose zero set is the range of a stable(α) subordinator. An interpretation can also be given in terms of a discrete renewal process derived from a stable(α) subordinator, as indicated in [351] and [440].

• ($\beta = 0, 0 < \alpha < 1, \theta > 0$). This generalization of the previous case can be represented in an obviously equivalent way as follows:

$$(0,\theta) \quad \underset{(0,\theta/\alpha)\text{-COAG}}{\overset{(\alpha,0)\text{-FRAG}}{\rightleftarrows}} \quad (\alpha,\theta). \tag{5.26}$$

This presentation emphasises the construction of PD(α, θ) as PD$(0, \theta)$ fragmented by PD$(\alpha, 0)$, which was derived already in Proposition 3.16.

• $(0 < \alpha < 1, 0 < \beta < 1, \theta = 0.)$ Now (5.24) reads

$$(\alpha, 0) \quad \underset{(\alpha, -\alpha\beta)\text{-FRAG}}{\overset{(\beta, 0)\text{-COAG}}{\rightleftarrows}} \quad (\alpha\beta, 0) \tag{5.27}$$

Thus we recover the basic identity (5.17) underlying the semigroup of the Bolthausen-Sznitman coalescent. But this time we get as well a recipe for construction of the time-reversed process, which has numerous applications. For instance:

Corollary 5.24. [357, Corollary 16] *Let $(X(t), t > 0)$ be the càdlàg $\mathcal{P}_1^{\downarrow}$-valued Bolthausen-Sznitman coalescent such that $X(t)$ has PD$(e^{-t}, 0)$ distribution, with $X(t)$ represented as the frequencies of $\Pi(t)$, where $(\Pi(t), t \geq 0)$ is an exchangeable $\mathcal{P}_{\mathbb{N}}$-valued coalescent process such that $\Pi(t)$ is an $(e^{-t}, 0)$ partition of $\mathbb{N}$. Let $\tilde{X}_1(t)$ be the frequency of the component of $\Pi(t)$ containing 1, so $\tilde{X}_1(t)$ is for each t a size-biased pick from the parts of $X(t)$. Then*

$$(\tilde{X}_1(t), t \geq 0) \stackrel{d}{=} (\Gamma_{1-\exp(-t)}/\Gamma_1, t \geq 0)$$

where Γ is a gamma subordinator. Equivalently, if ρ is the sequence of ranked jumps of $(\tilde{X}_1(t), t > 0)$, and T_i is when ρ_i occurs, then ρ has PD$(0,1)$ distribution and the T_i are i.i.d. exponential(1) independent of ρ.

See [357] for the proof of this and other properties of the Bolthausen-Sznitman coalescent which follow from the duality relation (5.27).

Notes and comments

It seems very difficult to describe the semigroup of the Bolthausen-Sznitman coalescent on $\mathcal{P}_1^{\downarrow}$ much more explicitly than has been done here. For the particular entrance law with the state at time t distributed according to PD$(e^{-t}, 0)$, some descriptions can be read from [371]. But there are no known extensions of these formulae to the case of a given initial state in $\mathcal{P}_1^{\downarrow}$. For a more complete development of ideas in this chapter, see the forthcoming book *Random fragmentation and coagulation processes* by Jean Bertoin, to be published by Cambridge University Press.

6

Random walks and random forests

This chapter is inspired by the following quotation from Harris's 1952 paper [193, §6]:

> *Walks and trees.* Random walks and branching processes are both objects of considerable interest in probability theory. We may consider a random walk as a probability measure on sequences of steps-that is, on "walks". A branching process is a probability measure on "trees". The purpose of the present section is to show that *walks and trees are abstractly identical objects* and to give probabilistic consequences of this correspondence. The identity referred to is non-probabilistic and is quite distinct from the fact that a branching process, as a Markov process, may be considered in a certain sense to be a random walk, and also distinct from the fact that each step of the random walk, having two possible directions, represents a twofold branching.

This *Harris correspondence* between walks and trees has been developed and applied in various ways, to enrich the theories of both random walks and branching processes. The chapter is organized as follows:

6.1. Cyclic shifts and Lagrange inversion This section presents a well-known probabilistic interpretation of the Lagrange inversion formula in terms of hitting times of random walks.

6.2. Galton-Watson forests The Lagrange inversion formula appears also in the theory of Galton-Watson branching processes. This is explained by Harris's correspondence between random walk paths on the one hand, and trees or forests on the other.

6.3. Brownian asymptotics for conditioned, Galton-Watson trees The Harris correspondence leads to results due to Aldous and Le Gall, according to which the height profiles of suitably conditioned Galton-Watson trees and forests converge weakly to Brownian excursion or reflecting Brownian bridge, as the number n of vertices in the tree or forest tends to ∞.

6.4. Critical random graphs and the multiplicative coalescent Aldous developed these ideas to obtain the asymptotic behaviour of component sizes of the Erdős-Rényi random graph process $\mathcal{G}(n,p)$ in the critical regime $p \approx 1/n$ as $n \to \infty$. The *multiplicative coalescent process* governs mergers of connected components in the random graph process as the parameter p increases.

6.1. Cyclic shifts and Lagrange inversion

For a sequence $\mathbf{x} := (x_1, \ldots, x_n)$ the *walk with steps* $\mathbf{x}$ is the sequence $s_0 = 0, s_1, \ldots, s_n$ with $s_j := \sum_{i=1}^{j} x_i$. Say the walk *first hits b at time n* if $s_i \neq b$ for $i < n$ and $s_n = b$. For $i \in [n]$ let $\mathbf{x}^{(i)}$ denote the ith *cyclic shift* of $\mathbf{x}$, that is the sequence of length n whose jth term is x_{i+j} with $i + j$ mod n. The following elementary lemma is a useful variant of the classical ballot theorem [409]:

Lemma 6.1. [409],[430, §3] *Let $\mathbf{x} := (x_1, \ldots, x_n)$ be a sequence with values in $\{-1, 0, 1, 2, \ldots\}$, and sum $-k$ for some $1 \leq k \leq n$. Then there are exactly k distinct $i \in [n]$ such that the walk with steps $\mathbf{x}^{(i)}$ first hits $-k$ at time n.*

Proof. Consider $i = m$, the least i such that $s_i = \min_{1 \leq j \leq n} s_j$ to see there is at least one such i. By first replacing $\mathbf{x}$ by $\mathbf{x}^{(m)}$, it may be assumed that the original walk first hits $-k$ at time n. But in that case, the walk with steps $\mathbf{x}^{(i)}$ first hits $-k$ at time n if and only if i is one of the k strict descending ladder indices of the walk, meaning the original walk first hits $-\ell$ at time i for some $1 \leq \ell \leq k$. □

Kemperman's formula Suppose now that a sequence of non-negative integer random variables $\mathbf{X} := (X_1, \ldots, X_n)$ is *cyclically exchangeable*, meaning that $\mathbf{X}^{(i)} \stackrel{d}{=} \mathbf{X}$ for each $1 \leq i \leq n$. Define

$$S_j := \sum_{i=1}^{j} X_i, \tag{6.1}$$

$$T_{-k} := \inf\{j > 0 : S_j - j = -k\} = \inf\{j > 0 : S_j = j - k\}, \tag{6.2}$$

Then, as observed by Takács [409], Lemma 6.1 implies *Kemperman's formula*,

$$\mathbb{P}(T_{-k} = n) = \frac{k}{n}\mathbb{P}(S_n = n - k) \qquad (1 \leq k \leq n). \tag{6.3}$$

Kemperman [234],[235, (7.15)] gave this formula for independent steps X_i with some arbitrary common distribution $(p_0, p_1, \ldots)$ on the non-negative integers. Continuing the discussion for independent and identically distributed X_i, let

$$g(z) := \sum_{j=0}^{\infty} p_j z^j = \mathbb{E}(z^{X_i}) \quad (|z| < 1) \tag{6.4}$$

be the probability generating function of the X_i, and for $k = 1, 2, \ldots$ let

$$h_k(z) := \sum_{n=1}^{\infty} \mathbb{P}(T_{-k} = n) z^n = \mathbb{E}(z^{T_{-k}}). \tag{6.5}$$

Because T_{-k} is the first passage time to $-k$ of the walk $(S_n - n)_{n=0,1,2,\ldots}$ which can move downward at most 1 at each step, T_{-k} is the sum of k independent copies of T_{-1}. Thus

$$h_k(z) = h(z)^k \tag{6.6}$$

where $h(z) := h_1(z)$. Moreover, by conditioning on X_1, the generating function h of T_{-1} solves

$$h(z) = zg(h(z)). \tag{6.7}$$

In particular, it is well known [192] that the hitting probability

$$\mathbb{P}(T_{-k} < \infty) = h_k(1) = q^k \tag{6.8}$$

where $q := h_1(1)$ is the least non-negative root of $q = g(q)$. So $q = 1$ or $q < 1$ according as $\mu \leq 1$ or $\mu > 1$, where $\mu := \sum_i ip_i$ and it is assumed that $p_1 < 1$. This brings us to:

Theorem 6.2. (Lagrange inversion formula [103]) *Let $g(\cdot)$ be analytic in a neighbourhood of 0 with $g(0) \neq 0$. Then the equation $h(z) = zg(h(z))$ has a unique analytic solution $h(\cdot)$ in a neighbourhood of 0 whose expansion in powers of z is such that*

$$[z^n]h(z)^k = \frac{k}{n}[z^{n-k}]g(z)^n. \tag{6.9}$$

Remark. While stated here in an analytic form, it is known that the Lagrange inversion formula can be regarded as an identity of formal power series. The formula has numerous variants and generalizations. See [407, §5.4] and papers cited there.

Sketch of proof. [380, 430] It is quite easy to establish existence and uniqueness, and to see that if $g(z) := \sum_{n=0}^{\infty} p_n z^n$, then both sides of (6.9) are polynomials in $p_0, \ldots, p_n$. In view of (6.6) and (6.7), Kemperman's formula (6.3) yields (6.9) for arbitrary non-negative $p_0, \ldots, p_n$ subject to $\sum_{i=0}^{n} p_i \leq 1$. The general conclusion then follows by polynomial continuation. □

To make the previous proof more explicit, it is immediate from the probabilistic interpretation that both $[z^n]h(z)^k$ and $[z^{n-k}]g(z)^n$ are polynomials in $p_0, \ldots, p_n$ with non-negative integer coefficients, which can be interpreted as follows. On the right side of (6.9), the coefficient of $\prod_i p_i^{n_i}$ in $[z^{n-k}]g(z)^n$ is

$$\#_{n_0,\ldots,n_n} := \binom{n}{n_0, \ldots, n_n} 1(\Sigma_i n_i = n \text{ and } \Sigma_i in_i = n-k) = \tag{6.10}$$

$$\#\{\text{lattice paths from } (0,0) \text{ to } (n,-k) \text{ with } n_i \text{ steps of size } i-1, 0 \leq i \leq n\}. \tag{6.11}$$

Whereas on the left side of (6.9), the coefficient of $\prod_i p_i^{n_i}$ in $[z^n]h(z)^k$ is the number of such paths which first hit $-k$ at time n, say $\#_{n_0,\ldots,n_n}\{\text{first hit } -k \text{ at } n\}$. So (6.9) reduces to the fact that the ratio of these two numbers is

$$\frac{\#_{n_0,\ldots,n_n}\{\text{first hit } -k \text{ at } n\}}{\#_{n_0,\ldots,n_n}} = \frac{k}{n} \tag{6.12}$$

for every choice of non-negative integers $n_0, \ldots, n_n$ with

$$\sum_i n_i = n \text{ and } \sum_i in_i = n-k.$$

This combinatorial fact (6.12) can be seen directly from Lemma 6.1.

Summary The Lagrange inversion formula (6.9) can be interpreted either probabilistically or combinatorially in terms of lattice paths. The key factor of k/n, appearing in the Lagrange inversion formula and its combinatorial equivalent (6.12), is interpreted by Kemperman's formula (6.3) as the conditional probability that the walk $(S_j - j)_{j=0,1,2,...}$ first hits $-k$ at step n given that $S_n - n = -k$, for any sequence of partial sums S_j of exchangeable non-negative integer valued random variables. The same factor of k/n appears, but is not so easy to interpret, in a number of other combinatorial expressions of the Lagrange formula presented in the exercises of the next section.

Exercises

6.1.1. (Discrete form of Vervaat's transformation) [355] Let $S_j := X_1 + \cdots + X_j$ where $(X_1, \ldots, X_n)$ is cyclically exchangeable with values in $\{-1, 0, 1, \ldots\}$, and $S_n = -1$. Let $M_n := \min\{i : S_i = \min_{1 \le j \le n} S_j\}$. Then

- M_n has uniform distribution on $\{1, \ldots, n\}$;
- the *cyclically shifted walk* with steps $(X_{M_n + j}, 1 \le j \le n)$, with $M_n + j$ understood mod n, is distributed like the original walk given it first hits -1 at time n;
- the cyclically shifted walk is independent of M_n.

6.1.2. (Vervaat's transformation) [425, 57] Let $S_j := X_1 + \cdots + X_j$ for independent and identically distributed integer-valued X_i with mean 0 and variance 1. Deduce from the previous exercise and the conditioned forms of Donsker's theorem (0.2) and (0.5) that if μ is the a.s. unique time at which a Brownian bridge $B^{\rm br}$ attains its minimum, then the process $(B^{\rm br}_{\mu+t} - B^{\rm br}_{\mu}, 0 \le t \le 1)$, with $\mu + t$ understood mod 1, is a standard Brownian excursion, independent of μ, which has uniform distribution on $[0,1]$. See [355] for a more elementary justification of the passage to the limit, using almost sure instead of weak convergence.

6.1.3. (Vervaat's transformation of a Lévy bridge) [95, 309] Generalize the result of the previous exercise to bridges and excursions derived from a suitable Lévy process with no negative jumps instead of Brownian motion.

Notes and comments

See also [96],[91],[48] for further variations of Vervaat's transformation. The continuous analog of Kemperman's formula (6.3) is *Kendall's formula* [237]

$$da\, \mathbb{P}(T_a \in dt) = \frac{a}{t} \mathbb{P}(X_t \in da) dt$$

for T_a the first hitting time of $a > 0$ for a Lévy process (X_t) with no positive jumps. See [74] and papers cited there.

6.2. Galton-Watson forests

It is well known to combinatorialists [380, 266, 101, 407] that the enumerations of lattice paths related to the Lagrange inversion formula, like (6.11)-(6.12), can also be expressed by suitable bijections as enumerations of various sets of trees and forests.

The term *forest* will be used here for a *finite rooted forest*, that is a directed graph $F \subseteq V \times V$ with a finite set of vertices V, such that each connected component of F is a tree with edges directed toward some root vertex. A forest with vertex set $V = V(F)$ is said to be *labeled by* V. For vertices v and w of a forest F write $v \xrightarrow{\mathbf{f}} w$, to show that (v, w) is a directed edge of F, and say *v is the child of w*, or *w is the parent of v*. Note that the direction of edges is from child to parent. In a *plane forest* F with k component trees, the set of roots of the tree components is ordered, as is the set of children of w for each vertex w of F. Regard a plane forest with k root vertices as a collection of family trees, one for each of k initial individuals, with each vertex in the forest corresponding to an individual, and with the order of the roots and the orders of children corresponding to the order of birth of individuals.

A plane forest is often depicted without labels as on the top left panel of Figure 6.1, and called an *unlabeled plane forest.* However, there is a natural way to identify each vertex of a plane forest by a finite sequence of non-negative integers which indicates the location of the vertex in the forest. So, following the convention of [192] for labeling family trees, the set of vertices $V(F)$ of a plane forest F can always be identified as a subset of the set of all finite sequences of integers, as illustrated in the top right panel of Figure 6.1. For a plane forest F with n vertices, two useful relabelings of $V(F)$ by $[n]$ are provided by the *depth-first* and *breadth-first* searches of $V(F)$, whose definition should be obvious from the examples in the lower panels of Figure 6.1.

Following Otter [334] and subsequent authors [192, 248, 125], regard a Galton-Watson process started with k individuals, with offspring distribution $(p_0, p_1, \ldots)$, as generating a collection of k family trees, which combine to form a *random family forest* $\mathcal{F}_k$. Let $\#\mathcal{F}_k$ be the *total progeny* of the branching process, meaning the number of vertices of $\mathcal{F}_k$. On the event $(\#\mathcal{F}_k < \infty)$ the random family forest $\mathcal{F}_k$ can be defined in an elementary way as a random element of the countable set $\mathbf{F}$ of all plane forests. The *distribution of* $\mathcal{F}_k$ is then the sub-probability distribution on $\mathbf{F}$ defined by the formula [334]

$$\mathbb{P}(\mathcal{F}_k = F) = \prod_{v \in V(F)} p_{c(v,F)} = \prod_{i \geq 0} p_i^{n_i(F)} \qquad \forall k \geq 1,\ F \in \mathbf{F}_k^{plane} \tag{6.13}$$

with the following notation

- $V(F)$ is the set of vertices of F;
- $c(v, F) := \#\{w : w \xrightarrow{\mathbf{f}} v\}$ is the *number of children* or *in-degree* of v in the forest F;

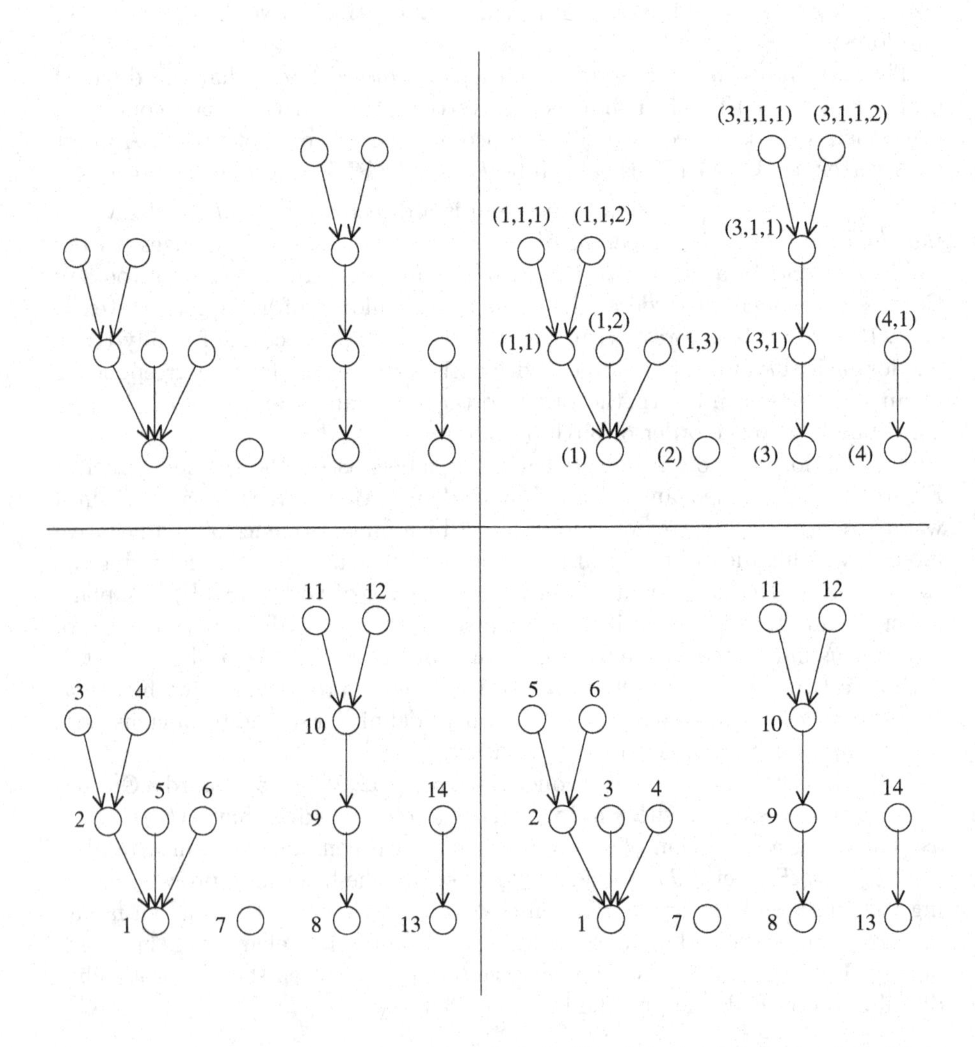

Figure 6.1: A plane forest, variously labeled. Top left: no labels needed. Top right: natural labeling of vertices by finite sequence. Bottom left: labeling by depth-first search. Bottom right: labeling by breadth-first search.

- $n_i(F) := \sum_{v \in V(F)} 1(c(v,F) = i)$ is the number of vertices of F with i children;
- $\mathbf{F}_k^{plane}$ is the set of plane forests with k root vertices.

The distribution of $\#\mathcal{F}_k$ induced by this distribution of $\mathcal{F}_k$ on $\mathbf{F}_k^{plane}$, with total mass $\mathbb{P}(\#\mathcal{F}_k < \infty) \leq 1$, was found by Otter [334] and Dwass [127]. Combining their result with Kemperman's formula (6.3), for $1 \leq k \leq n$

$$\mathbb{P}(\#\mathcal{F}_k = n) = \mathbb{P}(T_{-k} = n) = \frac{k}{n}\mathbb{P}(S_n = n - k) \tag{6.14}$$

where T_{-k} as before is the first hitting time of $-k$ by the walk $(S_n - n)_{n=0,1,\ldots}$ with increments $X_i - 1$ for a sequence of independent offspring variables X_i with distribution $(p_0, p_1, \ldots)$. Thus

$$\#\mathcal{F}_k \stackrel{d}{=} T_{-k} \text{ for each } k = 1, 2, \ldots. \tag{6.15}$$

One way to see (6.14) is to argue that the sequence of probability generating functions of $\#\mathcal{F}_k$ must solve equations (6.6) and (6.7), which according to Lagrange is solved uniquely by the sequence of generating functions h_k of T_{-k}. But more insight is gained from the following bijective proof of (6.15).

Bijection between plane forests and lattice walks The following lemma is well known and easily checked. Harris [193] gave it for just a single tree, but the extension to forests is immediate:

Lemma 6.3. *Given a plane forest F of k trees with n vertices, let x_i be the number of children of vertex i of F in order of depth-first search. This coding of F sets up a bijection*

$$F \leftrightarrow (x_1, \ldots, x_n) \tag{6.16}$$

between $\mathbf{F}_k^{plane}$ and sequences of non-negative integers $(x_1, \ldots, x_n)$ such that the lattice walk with steps $x_i - 1$ first reaches $-k$ at time n. Moreover, if the trees of the forest are of sizes $n_1, \ldots, n_k$, then for each $1 \leq i \leq k$, the walk first reaches $-i$ at the time $n_1 + \cdots + n_i$ when the depth-first search of the ith tree is completed.

The same is true with breadth-first instead of depth-first search of each tree of F. Call the lattice walk with increments $x_i - 1$ so associated with a forest F, the *depth-first walk* or the *breadth-first walk* as the case may be. This transformation from forests to walks has a well known interpretation in terms of queuing theory [236, 409, 410]. Variations of the bijection then correspond to various queue disciplines (last in – first out, first in – first out, etc).

To put the lemma in probabilistic terms, let X_i' be the number of children of the ith individual in the depth-first search order of a Galton-Watson family forest $\mathcal{F}_k$ started with k individuals. Then provided $\#\mathcal{F}_k < \infty$ the sequence $(X_1', \ldots, X_{\#\mathcal{F}_k}')$ determines the plane forest $\mathcal{F}_k$ uniquely, and there is the following refinement of (6.15):

$$(X_1', \ldots, X_{\#\mathcal{F}_k}')1(\#\mathcal{F}_k < \infty) \stackrel{d}{=} (X_1, \ldots, X_{T_{-k}})1(T_{-k} < \infty) \tag{6.17}$$

where on the right side the X_i are independent offspring variables, and T_{-k} is the first time j that $\sum_{i=1}^{j}(X_i - 1) = -k$. Keep in mind that $\mathcal{F}_k$ and the infinite sequence $X_1, X_2, \ldots$ might be defined on different probability spaces. However, by use of the bijection, it is clear that provided $p_1 < 1$ and $\sum_i ip_i \leq 1$, so that $P(\#\mathcal{F}_k < \infty) = P(T_{-k} < \infty) = 1$, there is defined on the same probability space as $X_1, X_2, \ldots$ an a.s. unique sequence of Galton-Watson forests $\mathcal{F}_k, k = 1, 2, \ldots$ such that (6.17) holds with almost sure equality instead of equality in distribution. Then $\mathcal{F}_j$ is the forest formed by the first j trees of $\mathcal{F}_k$ for each $j < k$. The jth tree of $\mathcal{F}_k$ for every $j < k$ is the unique tree whose depth-first walk has steps $X_{T_{-(j-1)}+i} - 1$ for $1 \leq i \leq T_{-j} - T_{-(j-1)}$.

Exercises

6.2.1. (Enumeration of plane forests by type) [133, 137, 334, 379] The *type* of a forest F is the sequence of non-negative integers (n_i), where n_i is the number of vertices of F with i children. Let $1 \leq k \leq n$ and let (n_i) be a sequence of non-negative integers with

$$\sum_i n_i = n \text{ and } \sum_i in_i = n - k. \tag{6.18}$$

A forest of type (n_i) has n vertices and $n - k$ non-root vertices, hence k root vertices and k tree components. For $1 \leq k \leq n$ and (n_i) subject to (6.18) the number $N^{plane}(n_0, n_1, \ldots)$ of plane forests of type (n_i) with k tree components and n vertices is

$$N^{plane}(n_0, n_1, \ldots) = \frac{k}{n}\binom{n}{n_0, \ldots, n_n}. \tag{6.19}$$

6.2.2. (Enumeration of labeled forests by type) [407, Cor. 3.5], [354, Th. 1.5] For $1 \leq k \leq n$ and (n_i) subject to (6.18), the number $N^{[n]}(n_0, n_1, \ldots)$ of forests labeled by $[n]$ of type (n_i) with k tree components and n vertices is

$$N^{[n]}(n_0, n_1, \ldots) = \frac{k}{n}\binom{n}{k}\frac{(n-k)!}{\prod_{i \geq 0}(i!)^{n_i}}\binom{n}{n_0, \ldots, n_n}. \tag{6.20}$$

6.2.3. (Enumeration of labeled forests by numbers of children) [356], [407, Thm. 3.4] For all sequences of non-negative integers $(c_1, \ldots, c_n)$ with $\sum_i c_i = n - k$ the number $N(c_1, \ldots, c_n)$ of forests F with vertex set $[n]$ in which vertex i has c_i children for each $i \in [n]$ (and hence F has k tree components) is

$$N(c_1, \ldots, c_n) = \frac{k}{n}\binom{n}{k}\binom{n-k}{c_1, \ldots, c_n}. \tag{6.21}$$

6.2.4. (Cayley's multinomial theorem) The enumeration (6.21) amounts to the following identity of polynomials in n commuting variables $x_i, 1 \leq i \leq n$:

$$\sum_{F \in \mathbf{F}_{k,n}} \prod_{i=1}^{n} x_i^{c(i,F)} = \frac{k}{n}\binom{n}{k}(x_1 + \cdots + x_n)^{n-k} \tag{6.22}$$

where the sum is over the set $\mathbf{F}_{k,n}$ of all forests with k tree components labeled by $[n]$, and $c(i, F)$ is the number of children of i in the forest F. Take the x_i to be identically 1 in (6.22) to recover the well known enumeration

$$\#\mathbf{F}_{k,n} = k\binom{n}{k}n^{n-k-1}. \tag{6.23}$$

This is equivalent to Cayley's [88] formula

$$\#\{\text{ forests with root set } [k] \text{ and vertex set } [n]\ \} = kn^{n-k-1} \tag{6.24}$$

which was derived in (1.56). In particular, for $k = 1$ the number of rooted trees labeled by $[n]$ is n^{n-1}. Equivalently, the number of unrooted trees labeled by $[n]$ is n^{n-2}. For various approaches to these formulae of Cayley, see [315, 356, 378, 402, 407, 411, 383].

Notes and comments

This section is based on the work of Harris [193]. The exercises were suggested by the treatment of tree enumerations in Stanley's book [407]. Solutions can be found in [354]. See also Chapter 10 for another approach to Cayley's multinomial theorem.

6.3. Brownian asymptotics for conditioned Galton-Watson trees

Recall that $\mathbf{F}_{k,n}^{plane}$ is the set of all plane forests of k trees with a total of n vertices. For $k = 1, 2, \ldots$ and $0 < p < 1$ let $\mathcal{G}_{k,p}$ be a Galton-Watson forest of k trees with the geometric(p) offspring distribution $p_i := p(1-p)^i$. Since for each $F \in \mathbf{F}_{k,n}^{plane}$ the total number of children of all vertices of F is $\sum_{v \in V(F)} c(v, F) = n - k$, the general product formula (6.13) gives

$$\mathbb{P}(\mathcal{G}_{k,p} = F) = \prod_{v \in V(F)} p(1-p)^{c(v,F)} = p^n(1-p)^{n-k} \quad \forall\ F \in \mathbf{F}_{k,n}^{plane}. \tag{6.25}$$

Hence, as observed by Harris [193], the conditional distribution of $\mathcal{G}_{k,p}$ given $(\#\mathcal{G}_{k,p} = n)$ is uniform on $\mathbf{F}_{k,n}^{plane}$. If $S_{n,p}$ is the sum of n independent geometric(p) random variables, then by the general formula (6.14) for the distribution of the size of a Galton-Watson forest of k trees, and the negative binomial formula [150, VI.8] for the distribution of $S_{n,p}$,

$$\mathbb{P}(\#\mathcal{G}_{k,p} = n) = \frac{k}{n}\mathbb{P}(S_{n,p} = n - k) = \frac{k}{n}\binom{2n-k-1}{n-k}p^n(1-p)^{n-k}. \tag{6.26}$$

Compare (6.25) and (6.26) to see that the number of plane forests of k trees with n vertices is

$$\#\mathbf{F}_{k,n}^{plane} = \frac{k}{n}\binom{2n-k-1}{n-k}. \tag{6.27}$$

In particular, for $k = 1$, the number of plane trees with n vertices is

$$\#\mathbf{F}_{1,n}^{plane} = \frac{1}{n}\binom{2n-2}{n-1} \tag{6.28}$$

which is the $(n-1)$th *Catalan number* [78, 197]. This is also the number of *lattice excursions of length* $2n$, that is sequences $(s_j, 0 \le j \le 2n)$ where $s_0 = s_{2n} = 0$, and $s_j > 0$ and $s_{j+1} - s_j \in \{-1, +1\}$ for all $0 \le j \le 2n-1$. As observed by Harris [193], there is a natural bijection between $\mathbf{F}_{1,n}^{plane}$ and the set of lattice excursions of length $2n$. Given a plane tree with n vertices, start from the root and traverse the plane tree as follows. At each step move away from the root along the first edge that has not been walked on yet. If this is not possible then step back along the edge leading towards the root. A walk with steps of ± 1 is obtained by plotting the height at each step. Appending a $+1$ step at the beginning and a -1 step at the end gives a lattice excursion of $2n$ steps, as illustrated by Figure 6.2.

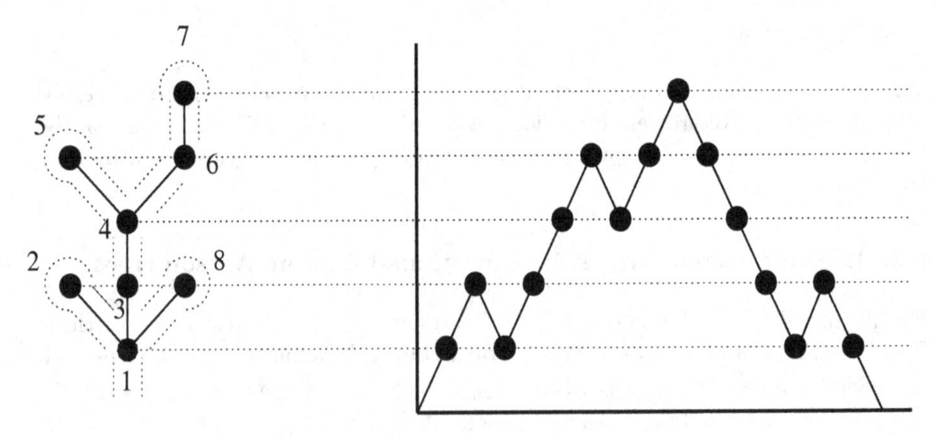

Figure 6.2: Harris walk for a finite tree.

This bijection extends to one between $\mathbf{F}_{k,n}^{plane}$ and the set of non-negative lattice walk paths from $(0,0)$ to $(0,2n)$ with increments of ± 1 and exactly k returns to 0.

Let $\mathcal{T}$ denote a Galton-Watson tree with some offspring distribution (p_i), and let $\mathcal{T}_n$ denote the random tree with n vertices defined by conditioning $\mathcal{T}$ to have n vertices. To abbreviate, call $\mathcal{T}_n$ a GW(n) *tree with offspring distribution* (p_i). Such random trees are also called *simply generated trees* [305]. It follows immediately from (6.13) that the distribution of a GW(n) tree is the same for offspring distribution (p_i) as for $(p_i\theta^i/g(\theta))$ for arbitrary $\theta > 0$ such that $g(\theta) := \sum_i p_i \theta^i < \infty$. Consequently, in consideration of GW(n) trees there is no loss of generality in supposing that the offspring distribution has mean 1. According to the previous discussion, the Harris walk associated with a GW(n) tree with geometric offspring distribution is the unsigned excursion of a simple symmetric random walk conditioned to have length $2n$. By the conditioned form of Donsker's theorem (0.5), a suitable normalization of this uniform lattice excursion converges in distribution to a standard Brownian excursion B^{ex}. For GW(n) trees with other offspring distributions, the Harris walk is no longer the excursion of a Markovian random walk. Rather it is some non-uniformly distributed lattice excursion with increments of ± 1 and a rather complicated dependence structure. Nonetheless, according to the following theorem, with suitable scaling the asymptotic behaviour of Harris walks of large Galton-Watson trees is the same, no matter what the offspring distribution with finite variance.

Theorem 6.4. Aldous [5, 6] *Let $\mathcal{T}_n$ be a GW(n) tree, with offspring distribution with mean* 1 *and variance $\sigma^2 \in (0,\infty)$. Let $H_n(k), 0 \le k \le 2n$ be the Harris walk associated with $\mathcal{T}_n$. Then as $n \to \infty$ through possible sizes of the unconditioned Galton-Watson tree,*

$$(H_n(2nu)/\sqrt{n}, 0 \le u \le 1\}) \xrightarrow{d} (2\sigma^{-1} B_u^{\mathrm{ex}} : 0 \le u \le 1) \tag{6.29}$$

where B^{ex} is the standard Brownian excursion and $\xrightarrow{d}$ is the usual weak convergence of processes in $C[0,1]$.

Since numerous features of $\mathcal{T}_n$ are encoded as continuous functionals of the associated Harris walk, the asymptotic distributions of these features of trees can be read from the distribution of the corresponding functional of Brownian excursion. For instance, if $\overline{H}(\mathcal{T})$ denotes the maximum height above the root of any vertex of a tree $\mathcal{T}$, so $\overline{H}(\mathcal{T}_n) = \max_{0\le u\le 1} H_n(2nu)$, then we read from (6.29) that

$$\overline{H}(\mathcal{T}_n)/\sqrt{n} \xrightarrow{d} 2\sigma^{-1} \max_{0\le u\le 1} B_u^{\mathrm{ex}}. \tag{6.30}$$

See [60] for a review of properties of the distribution of $\max_{0\le u\le 1} B_u^{\mathrm{ex}}$, whose Mellin transform is related to the Riemann zeta function.

Marckert and Mokkadem [293] derive Theorem 6.4, under further moment conditions, by showing that for large n the normalized Harris path on the left side of (6.29) is with high probability close in $C[0,1]$ to

$$(2\sigma^{-2}(S_{nu} - nu)/\sqrt{n}, 0 \le u \le 1 \,|\, T_{-1} = n) \tag{6.31}$$

where $(S_k - k, 1 \le k \le T_{-1})$ is the depth-first walk of $\mathcal{T}_n$, as in (6.17). The process in (6.31) converges in distribution to $2\sigma^{-1}B^{\mathrm{ex}}$ by the conditioned form of Donsker's theorem (0.5). The appearance of the factor $2/\sigma^2$ in approximating the Harris walk by the normalized depth-first walk is nicely explained by Bennies-Kersting [33]. Besides the geometric case discussed above, where the offspring distribution has standard deviation $\sigma = \sqrt{2}$, and Theorem 6.4 reduces to the conditioned form of Donsker's theorem (0.5), two other important examples are as follows:

Poisson branching $p_i = e^{-1}/i!$, $\sigma = 1$. As indicated by Kolchin and Aldous, a Poisson-GW(n) tree $\mathcal{T}_n$ may be constructed as follows, from $\mathcal{U}_n$ with uniform distribution on the set of n^{n-1} rooted trees labeled by $[n]$. First give the children of each vertex of $\mathcal{U}_n$ the order they acquire from the usual order on $[n]$, (or, equivalently, independent random orders), then ignore the labels to obtain a plane tree. See [354, §7] for further discussion. This transformation allows numerous asymptotic results for the uniform labeled tree $\mathcal{U}_n$ to be deduced from corresponding results for the non-uniform Poisson-GW(n) tree $\mathcal{T}_n$, which can be read from Theorem 6.4. For instance, the maximum height $\overline{H}(\mathcal{U}_n)$ is by construction identical to $\overline{H}(\mathcal{T}_n)$, whose asymptotic distribution is given by (6.30) with $\sigma = 1$.

Binary branching [180] $p_0 = p_2 = \frac{1}{2}$, $\sigma = 1$. Now $\mathcal{T}_n$ is a uniform ordered binary rooted tree on n vertices. Note that n must be odd. Similarly, for any $m = 2, 3, \ldots$ one can realize a uniform ordered m-ary rooted tree on n vertices as a conditioned Galton-Watson tree.

Concatenated Harris walks A result similar in spirit to Theorem 6.4 can be obtained by considering a sequence of independent and identically distributed critical Galton Watson trees. By concatenating the individual Harris walks (resp depth-first walks) one gets an infinite walk which upon scaling converges to the reflected Brownian motion.

Theorem 6.5. Le Gall [275] *Let $(H(t), t \geq 0)$ be the continuous path obtained by concatenation of the Harris walks of an infinite independent and identically distributed sequence of critical GW trees with finite non-zero offspring variance σ^2. Then as $n \to \infty$*

$$(H(2nt)/\sqrt{n}, t \geq 0) \xrightarrow{d} \left(\frac{2}{\sigma}|B_t|, t \geq 0\right) \tag{6.32}$$

in the sense of weak convergence in $C[0,\infty)$, where B is a standard Brownian motion.

The proof of this theorem again proceeds by using standard results for depth-first walk and then relating the Harris walk to the depth-first walk. It is convenient to use another walk associated with trees called the height process as an intermediary in this comparison.

The height process For a rooted tree labelled by $[n]$ in depth-first order, the associated height process is defined by

$$\mathcal{H}_i = \text{height of vertex } i+1 \text{ for } i = 0, 1 \ldots, n-1$$

For a sequence of GW trees $\mathcal{T}^{(1)}, \mathcal{T}^{(2)}, \ldots$ denote the associated(concatenated) Harris walk, depth-first walk and height process by H_n, Y_n and $\mathcal{H}_n$, $n \geq 0$, where values of the processes at non-integer times are defined by linear interpolation. The following lemma is well known:

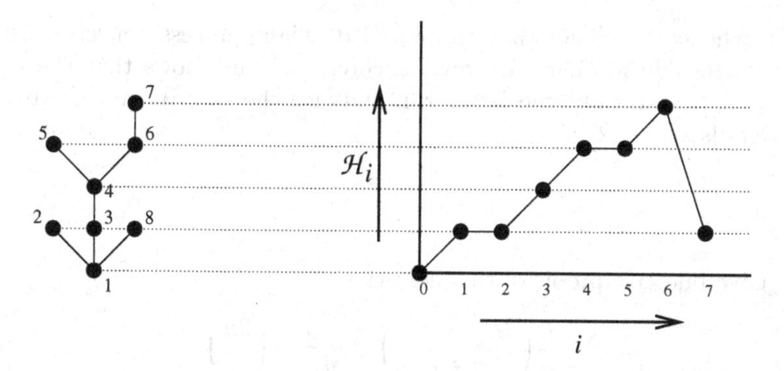

Figure 6.3: The height process for a rooted labeled tree of 8 vertices.

Lemma 6.6. *Let $\tau_1, \tau_2, \ldots$ be the ascending weak ladder points of the depth-first walk Y. That is, $\tau_0 = 0$ and $\tau_i = \inf\{n > \tau_{i-1} : Y_n = M_n\}$ where $M_n = \sup_{0 \le k \le n} Y_k$. Then $Y_{\tau_{i+1}} - Y_{\tau_i}, i = 0, 1, \ldots$ are independent and identically distributed, and*

$$\mathbb{P}(Y_{\tau_1} = k) = \sum_{i=k+1}^{\infty} p_i; \qquad \mathbb{E}(Y_{\tau_1}) = \frac{\sigma^2}{2}$$

Observing that $M_n = \sum_{i=1}^{K_n} (Y_{\tau_i} - Y_{\tau_{i-1}})$ where

$$K_n = \#\{k \in \{1, \ldots, n\} : Y_k = M_k\},$$

and using Lemma 6.6, we deduce from the law of large numbers that

$$\frac{M_n}{K_n} \xrightarrow{\text{a.s}} \mathbb{E}(Y_{\tau_1}) = \frac{\sigma^2}{2} \tag{6.33}$$

as $n \to \infty$. Now for any fixed n the time reversed walk $\{Y_n - Y_{n-i}\}_{i=0}^n$ has the same distribution as $\{Y_i\}_{i=0}^n$. From Exercise 6.3.2 below it follows that the corresponding M_n and K_n for the reversed walk are precisely $Y_n - \inf_{0 \le k \le n} Y_k$ and $\mathcal{H}_{n+1}$. By Donsker's theorem, for any $t_1, \ldots, t_r$

$$\begin{aligned} \frac{1}{\sigma\sqrt{n}} \left(Y_{[nt_k]} - \inf_{0 \le i \le [nt_k]} Y_i \right)_{1 \le k \le r} &\xrightarrow{d} \left(B_{t_k} - \inf_{0 \le u \le t_k} B_u \right)_{1 \le k \le r} \\ &\stackrel{d}{=} (|B_{t_k}|)_{1 \le k \le r} \end{aligned}$$

Hence by using (6.33) for the reversed walk, we get

$$\frac{\sigma}{2\sqrt{n}} (\mathcal{H}_{[nt_1]}, \ldots, \mathcal{H}_{[nt_r]}) \xrightarrow{d} (|B_{t_1}|, \ldots, |B_{t_r}|).$$

This shows convergence of finite dimensional distributions of the height process to the reflected Brownian motion. One can further show tightness as in [275], whence it follows that the rescaled height process converges to the reflected Brownian motion. To prove Theorem 6.5 one shows that the height process and Harris walk (the latter run at twice the speed) are close to each other. Details are in [275].

Exercises

6.3.1. Give bijective proofs of the identities

$$\sum_{k=1}^{n} \frac{k}{n} \binom{2n-k-1}{n-k} = \frac{1}{n+1}\binom{2n}{n} \tag{6.34}$$

where the sum is the nth Catalan number, and

$$\sum_{k=1}^{n} \frac{k}{n} \binom{2n-k-1}{n-k} 2^k = \binom{2n}{n}. \tag{6.35}$$

6.3.2. (Depth-first and height processes)[277][33]
(a) Show that the depth-first walk of a tree is given by $Y_{n-1} = \sum_v R_v$, where R_v is the number of younger siblings of v and the sum is over all ancestors v of n(including $v = n$).
(b) Show that the height process $\mathcal{H}$ of a sequence of finite trees is related to the depth-first walk Y by

$$\mathcal{H}_n = \#\{k \in \{0,1,\ldots,n-1\} : Y_k = \inf_{k \le j \le n} Y_j\}$$

(Hint: Show that $Y_k = \inf_{k \le j \le n} Y_j$ if and only if k is an ancestor of $n+1$)

6.3.3. (Harris's embedding in Brownian motion of inhomogeneous random walks and associated geometric branching processes) [193] Show how a random walk on the non-negative integers started at $k > 0$, with transition probabilities p_i for $i \to i+1$, and $1 - p_i$ for $i \to i-1$, $i > 0$, and absorption at 0, can be embedded in the path of a Brownian motion started at a suitable $z_k > 0$ and stopped when it first reaches 0. As Harris observed, the path of such a walk, started at 1 and stopped when it first reaches 0, is bijectively equivalent to the family tree of a branching process, starting with 1 individual in generation 0, in which each individual in the $(k-1)$th generation has probability $p_k^r q_k$ of having exactly r children, $r = 0, 1, 2, \ldots$. Each birth in the $(k-1)$th generation of the branching process corresponds to a transition of the walk from k to $k+1$, which corresponds to an upcrossing of the Brownian path from z_k to z_{k+1} before the path first reaches 0. Thus the random plane tree generated by such a branching process, with geometric offspring distribution with parameter q_k in the kth generation, is embedded in the Brownian path in a natural way. This idea was further exploited by Knight [254] to derive the Ray-Knight theorems, whereby portions of the local time process $(L_T^x(B), x \in \mathbb{R})$ are described in terms of various continuous-state branching processes, commonly represented as squares of Bessel processes, as discussed in Chapter 8.

Some refinements of the convergence of Harris walks Consider critical GW trees with offspring variance $\sigma^2 \in (0, \infty)$.

6.3.4. In the setting of Theorem 6.5, let $N(t)$ be the number of zeros of $H(\cdot)$ in $(0, t]$, which is the number of complete trees encoded by the concatenated walk up to time t. Then (6.32) holds jointly with

$$(N(2nt)/\sqrt{n}, t \geq 0) \stackrel{d}{\rightarrow} (\sigma L_t, t \geq 0) \tag{6.36}$$

where $(L_t, t \geq 0)$ is the usual local time process of B at 0, normalized so that $L_t \stackrel{d}{=} |B_t|$ for each $t > 0$.

6.3.5. [17] Let $\mathcal{F}_{n,k}$ be a forest of k critical GW trees conditioned to have total size n. Let $H_{n,k}(t), 0 \leq t \leq 2n$ be the Harris path associated with $\mathcal{F}_{n,k}$, by concatenation of the Harris paths of the k trees. It is to be anticipated from (6.32) and (6.36) that as $n \to \infty$ and k varies with n

$$(H_{n,k}(2nu)/\sqrt{n}, 0 \leq u \leq 1\} \stackrel{d}{\rightarrow} (2\sigma^{-1}|B_\ell^{\mathrm{br}}(u)| : 0 \leq u \leq 1) \text{ if } \frac{k}{\sigma\sqrt{n}} \to \ell \tag{6.37}$$

where B_ℓ^{br} is Brownian bridge B^{br} conditioned on $L_1^0(B^{\mathrm{br}}) = \ell$, as defined in Lemma 4.10. Show this is true for Poisson offspring distribution, with $\sigma = 1$.

6.3.6. (Problem) Does (6.37) hold for every critical aperiodic offspring distribution with $\sigma^2 \in (0, \infty)$?

6.3.7. (Problem) In the same vein, let $(H(t), t \geq 0)$ be the continuous path obtained by concatenation of the Harris walks of an infinite independent and identically distributed sequence of critical GW trees. Then it is to be anticipated that as $n \to \infty$

$$(H(2nu)/\sqrt{n}, 0 \leq u \leq 1) \text{ given } H(2n) = 0 \stackrel{d}{\rightarrow} \left(\frac{2}{\sigma}|B^{\mathrm{br}}(u)|, 0 \leq u \leq 1\right), \tag{6.38}$$

where the event $H(2n) = 0$ is the event that after n steps the depth-first search of the forest completes the search of some tree. This was argued in [17], jointly with the obvious variant of (6.36), in the Poisson case. Is it true for every critical aperiodic offspring distribution with $\sigma^2 \in (0, \infty)$?

Notes and comments

See Duquesne [118] and Marckert and Mokkadem [293, 294] for some further developments.

6.4. Critical random graphs and the multiplicative coalescent

Aldous [9] used a correspondence between trees and walks to study the asymptotic behaviour of component sizes in the Erdős-Rényi random graph process

$\mathcal{G}(n,p)$ in the critical regime $p \approx 1/n$ as $n \to \infty$. Recall that $\mathcal{G}(n,p)$ is the model of an undirected random graph on $[n]$ in which each of $\binom{n}{2}$ undirected edges is present with probability p, independently. The random partition of $[n]$ generated the connected components of $\mathcal{G}(n,p)$ is clearly exchangeable. But the asymptotic theory of exchangeable partitions is of little use in this example, because the interesting behaviour in the critical regime lies beyond the reach of Kingman's theory. The key to understanding this behaviour is the following algorithm for constructing the connected components of $\mathcal{G}(n,p)$ in the size-biased order of their least elements.

The breadth first walk [9, §1.3] Start by considering a fixed undirected graph G with vertex set $[n]$. So G is a subset of the set of $\binom{n}{2}$ subsets of $[n]$ of size 2. Construct a *rooted ordered spanning forest of* G, say $F^{[n]}(G)$, with vertices labeled by $[n]$, as follows. The root vertices (or zeroth generation of the forest) are the least elements of components of G, with their natural order from $[n]$. The children of a particular root vertex v are the vertices w such that there is an edge from v to w in G. These children, which form the first generation of the forest, are ordered firstly by the order of their parents, and secondarily by their natural order from $[n]$. In general, for $m \geq 0$, given that the first m generations of the forest have been defined and the vertices of the mth generation have been ordered, the $(m+1)$th generation is the set of all vertices w such that there is an edge from w to some v in the mth generation, and w is not in generations $0, \ldots, m$. Then w is the child of the first such v with respect to the order on the mth generation, and the vertices in the the $(m+1)$th generation are ordered firstly by the order of their parents, and secondarily by their natural order from $[n]$. The *breadth-first search of* $[n]$ *induced by* G is the permutation $\beta_G : [n] \to [n]$ where $\beta_G(i)$ is the ith vertex in the list of vertices of $F^{[n]}(G)$, starting with the zeroth generation, then the first generation, then the second generation, and so on, where the vertices of the mth generation are put in the order they were given in the construction of $F^{[n]}(G)$. Put another way, $\beta_G(i)$ is the label of the ith vertex of $F^{[n]}(G)$ visited in the usual breadth-first search of $F^{[n]}(G)$ regarded as a plane tree by ignoring the vertex labels. The *breadth first walk derived from* G is the breadth first walk derived from $F^{[n]}(G)$ and β_G. See [9, Fig. 1]. To be precise, the walk is $(w_0, w_1, \ldots, w_n)$, where $w_0 = 0$, and for $1 \leq i \leq n$

$$w_i - w_{i-1} = c_i - 1$$

where c_i is the number of children of the vertex labeled $\beta_G(i)$ in $F^{[n]}(G)$. Consequently, by the variant of Lemma 6.3 for breadth-first search, the component sizes of G, in order of least elements, say $\tilde{N}_1, \tilde{N}_2, \ldots$, can be recovered from its breadth first walk as

$$\tilde{N}_j = T_{-j} - T_{-(j-1)}$$

where

$$T_{-j} := \min\{i : w_i = -j\}.$$

To discuss the distribution of the walk induced by a random graph, one more concept is needed. In the breadth-first search of $[n]$ induced by G, after i steps, call a vertex v *marked at stage* i if v is not in the set $\{\beta_G(j), 1 \leq j \leq i\}$ of vertices searched in the first i steps, but v is a child of one of these vertices in $F^{[n]}(G)$, meaning there is an edge of G joining v to $\{\beta_G(j), 1 \leq j \leq i\}$. Let M_i denote the number of marked vertices at stage i. Note that M_i is determined by the first i steps of the walk. Also $M_i \leq N_{h(i)}(T_i) + N_{h(i)+1}(T_i)$, where T_i is the tree component of $\beta_G(i)$ in $F^{[n]}(G)$, $h(i)$ is the height of $\beta_G(i)$ in T_i, and $N_j(T_i)$ is the number of vertices of T_i at height j. Thus M_i is at most the number of vertices in two slices through a tree component of $F^{[n]}(G)$, which turns out to be negligible relative to i in the asymptotic regimes discussed here. It follows from these definitions and the definition of $\mathcal{G}(n,p)$ that the dynamics of the breadth-first walk derived from $\mathcal{G}(n,p)$ can be described as follows:

Lemma 6.7. *Fix n. Let $(W_1, \ldots, W_n)$ be the breadth-first walk derived from $\mathcal{G}(n,p)$. Then for each $i \geq 1$,*

$$W_{i+1} - W_i \text{ given } (W_1, \ldots, W_i) \stackrel{d}{=} \text{BINOMIAL}(n - i - M_i, p) - 1 \tag{6.39}$$

where BINOMIAL(m,p) *denotes a binomial variable with parameters m and p, and M_i is the number of marked vertices after i steps, which is some function of $(W_1, \ldots, W_i)$.*

Suppose now that $n \to \infty$ and $p = \lambda/n \to 0$. Then it is clear that for some large finite number of steps the numbers $W_{i+1} - W_i + 1$ of children of vertices of $F^{[n]}(G)$ in order of breadth-first search will be approximately independent Poisson(λ) variables. Thus one easily obtains:

Lemma 6.8. *Let $\mathcal{T}_1(n,p)$ be the plane tree derived from the subtree rooted at 1 in the forest $F^{[n]}(G)$ derived from G by breadth-first search, for G distributed according to $\mathcal{G}(n,p)$. Let $\lambda \in (0,\infty)$ and $h > 0$. Then as $n \to \infty$, the restriction of $\mathcal{T}_1(n, \lambda/n)$ to levels below height h, converges in distribution to the restriction to levels below h of a Galton-Watson tree with Poisson(λ) offspring distribution.*

Recall that the limiting Poisson-Galton-Watson (λ) tree is sub-critical, super-critical, or critical according to whether $\lambda < 1$, $\lambda > 1$ or $\lambda = 1$. There is the following well known result:

Theorem 6.9. (Erdős-Rényi[134]) *As $n \to \infty$ with $p = \lambda/n$ for some $\lambda \in (0,\infty)$,*
If $\lambda < 1$, then the largest component of $\mathcal{G}(n,p)$ has size $O(\log n)$.
If $\lambda > 1$, then the largest component has size $(1 + o(1))P_\lambda n$ where P_λ with $\exp(-\lambda P_\lambda) = 1 - P_\lambda$ *is the probability that a Poisson-Galton-Watson(λ) tree is infinite in size, while the second largest component is of size at most $O(\log n)$.*
If $\lambda = 1$, then the sizes of the largest and second largest components are both of order $n^{2/3}$.

Let $\Pi_n(p)$ be the partition of $[n]$ determined by the components of $\mathcal{G}(n,p)$. Note that as $n \to \infty$ with $p = \lambda/n$ for $\lambda > 1$, the limit in distribution of $\Pi_n(\lambda/n)$ is the partition of $\mathbb{N}$ with a single block whose frequency is $P_\lambda \in (0,1)$, and all other components singletons. So this is an instance where an exchangeable partition with improper frequencies arises naturally as a limit. But in the critical case, the limit in distribution of $\Pi_n(1/n)$ is the trivial partition into singletons. To obtain interesting limits in distribution around the critical stage $p \approx 1/n$, another scaling is necessary.

To this end, consider $\mathcal{G}(n, p_n(r))$ with $p_n(r) := 1/n + r/n^{4/3}$, for $r \in \mathbb{R}$. From Lemma 6.7, we find that after the ith step of the breadth-first walk, as $n \to \infty$

$$\begin{aligned} \mathbb{E}(W_{i+1} \,|\, W_1 \dots W_i) &= -1 + p_n(r)(n - i - M_i) \\ &\approx -1 + (1/n + r/n^{4/3})(n - i) \qquad (6.40) \\ &= \frac{r - in^{-2/3}}{n^{1/3}}. \qquad (6.41) \end{aligned}$$

where the approximation is just to ignore the M_i marked vertices, and similarly

$$\mathrm{Var}(W_{i+1} \,|\, W_1 \dots W_i) \approx 1.$$

Now measure time in units of $n^{2/3}$ and component sizes in units of $n^{1/3}$, to obtain, at rescaled time s, a conditional mean drift rate of $r - s$, and a conditional variance rate that remains 1. Then we arrive at:

Theorem 6.10. Aldous [9, Theorem 3] *Let $(B(t), t \geq 0)$ be standard Brownian motion. Set*

$$B^r(t) := B(t) + rt - t^2/2 \text{ for } t \geq 0$$

so $B^r(0) = 0$ and $dB^r(t) = dB(t) + (r - t)dt$, so B^r is a Brownian motion with drift $r - t$ at time t. For $r \in \mathbb{R}$ let $W_n^r(i), 0 \leq i \leq n$, with linear interpolation between integer values i, be the breadth-first walk which is associated with $\mathcal{G}(n, 1/n + r/n^{4/3})$. Then

$$n^{-1/3}(W_n^r(n^{2/3}t), t \geq 0) \xrightarrow{d} (B^r(t), t \geq 0) \qquad (6.42)$$

in $C[0, \infty)$.

Sketch of proof. A minor perturbation of the breadth-first walk is presentable as the sum of a martingale and a bounded variation process; the former rescales to satisfy standard hypotheses for a martingale to converge in distribution to B, while the bounded variation term rescales to give the drift. See [9] for details. □

Given that component sizes of $\mathcal{G}(n, p_n(r))$ are coded as the gaps between successive drops to new lows of the corresponding walk, it is to be expected that the distribution of the ranked component sizes should approximate the excursions of the limit process. This is expressed rigorously as follows:

Theorem 6.11. Aldous [9, Corollary 2] *For each fixed $r \in \mathbb{R}$, the sequence of ranked component sizes of $\mathcal{G}(n, 1/n + r/n^{4/3})$ converges in distribution in $l^2_\downarrow$, the l_2-normed space of non-increasing sequences:*

$$n^{-2/3}(N^{\downarrow, r}_{n,j}, j \geq 1) \xrightarrow{d} (X^{(\infty)}_j(r), j \geq 1) \qquad (6.43)$$

where the limit $(X_j^{(\infty)}(r), j \geq 1)$ *has the same distribution as the sequence of ranked lengths of excursions away from* 0 *of the reflected process* $B^r - \underline{B}^r$, *where* B^r *is the Brownian motion with drift* $r-t$ *at time* t, *and* $\underline{B}^r(t) := \inf_{0 \leq s \leq t} B^r(s)$.

The multiplicative coalescent This discussion becomes more interesting if we take the dynamic view of $(\mathcal{G}(n,p), 0 \leq p \leq 1)$ as a process in which edges are born as the parameter p increases. If the edges are assumed to be born at independent exponential(1) times, then $\mathcal{G}(n,p)$ describes the state of the graph at time t with $p = 1 - e^{-t}$. Recall Definition 5.2 of a $\mathcal{P}_{[n]}$-valued, K-coalescent, called a *multiplicative coalescent* for $K(x,y) = xy$.

Lemma 6.12. [79] *The process* $(\Pi_n(t), t \geq 0)$, *where* $\Pi_n(t)$ *is the partition of* $[n]$ *defined by connected components of* $\mathcal{G}(n, 1-e^{-t})$, *is a* $\mathcal{P}_{[n]}$*-valued multiplicative coalescent, with mass defined by counting.*

Proof. If a graph has two particular components of sizes a and b, there are ab possible edges which when added to the graph would connect the two components. Combined with standard properties of independent exponential variables, this yields the conclusion. □

To compare the evolution of component sizes of $(\Pi_n(t), t \geq 0)$ for different values of n, consider the process $X^{(n)}(t)$ of ranked component sizes

$$X^{(n)}(t) := (N_{n,j}^{\downarrow}(t), j \geq 1) \tag{6.44}$$

where $N_{n,j}^{\downarrow}(t)$ is the size of the jth largest component of $\Pi_n(t)$ derived from $\mathcal{G}(n, 1-e^{-t})$. Then by Lemma 6.12, for each n the process $(X^{(n)}(t), t \geq 0)$ is a ranked multiplicative coalescent, with initial state a vector of n entries 1 padded by zeros. The ranked multiplicative coalescent has the following obvious *scaling property*: if $(X(t), t \geq 0)$ is a ranked multiplicative coalescent, then so is $(cX(c^2 t), t \geq 0)$ for arbitrary $c > 0$. In particular, the process

$$(n^{-2/3} X^{(n)}(t/n^{4/3}), t \geq 0) \tag{6.45}$$

is a ranked multiplicative coalescent with initial state a vector of n entries $n^{-2/3}$ padded by zeros. Theorem 6.11 suggests that to capture the evolution of component sizes $X^{(n)}(t)$ around the critical time $t \sim n^{-1}$, the process (6.45) should be considered at time $t = n^{1/3} + r$ for $r \geq -n^{1/3}$. This process is a ranked multiplicative coalescent, with time parameter set $[-n^{1/3}, \infty)$, and initial state at time $-n^{1/3}$ a vector of n entries $n^{-2/3}$ padded by zeros. This brings us to the following refinement of Theorem 6.11

Theorem 6.13. Aldous [9, Corollary 24] *Let* $(X^{(n)}(t), t \geq 0)$ *be the ranked multiplicative coalescent defined by the component sizes of the random graph process* $\mathcal{G}(n, 1-e^{-t})$ *for* $t \geq 0$, *regarded as a process with values in* $l^2_{\downarrow}$. *Then as* $n \to \infty$

$$(n^{-2/3} X^{(n)}(1/n + r/n^{4/3}), r \geq -n^{1/3}) \xrightarrow{d} (X^{(\infty)}(r), r \in \mathbb{R}) \tag{6.46}$$

in the sense of convergence of finite-dimensional distributions, where the limit is the $l^2_\downarrow$-valued multiplicative coalescent process such that the distribution of $X^{(\infty)}(r)$ is that of the ranked lengths of excursions away from 0 of the reflected process $B^r - \underline{B}^r$, where B^r is the Brownian motion with drift $r - t$ at time t.

Proof. This follows easily from the above discussion, once it is checked that the transition kernel $P_t(x, \cdot)$ of the ranked multiplicative coalescent on $\mathcal{P}^{\downarrow}_{\text{finite}}$ admits a unique extension to $l^2_\downarrow$, such that $P_t(x, \cdot)$ is weakly continuous in x for each fixed t. But that is much harder to prove than might be expected: see [9, §4.2]. □

The process $(X^{(\infty)}(r), r \in \mathbb{R})$ defined by (6.46) is the *standard multiplicative coalescent.* A process with time-parameter set $\mathbb{R}$ may be called *eternal.* Theorem 6.13 raises the question of whether there exist other eternal multiplicative coalescents besides shifts of the standard one. Indeed there are many of them. From the standard multiplicative coalescent one can construct other multiplicative coalescents, the simplest of which is obtained by lumping together suitably-chosen components to form a distinguished component of the new multiplicative coalescent. It turns out [13], but is technically hard to prove, that the most general extreme multiplicative coalescent can be obtained by such lumping procedures and a weak convergence construction.

Exercises

6.4.1. Let $\Pi(n, p)$ be the partition generated by components of $\mathcal{G}(n, p)$ and let $c(n, p) := \mathbb{P}(\Pi(N, p) = \{[n]\})$, that is the probability that $\mathcal{G}(n, p)$ is connected.

- Give an explicit formula for the EPPF of $\Pi(n, p)$ in terms of $c(j, p)$ for $1 \le j \le n$.
- Deduce that $\Pi(n, p)$ is not a Gibbs partition.
- Are the partitions $\Pi(n, p)$ consistent in distribution as n varies?
- Describe the weak limit of $\Pi(n, p)$ as $n \to \infty$ for fixed p.
- Describe the weak limit of $\Pi(n, \lambda/n)$ as $n \to \infty$ for fixed λ.

6.4.2. [70] Prove Lemma 6.8. Deduce that if $\lambda \le 1$, then as $n \to \infty$ the distribution of $\mathcal{T}_1(n, \lambda/n)$ converges to that of a Poisson-Galton-Watson (λ) tree. Show also that if $\lambda > 1$, and $o(n)$ is some arbitrary sequence tending to infinity with $o(n)/n \to 0$, then the conditional distribution of $\mathcal{T}_1(n, \lambda/n)$, given that it has at most $o(n)$ vertices, converges to that of a Poisson-Galton-Watson ($\hat{\lambda}$) tree, for some $\hat{\lambda} < 1$, called the *conjugate of* λ, whose value should be determined.

6.4.3. (A tree-growth process) For each fixed n, regard $\mathcal{T}_1(n, \lambda/n)$ as a tree-valued process indexed by $\lambda \ge 0$, with the convention that this process jumps to some terminal state † as soon as its number of vertices exceeds $o(n)$, for some $o(n) \to \infty$ with $o(n)/n \to 0$. Show that this process has a limit in distribution $(\mathcal{T}(\lambda), \lambda \geq 0)$ as $n \to \infty$, which is an inhomogeneous tree-valued Markov chain studied in [4, 19]. Describe the transition mechanism of the limiting tree-growth process as explicitly as possible. In particular, find the distribution of the *ascension time* $A := \inf\{\lambda : \mathcal{T}(\lambda) = \dagger\}$. Intuitively, this is the asymptotic distribution of the time it takes for a single vertex in $\mathcal{G}(n, p)$ to be a giant component whose size is $O(n)$. Show also that $\mathcal{T}(A-)$ is a.s. finite, and find its distribution. See [19] for further analysis, and generalizations.

6.4.4. [9, §5.1] As an immediate corollary of Theorem 6.11, the excursion lengths of $B^r - \underline{B^r}$ are square summable. Derive this directly, by analysis of Brownian paths.

6.4.5. [9, §1.5] Fix $x \in l^2_\downarrow$. Let $\varepsilon_{i,j}$ be independent exponential (1) variables indexed by $i, j \in \mathbb{N}$ with $i < j$. Let $\mathcal{G}(x, t)$ be the graph on $\mathbb{N}$ with an edge (i, j) if and only if $\varepsilon_{i,j} \leq t x_i x_j$, and let $X(x, t)$ be the vector of ranked x-masses of connected components of $\mathcal{G}(x, t)$. Then $(X(x, t), t \geq 0)$ is a realization of the $l^2_\downarrow$-valued ranked multiplicative coalescent with initial state x.

6.4.6. (Problem) Theorems 6.11 and 6.13 strongly suggest that there exists a $C[0, \infty)$-valued process $(B^r, r \in \mathbb{R})$ such that B^r is a Brownian motion with drift $r - t$ at time t, and if $X^{(\infty)}(r)$ is the sequence of ranked lengths of excursions of $B^r - \underline{B^r}$, then $(X^{(\infty)}(r), r \in \mathbb{R})$ is the standard multiplicative coalescent. However, the results of Aldous [9] do not even establish existence of a joint distribution of B^r and B^s for $r \neq s$, let alone the finite-dimensional distributions or path-properties of such a process. Does it exist, and if so what can be said about it?

Notes and comments

I learned everything in this section from David Aldous. The book [216] surveys the theory of random graphs. See also [215] and [9] for more detailed information about the birth of the giant component around time $t \approx 1/n$. There are striking parallels between the theory of the multiplicative coalescent described here, and that of the additive coalescent discussed in Chapter 10. Both processes arise naturally from random graphs, their combinatorial structure is related to random trees, and they admit eternal versions whose entrance laws are related to the lengths of excursions of Brownian motion. Another important similarity, which is the key to many deeper results, is that there is an essentially combinatorial construction of the coalescent process with an infinite number of initial masses subject to appropriate conditions: see Exercise 6.4.5 and Corollary 10.6. See [18] for a more technical comparison of the similarities and differences between these two processes.

7

The Brownian forest

The Harris correspondence between random walks and random trees, reviewed in Section 6.3, suggests that a continuous path be regarded as encoding some kind of infinite tree, with each upward excursion of the path corresponding to a subtree. This idea has been developed and applied in various ways by Neveu-Pitman [324, 323], Aldous [5, 6, 7] and Le Gall [271, 272, 273, 275]. This chapter reviews this circle of ideas, with emphasis on how the Brownian forest can be grown to explore finer and finer oscillations of the Brownian path, and how this forest growth process is related to Williams' path decompositions of Brownian motion at the time of a maximum or minimum.

7.1. Plane trees with edge-lengths This section introduces the notion of a finite plane tree with edge-lengths, and shows how the Harris correspondence, between combinatorial plane trees and their Harris paths, extends nicely to plane trees with edge-lengths.

7.2. Binary Galton-Watson trees A particular parametric family of Galton-Watson trees with edge-lengths and binary branching is related to the uniform distribution on the set of n-leaf reduced plane trees with a given length.

7.3. Trees in continuous paths The notion of a tree embedded in a continuous function generalizes the correspondence between a reduced plane tree with edge-lengths and its Harris path.

7.4. Brownian trees and excursions Brownian excursion theory shows how the infinite tree embedded in a Brownian path may be identified as a forest of subtrees, the *Brownian forest.* The distribution of these subtrees is determined by Aldous's description of the tree in a standard Brownian excursion.

7.5. Plane forests with edge-lengths Such forests are naturally embedded in a continuous path by a suitable sequence of sampling times. The entire forest of Brownian trees may be regarded as a kind of projective limit of such forests.

7.6. Sampling at downcrossing times and 7.7. Sampling at Poisson times These two sections present constructions of the Brownian forest based on sampling the Brownian path at two differently defined sequences of stopping times. In each case, the trees in the forest can be recognized as critical binary Galton-Watson trees with exponentially distributed edge-lengths, and increasing a parameter θ corresponds to growth of the forest to explore finer and finer oscillations of the Brownian path. Both forest-valued processes turn out to be Markovian, but their dynamics are different.

7.8. Path decompositions This section explains how the structure of these forest-valued processes is related to Williams' path decompositions for Brownian paths at the time of a maximum or minimum.

7.9. Further developments This is a brief survey of recent work on continuum random trees.

7.1. Plane trees with edge-lengths

This section describes in combinatorial and geometric terms the kinds of random trees which turn out to be naturally embedded in Brownian paths. The reader is assumed to be familiar with basic notions of plane trees and forests, such as depth-first search, as summarized in Section 6.2 .

Definition 7.1. A *finite plane tree with edge-lengths* is a pair

$$\mathcal{T} = (\text{SHAPE}(\mathcal{T}), \text{LENGTHS}(\mathcal{T}))$$

where $\text{SHAPE}(\mathcal{T})$, the *combinatorial shape of* $\mathcal{T}$, is a plane tree with a finite number of vertices, as in Section 6.2 and if $\text{SHAPE}(\mathcal{T})$ has m edges then $\text{LENGTHS}(\mathcal{T})$ is a sequence of m strictly positive numbers, to be interpreted as the lengths of edges of $\text{SHAPE}(\mathcal{T})$ in order of depth-first search.

From now on, the term *plane tree* will be used for a finite plane tree $\mathcal{T}$ with edge-lengths. A *combinatorial plane tree* such as $\text{SHAPE}(\mathcal{T})$ is treated as a plane tree, all of whose edge-lengths equal 1. A plane tree $\mathcal{T}$ can be represented, as in Figure 7.1, as a subset of the plane $\mathbb{R}^2$, formed by a union of line segments.

In such a *graphical representation of* $\mathcal{T}$ $\mathcal{T}$ is identified with a subset of $\mathbb{R}^2$. The set of vertices of $\text{SHAPE}(\mathcal{T})$ is then identified with a corresponding subset of $\mathcal{T}$, the *set of vertices of* $\mathcal{T}$, in such a way that each edge of $\text{SHAPE}(\mathcal{T})$ with length ℓ corresponds to a line segment of length ℓ joining corresponding vertices of $\mathcal{T}$. These line segments, regarded as subsets of $\mathcal{T} \subseteq \mathbb{R}^2$, are the *edges of* $\mathcal{T}$. Two edges of $\mathcal{T}$ may intersect only at a vertex of $\mathcal{T}$, and $\mathcal{T}$ is the union of its finite collection of edges.

Of course, each plane tree has many graphical representations. But it is easily verified that various definitions made below in terms of a graphical representation do not depend on the choice of representation. So the same symbol $\mathcal{T}$ may be used for a plane tree with edge-lengths, or one of its graphical representations, as convenient.

The correspondence between a combinatorial plane tree and its Harris path has a natural extension to plane trees $\mathcal{T}$ with edge-lengths. Let $\text{LENGTH}(\mathcal{T})$ be the total length of all edges of $\mathcal{T}$. Assuming that $\mathcal{T}$ is graphically represented as a

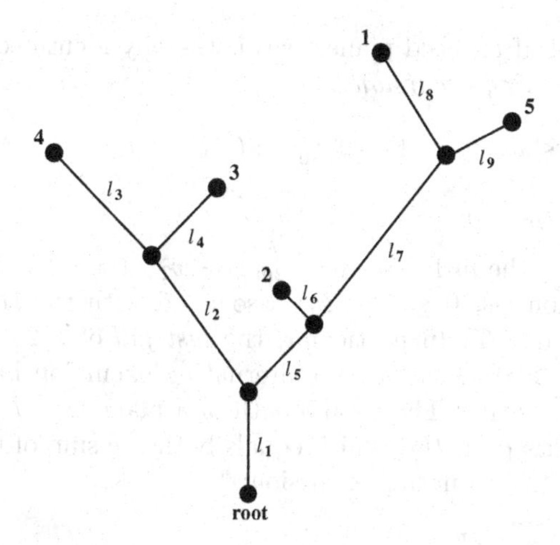

Figure 7.1: Graphical representation of a finite rooted plane tree with 5 leaves labeled by $\{1, 2, 3, 4, 5, \}$.

subset of the plane, there is a unique continuous map $\sigma_{\mathcal{T}} : [0, 2\text{LENGTH}(\mathcal{T})] \to \mathcal{T}$, the *depth-first search of* $\mathcal{T}$, which runs along the edges of $\mathcal{T}$ at unit speed, starting at the root, in order of depth-first search, in such a way that each edge of $\mathcal{T}$ is traversed twice, first moving away from the root, then later moving back towards the root. Define the *Harris path of* $\mathcal{T}$ to be the continuous function $H_{\mathcal{T}}(s), 0 \le s \le 2\text{LENGTH}(\mathcal{T})$, which gives the distance from the root at time s of the depth-first search of $\mathcal{T}$ at unit speed, with distance measured along edges of $\mathcal{T}$. More formally, the graphical representation of $\mathcal{T}$ is regarded as a *metric space* with metric $d_{\mathcal{T}}$, where $d_{\mathcal{T}}(x, y)$ for $x, y \in \mathcal{T}$ is the distance from x to y measured along edges of $\mathcal{T}$. Then

$$H_{\mathcal{T}}(s) := d_{\mathcal{T}}(\sigma_{\mathcal{T}}(0), \sigma_{\mathcal{T}}(s)) \qquad (0 \le s \le 2\text{LENGTH}(\mathcal{T})), \tag{7.1}$$

where $\sigma_{\mathcal{T}}(0)$ is the root vertex of $\mathcal{T}$. The following basic facts are easily verified:

- a function $H(u), 0 \le u \le t$, is the Harris path of some n-leaf plane tree $\mathcal{T}$ iff H is non-negative and continuous, with $H(0) = H(t) = 0$, and the graph of H is a union of $2n$ line segments of alternating slopes ± 1; then H has n local maxima, at the times s_i at which the depth-first search of $\mathcal{T}$ visits the leaves v_i of $\mathcal{T}$, $1 \le i \le n$.
- Corresponding to each such Harris path H, there is a unique n-leaf plane tree $\mathcal{T}$ which is *reduced*, meaning that $\mathcal{T}$ has no non-root vertices of degree 2.
- Such a Harris path H, with n local maxima, is uniquely specified by the sequence $(w_i, 1 \le i \le 2n-1)$, where $w_i > 0$ is the level at which the slope of H changes sign for the ith time.

- The set of n-leaf reduced plane trees is thereby identified with the set of *$2n$-step alternating excursions*

$$\{(w_i, 1 \leq i \leq 2n-1) \in \mathbb{R}_{>0}^{2n-1} : (-1)^{i-1}(w_i - w_{i-1}) > 0\} \tag{7.2}$$

where $w_0 = w_{2n} = 0$.

Call $w_{2m-1} - w_{2m-2}$ the mth *rise* and $w_{2m-1} - w_{2m}$ the mth *fall* of the $2n$-step alternating excursion $(w_i, 0 \leq i \leq 2n)$ associated with the Harris path H or the reduced plane tree $\mathcal{T}$. In particular, the *last fall* of a $2n$-step alternating excursion is $w_{2n-1} > 0$. So a $2n$-step alternating excursion has n rises and n falls, each strictly positive. The total length of a plane tree $\mathcal{T}$ is half the total variation of its Harris path $H_{\mathcal{T}}$, which equals both the sum of the rises and the sum of the falls of its alternating excursion:

$$\text{LENGTH}(\mathcal{T}) = \sum_{m=1}^{n}(w_{2m-1} - w_{2m-2}) = \sum_{m=1}^{n}(w_{2m-1} - w_{2m}). \tag{7.3}$$

Exercises

7.1.1. (Another coding of plane trees) An n-leaf reduced plane tree $\mathcal{T}$ is determined by the sequence of $2n-1$ distances

$$(d_{\mathcal{T}}(0, \ell_i), 1 \leq i \leq n) \text{ and } (d_{\mathcal{T}}(\ell_i, \ell_{i+1}), 1 \leq i \leq n-1) \tag{7.4}$$

where 0 is the root of $\mathcal{T}$, and ℓ_i is the ith leaf of $\mathcal{T}$ in order of depth-first search. The vector of $2n-1$ distances (7.4) is a linear transformation of the alternating excursion associated with $\mathcal{T}$, subject to constraints implied by (7.2).

7.2. Binary Galton-Watson trees

Call $\mathcal{T}$ a *planted binary plane tree* if the root of $\mathcal{T}$ has degree 1, and every other vertex of $\mathcal{T}$ is either a leaf or an *internal vertex* of degree 3. Every planted binary plane tree with n leaves has $n-1$ internal vertices and $2n-1$ edges. The number of different possible shapes of such a tree is the $(n-1)$th Catalan number

$$C_{n-1} = \frac{1}{n}\binom{2n-2}{n-1}.$$

See (7.27) for a derivation.

The following definition introduces a standard model for random binary trees, with a non-standard parameterization to be explained later.

Definition 7.2. For $0 \leq \lambda < \mu$, a *binary*(λ, μ) *tree*, denoted $\mathcal{G}_{\lambda,\mu}$, is a random planted binary plane tree such that the restriction of $\text{SHAPE}(\mathcal{G}_{\lambda,\mu})$ to levels

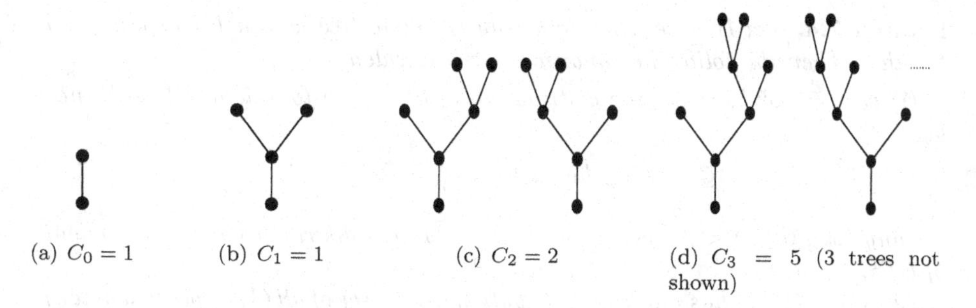

(a) $C_0 = 1$ (b) $C_1 = 1$ (c) $C_2 = 2$ (d) $C_3 = 5$ (3 trees not shown)

Figure 7.2: The shapes of some planted binary plane trees.

$1, 2, 3, \ldots$ is a Galton-Watson tree with a single individual at level 1 and offspring distribution

$$p_0 = \frac{\mu + \lambda}{2\mu} \text{ and } p_2 = \frac{\mu - \lambda}{2\mu},$$

and given $\text{SHAPE}(\mathcal{G}_{\lambda,\mu})$ is some planted binary tree with n leaves, the $2n - 1$ edge-lengths of $\mathcal{G}_{\lambda,\mu}$, enumerated by depth-first search, say, are independent exponential(2μ) variables, with density $2\mu e^{-2\mu x}$, $x \geq 0$.

The mean of the offspring distribution is $(\mu - \lambda)/\mu$, so binary(λ, μ) trees are critical or subcritical according to whether $0 = \lambda < \mu$ or $0 < \lambda < \mu$. The next Lemma presents an alternative construction of a binary(λ, μ) tree $\mathcal{G}_{\lambda,\mu}$ in terms of exponential variables with rates $\beta = \mu - \lambda$ and $\delta = \mu + \lambda$, which can be interpreted as birth rates and death rates in another well known construction of $\mathcal{G}_{\lambda,\mu}$ recalled in Exercise 7.2.1 . The non-standard parameterization of these trees by (λ, μ) instead of $(\beta, \delta) = (\mu - \lambda, \mu + \lambda)$ is made to simplify later discussion of associated random forests.

Lemma 7.3. *For $0 \leq \lambda < \mu$, a random reduced plane tree is a binary(λ, μ) tree iff the succession of rises and falls of its Harris path, excluding the last fall, is distributed like a sequence of independent exponential variables, with exponential$(\mu + \lambda)$ rises and exponential$(\mu - \lambda)$ falls, stopped one step before the sum of successive rises and falls first becomes negative.*

Proof. In the critical case $0 = \lambda < \mu$, this was pointed out in [271] and [324, p. 246]. See [271] for a proof in this case. The case $0 \leq \lambda < \mu$ can be handled by a change of measure relative to the critical case. □

For each fixed $t > 0$ and $n = 1, 2, \ldots$ there is a natural *uniform distribution* on the set of n-leaf reduced plane trees whose total length is t, which concentrates on planted binary trees. A random tree $\mathcal{B}^{n,t}$ with this uniform distribution can be described in three equivalent ways as follows:

Lemma 7.4. *Let $\mathcal{B}^{n,t}$ be a random reduced plane tree with n leaves and total length t. Then the following conditions are equivalent.*

(i) $\mathcal{B}^{n,t} \stackrel{d}{=} t\mathcal{B}^{n,1}$, *where the distribution of $\mathcal{B}^{n,1}$ with total length 1 is defined by*

$$(\mathcal{G}_{\lambda,\mu} \,|\, \mathcal{G}_{\lambda,\mu} \text{ has } n \text{ leaves }) \stackrel{d}{=} \frac{\Gamma_{2n-1}}{2\mu} \mathcal{B}^{n,1} \tag{7.5}$$

for any fixed $0 \leq \lambda < \mu$, where Γ_{2n-1} is a Gamma-distributed r.v. independent of $\mathcal{B}^{n,1}$.

(ii) SHAPE$(\mathcal{B}^{n,t})$ *has uniform distribution on the set of all C_{n-1} planted binary plane trees with n leaves, and independently of* SHAPE$(\mathcal{T}_n)$ *the sequence of edge-lengths of $\mathcal{B}^{n,t}$ has distribution proportional to $(2n-2)$-dimensional Lebesgue measure on sequences of $(2n-1)$ non-negative numbers with sum t. In particular, the sequence of edge-lengths of $\mathcal{B}^{n,1}$ with sum 1 has the Dirichlet distribution with $2n-1$ parameters equal to 1.*

(iii) *the $2n$-step alternating excursion, derived from the Harris path of $\mathcal{B}^{n,t}$, has distribution proportional to $(2n-2)$-dimensional Lebesgue measure on the set of all $2n$-step alternating excursions whose total rising length is t.*

Proof. Take (i) as the definition of $\mathcal{B}^{n,t}$. Then (ii) follows from Definition 7.2 and the fact that a binary Galton-Watson tree conditioned to have n leaves has the uniform distribution on the set of all n-leaf binary trees (see Gutjahr [180]). (iii) follows from Lemma 7.3. That (ii) and (iii) characterize the distribution of $\mathcal{B}^{n,t}$ is evident from the bijective equivalence of the various codings of reduced plane trees. □

The classification of $2n$-step alternating excursions according to the shape of their associated plane tree corresponds to a decomposition of the polytope of possible $2n$-step alternating excursions into C_{n-1} chambers of equal $(2n-2)$-dimensional volume. Various shapes of non-binary trees then correspond to various facets of the chambers. See [363] for more about polytopal subdivisions related to plane trees.

Exercises

7.2.1. (Lifeline representation of binary Galton-Watson trees) A population starts with 1 individual at time 0. Each individual has an exponential(δ) lifetime, and throughout its lifetime gives birth to new individuals according to a Poisson process with rate β, assumed independent of its lifetime. These offspring continue to reproduce independently of each other in the same manner, and so on. Assuming $\beta \leq \delta$, let $\mathcal{T}$ be the random family tree generated by this process, regarded as a random plane tree, drawn so that

- each vertex of $\mathcal{T}$ corresponds to the moment of a birth or death in the population;

- the lifespan of each individual is represented by a path in $\mathcal{T}$, its *lifeline*, from the vertex on its parent's lifeline representing its birth moment to a leaf representing its death moment;
- the lifeline of each individual in $\mathcal{T}$ branches to the right of the lifeline of its parent;
- each *edge* or *segment* of $\mathcal{T}$ represents the portion of the lifeline of some individual between some birth moment and the next birth or death moment along the lifeline of that individual;
- each segment-length represents the length of the corresponding time interval.

The segments of $\mathcal{T}$ are connected according to the binary Galton-Watson process in which each segment either terminates with probability $\delta/(\beta+\delta)$ or branches into two segments with probability $\beta/(\beta+\delta)$, and segments have independent exponential$(\beta+\delta)$ lifetimes. So $\mathcal{T}$ is a binary(λ,μ) tree with $\lambda = (\delta-\beta)/2$ and $\mu = (\delta+\beta)/2$. The collection of lifelines of individuals in $\mathcal{T}$ can also be regarded in an obvious way as a Galton-Watson process with geometric$(\delta/(\beta+\delta))$ offspring distribution with mass function $p_n = (\beta/(\beta+\delta))^n(\delta/(\beta+\delta))$, whose total progeny equals the number of leaves of the binary branching tree $\mathcal{T}$.

7.2.2. (The linear birth and death process) [150, p. 456] For $\mathcal{T}$a binary(λ,μ) tree with $\lambda = (\delta-\beta)/2$ and $\mu = (\delta+\beta)/2$ as above, let Z_t be the number of branches of $\mathcal{T}$ at distance t from the root. Then $(Z_t, t \geq 0)$ is a Markovian birth and death process on the non-negative integers with transition rates $i\beta$ for $i \to i+1$ and $i\delta$ for $i \to i-1$. In particular, in the critical case $\lambda = 0, \beta = \delta = \mu$,

$$\mathbb{P}(Z_t = 0) = \frac{\mu t}{1+\mu t}; \; \mathbb{P}(Z_t = n) = \frac{(\mu t)^{n-1}}{(1+\mu t)^{n+1}} \quad (n \geq 1). \tag{7.6}$$

Notes and comments

There are numerous natural generalizations of the binary branching trees considered in this section. See for instance Geiger [161, 162, 163, 164, 165].

7.3. Trees in continuous paths

Fix some graphical representation of a plane tree $\mathcal{T}$ with depth-first search path $\sigma : I \to \mathcal{T}$ and Harris path $H : I \to \mathbb{R}_{\geq 0}$, where $I := [0, 2\,\text{LENGTH}(\mathcal{T})]$. Then for $u, v \in I$ the height in $\mathcal{T}$ of the branch point at which the path from $\sigma(0)$ to $\sigma(u)$ diverges from the path from $\sigma(0)$ to $\sigma(v)$ is evidently

$$\underline{H}[u,v] := \inf_{t \in [u,v]} H(t). \tag{7.7}$$

Let

$$d_H(u,v) := (H(u) - \underline{H}[u,v]) + (H(v) - \underline{H}[u,v]) \quad (u, v \in I). \tag{7.8}$$

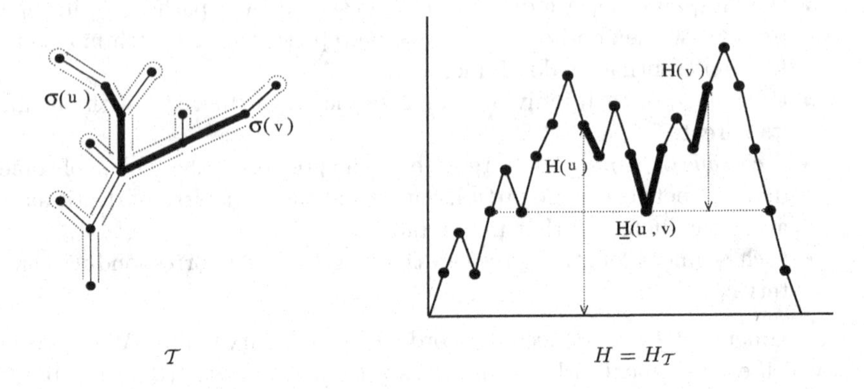

Figure 7.3: Depth-first search and Harris paths for a finite plane tree with edge-lengths.

so the distance from $\sigma(u)$ to $\sigma(v)$ in $\mathcal{T}$ is

$$d_{\mathcal{T}}(\sigma(u), \sigma(v)) = d_H(u, v), \tag{7.9}$$

as illustrated by Figure 7.3.

For any subinterval I of $\mathbb{R}$, and any locally bounded function $H : I \to \mathbb{R}_{\geq 0}$ it is obvious that formula (7.8) defines a *pseudo-metric* on I. Identify t and t', and write $t \sim_H t'$ if $d_H(t, t') = 0$. Then the previous discussion can be summarized as follows:

Proposition 7.5. *Let $H = H_{\mathcal{T}}$ be the Harris path of some graphically represented plane tree $\mathcal{T}$, and let $I := [0, 2\,\text{LENGTH}(\mathcal{T})]$ be the domain of H. Then the depth-first search of $\mathcal{T}$ at unit speed defines an isometry between $(I/\sim_H, d_H)$ and $(\mathcal{T}, d_{\mathcal{T}})$.*

A metric space (M, d) is called a *tree* if for each choice of $u, v \in M$ there is a unique continuous path $\sigma_{u,v} : [0, d(u, v)] \to M$ which travels from u to v at unit speed, meaning $d(u, \sigma_{u,v}(t)) = t$ for $0 \leq t \leq d(u, v)$ and $\sigma_{u,v}(d(u, v)) = v$, and for any simple continuous path $f : [0, T] \to M$ with $f(0) = u$ and $f(T) = v$, the ranges of f and $\sigma_{u,v}$ coincide.

Definition 7.6. [272, 7] If $H \in C[I]$, the space of continuous functions from I to $\mathbb{R}$, then $(I/\sim_H, d_H)$ is a tree, to be denoted $\text{TREE}(H)$. If $I = [G, D]$ or $I = [G, \infty)$ the *root* of $\text{TREE}(H)$ is taken to be $\{t \in I : t \sim_H G\}$.

In view of Proposition 7.5, this notion of $\text{TREE}(H)$ for $H \in C[I]$ generalizes the correspondence between a reduced plane tree with edge-lengths $\mathcal{T}$ and its Harris path H. Refer to Figure 7.4 for an example.

For a subset S of I, and $H \in C[I]$, the *subtree of* $\text{TREE}(H)$ *spanned by* S, denoted $\text{SUBTREE}(H; S)$ is the union over $s, t \in S$ of the range of the path from s to t in $\text{TREE}(H)$, equipped with the tree metric d_H. Let $H \in C[I]$ where I has

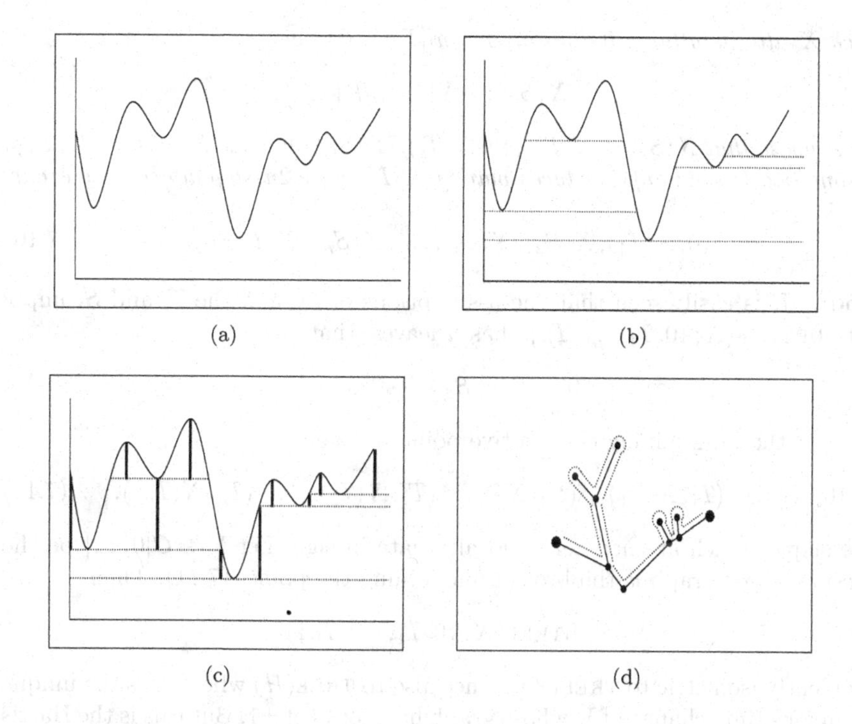

Figure 7.4: Construction of tree from a tame function, with a finite number of local extrema. The whole tree is evidently isometric to a finite plane tree with edge-lengths, as in the bottom right panel. Note from Definition 7.6 that the root of this tree is its left-most extremity, *not* the node corresponding to the minimal point. The angle between the two branches meeting at this node could be straightened out, and this node disregarded as a vertex, to make an isometric representation by a reduced plane tree.

left end 0. Assume for simplicity that $0 = H(0) \leq H(t)$ for all $t \in I$. Let $T_1 > 0$. Then $\text{SUBTREE}(H; \{0, T_1\})$ is isometric to a line segment of length $H(T_1)$. For $0 < T_1 < T_2$, let S_1 be a time at which H attains its minumum on $[T_1, T_2]$, and suppose to avoid degenerate cases that $H(S_1)$ is strictly less than both $H(T_1)$ and $H(T_2)$. Then $\text{SUBTREE}(H; \{0, T_1, T_2\})$ is isometric to a Y-shaped tree with 3 edges of lengths $H(S_1)$, $H(T_1) - H(S_1)$, and $H(T_2) - H(S_1)$ respectively, whose root and leaves may be identified, in clockwise order from the bottom, as 0, T_1 and T_2 respectively, while the junction point is identified with S_1. Continuing in this way, it is clear that for any finite sequence of *sample times* $T_1, \ldots, T_n$. the order and metric structure of $\text{SUBTREE}(H; \{0, T_1, \ldots, T_n\})$ will be encoded by some plane tree. This can be made precise as follows, with some simplifying assumptions to avoid annoying cases:

Lemma 7.7. *Let $X \in C[0, T]$ and suppose $X(t) \geq X(0) = 0$ for all $t \in [0, T]$. Let $0 < T_1 < \cdots < T_n \leq T$, and for $1 < m \leq n$ let S_i be a time in $[T_{i-1}, T_i]$ at* *which X attains attains its minimum on $[T_{i-1}, T_i]$, so*

$$X(S_i) := \underline{X}[T_{i-1}, T_i],$$

and suppose that $X(S_i) < X(T_{i-1}) \wedge X(T_i)$. Then SUBTREE$(X; \{0, T_1, \ldots, T_n\})$ *is isometric to the unique reduced plane tree $\mathcal{T}$ whose $2n$-step alternating excursion is*

$$(0, X(T_1), X(S_2), X(T_2), \ldots, X(S_n), X(T_n), 0). \tag{7.10}$$

Proof. It is easily seen that the assumptions on X and the T_i and S_i imply that SUBTREE$(X; \{0, T_1, \ldots, T_n\})$ has n leaves, that

$$0 < T_1 < S_2 < \cdots < T_n$$

and that the lines joining consecutive points in the list

$$(0, X(0)),\ (T_1, X(T_1)),\ (S_2, X(S_2),\ (T_2, X(T_2)), \ldots\ (T_n, X(T_n)). \tag{7.11}$$

have slopes which are non-zero and alternate in sign. Let $Y \in C[0, T_n]$ be the function whose graph is the broken-line joining the points (7.11). Then

$$\text{SUBTREE}(X; \{0, T_1, \ldots, T_n\})$$

is evidently isometric to TREE(Y), hence also to TREE(H) where H is the unique continuous time-change of Y whose graph has slopes of ± 1. But this is the Harris path whose $2n$-step alternating excursion is (7.10), and the conclusion follows by Proposition 7.5. □

Now consider the construction of TREE(X) for a random function $X \in C[I]$, such as a Brownian path restricted to I. Intuitively, TREE(X) should be regarded as a random metric space. Technically however, for purposes of measure theory, let us simply identify TREE(X) with the random element d_X of $C[I \times I]$. So for instance TREE$(X) \stackrel{d}{=}$ TREE(Y) means $d_X \stackrel{d}{=} d_Y$ which is the same as

$$(d_X(u, v), u, v \in I) \stackrel{d}{=} (d_Y(u, v), u, v \in I)$$

in the sense of equality of finite dimensional distributions. Now give the space of reduced plane trees the measurable structure it acquires when identified as a subset of $C[0, \infty)$ by using the Harris correspondence $\mathcal{T} \leftrightarrow H_{\mathcal{T}}$. Here $C[0, t]$ is regarded as a subset of $C[0, \infty)$, by stopping paths at time t. Then it is clear that SUBTREE$(X; \{0, T_1, \ldots, T_n\})$ may be regarded as a random reduced plane tree for arbitrary I-valued random variables T_i defined on the same probability space as a continuous path process $X = (X(t), t \in I)$.

7.4. Brownian trees and excursions

Let $B = (B(t), t \geq 0)$ be a standard Brownian motion. An obvious geometric feature of TREE(B) is that this tree has a unique infinite branch, namely the *descending ladder set* $\{t : B(t) = \underline{B}[0,t]\}$. Except on an event with probability zero, this branch of $\text{TREE}(B)$, regarded as a subset of $\mathbb{R}_{\geq 0}/\sim_B$, is traversed at unit speed by the map $\ell \to \{T_{\ell-}, T_\ell\}$ where

$$T_\ell := \inf\{t : B(t) < -\ell\}.$$

Let $I_\ell := [T_{\ell-}, T_\ell]$. Then $\text{TREE}(B)$ is the union over $\ell \geq 0$ of $\text{SUBTREE}(B; I_\ell)$, which is I_ℓ equipped with a tree structure isometric by a shift to $\text{TREE}(B[I_\ell])$ on $[0, \infty)$, where

$$B[I_\ell](t) := B((T_{\ell-} + t) \wedge T_\ell) \quad (t \geq 0).$$

Call the descending ladder set of B the *floor line*, to be identified with $\mathbb{R}_{\geq 0}$ by traversal at unit speed in $\text{TREE}(B)$, as above. Then $\text{TREE}(B)$ is identified by a *forest of subtrees attached to the floor line,* with a copy of $\text{TREE}(B[I_\ell])$ attached to the point on the floor at distance ℓ from the root. Thus to describe the structure of $\text{TREE}(B)$ it is enough to describe the forest of subtrees $(\text{TREE}(B[I_\ell]), \ell > 0)$.

Recall that if B^{ex} is the standard excursion of length 1, then a Brownian excursion $B^{\text{ex},t}$ of length t can be defined by Brownian scaling. The distance from u to v in $\text{TREE}(B^{\text{ex},t})$ is just $\sqrt{t}$ times the distance from u/t to v/t in B^{ex}, see Chapter 0.

Proposition 7.8. *Let $\mu_\ell := T_\ell - T_{\ell-}$, which is the length of I_ℓ. Then the forest of Brownian subtrees $(\text{TREE}(B[I_\ell]), \ell > 0)$ is such that*

$$\{(\ell, \mu_\ell, \text{TREE}(B[I_\ell])) : \ell > 0, \mu_\ell > 0\}, \tag{7.12}$$

is the set of points of a Poisson point process on $\mathbb{R}_{\geq 0} \times \mathbb{R}_{\geq 0} \times C([0,\infty)^2)$ whose intensity measure is

$$d\ell \, \frac{dt}{\sqrt{2\pi} t^{3/2}} \, \mathbb{P}(\text{TREE}(B^{\text{ex},t}) \in d\tau), \tag{7.13}$$

where $\text{TREE}(B^{\text{ex},t})$, the tree generated by a Brownian excursion of length t, is identified by its distance function as a random element of $C([0,\infty)^2)$.

Proof. According to the preceding discussion, $\text{TREE}(B[I_\ell])$ is a trivial tree with only one vertex unless $T_{\ell-} < T_\ell$, in which case $\text{TREE}(B[I_\ell])$ is the tree generated by the excursion of $B - \underline{B}$ over the interval I_ℓ. So the statement of the proposition is just a push-forward of the Lévy-Itô description of excursions of B above $\underline{B}$, which is reviewed in (0.17) . □

To complete a description of the structure of $\text{TREE}(B)$, regarded as an infinite forest of subtrees attached to the forest floor defined by the descending ladder set, it remains to describe $\text{TREE}(B^{\text{ex}})$. This is done by the following theorem. For ease of comparison with the work of Aldous [7], the theorem describes $\text{TREE}(2B^{\text{ex}}) = 2\,\text{TREE}(B^{\text{ex}})$, that is $\text{TREE}(B^{\text{ex}})$ with all distances doubled.

Theorem 7.9. Aldous [7]. *Let $U_1, U_2, \ldots$ be a sequence of independent uniform $[0,1]$ variables, independent of B^{ex}. For each $n = 1, 2, \ldots,$ let*

$$\mathcal{T}_n = \text{SUBTREE}(2B^{\mathrm{ex}}; \{0, U_1, \ldots, U_n\}), \tag{7.14}$$

regarded as a plane tree with edge-lengths, and let

$$\Theta_n := \text{LENGTH}(\mathcal{T}_n). \tag{7.15}$$

Then (i)

$$\Theta_n = \sqrt{2\sum_{i=1}^{n} \varepsilon_i} \tag{7.16}$$

for a sequence of independent exponential(1) *variables ε_i, and*

$$\mathcal{T}_n = \Theta_n \mathcal{B}^{n,1} \tag{7.17}$$

where $(\Theta_n, n \geq 1)$ and $(\mathcal{B}^{n,1}, n \geq 1)$ are independent, and $\mathcal{B}^{n,1}$ has uniform distribution on the set of planted binary plane trees whose total edge-length is 1*, as defined by Lemma* 7.4.

(ii) **(Poisson line-breaking construction)** *The distribution of the sequence of trees $(\mathcal{T}_n, n \geq 1)$ is determined by the prescription* (7.16) *of their lengths, and the following: for each $n \geq 1$, conditionally given $(\mathcal{T}_j, 1 \leq j \leq n)$, the new segment of length $\Theta_{n+1} - \Theta_n$, which is added to $\mathcal{T}_n$ to form $\mathcal{T}_{n+1}$, is attached at a point of $\mathcal{T}_n$ picked independently of the new segment length according to the uniform distribution on edges of $\mathcal{T}_n$, with equal probability to the left or the right, independently of the point of attachment and the new segment length.*

To rephrase the Poisson line-breaking construction: Let $0 < \Theta_1 < \Theta_2 < \ldots$ be the points of an inhomogenous Poisson process on $\mathbb{R}_{>0}$ of rate $t\,dt$. Break the line $[0, \infty)$ at points Θ_n. Grow trees $\mathcal{T}_n$ by letting $\mathcal{T}_1$ be a segment of length Θ_1, then for $n \geq 2$ attaching the segment $(\Theta_{n-1}, \Theta_n]$ as a "twig" attached to a random point of the tree $\mathcal{T}_{n-1}$ formed from the first $n-1$ segments. Aldous [7, §4.3] discovered this result via weak convergence of combinatorially defined random trees, as indicated in Exercise 7.4.8 and Exercise 7.4.9 . This combinatorial approach explains nicely why the Poisson line breaking construction of $(\mathcal{T}_n)$ implies (7.17), which is otherwise not very obvious. But Aldous's argument identifying the laws of these trees with those embedded in B^{ex} by (7.14) involved an invariance principle implied by the invariance principle for conditioned Harris walks Theorem 6.4 : Galton-Watson trees with Poisson offspring distribution or with geometric offspring distribution converge in distribution to the same limit "continuum tree"; this limit was identified with TREE$(2B^{\mathrm{ex}})$ in the geometric case via the Harris bijection (Exercise 7.4.15) while Poisson line-breaking arose in consideration of the Poisson offspring case (Exercise 7.4.9). Le Gall [273] gave another proof of (7.17)-(7.16) based on calculations with Itô's σ-finite excursion law. See also [275, Chapter 2]. A more elementary proof can be given as follows, by relating $\mathcal{T}_n$ derived from B^{ex} and $U_1, \ldots, U_n$ to the critical binary$(0, \frac{1}{2})$ Galton-Watson tree with exponential edge-lengths conditioned to

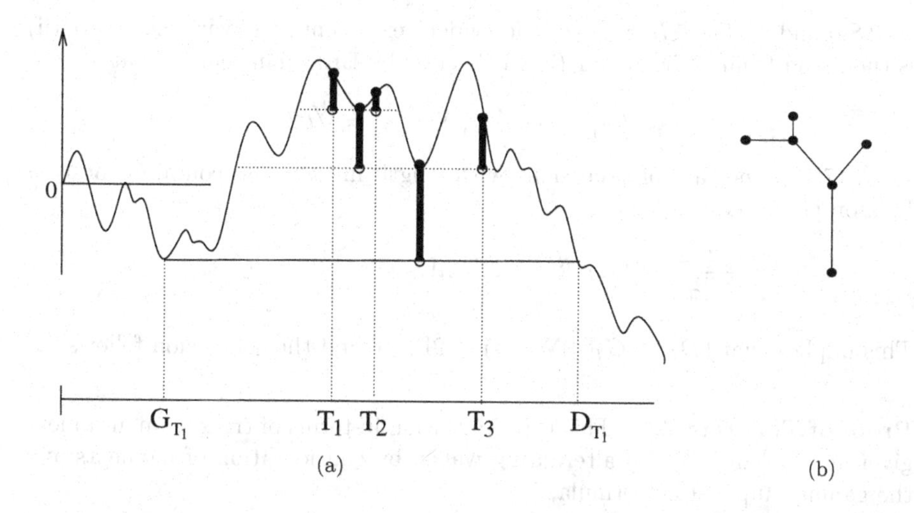

Figure 7.5: SUBTREE$(2B; \{G_{T_1}, T_1, \ldots, T_N\})$, as in Lemma 7.10.

have n leaves. Let Γ_r denote a gamma variable with parameter $r > 0$. Refer to Figure 7.5.

Lemma 7.10. *Let $T_1 < T_2 < \ldots$ be the times of points of a Poisson process with rate $\frac{1}{2}$ on $\mathbb{R}_{\geq 0}$, assumed independent of B. Let $[G_{T_1}, D_{T_1}]$ be the excursion interval of $B - \underline{B}$ straddling T_1, and let*

$$\mathcal{G} := \text{SUBTREE}(2B; \{G_{T_1}, T_1, \ldots, T_N\}) \text{ where } N := \sum_{i=1}^{\infty} 1(T_i \leq D_{T_1}) \quad (7.18)$$

is the number of T_i which happen to fall in $[G_{T_1}, D_{T_1}]$. Then
(i) [399, 200] *The tree $\mathcal{G}$ is a critical binary Galton-Watson tree, starting with one initial segment, in which each segment branches into either* 0 *or* 2 *segments with equal probability, and the segment lengths are independent exponential*(1) *variables.*
(ii) [355]

$$(\mathcal{G} \,|\, N = n) \stackrel{d}{=} \Gamma_{2n-1} \mathcal{B}^{n,1} \quad (7.19)$$

where Γ_{2n-1} is independent of $\mathcal{B}^{n,1}$, which has uniform distribution on planted binary trees with n leaves and total edge-length 1.
(iii) [355]

$$(\mathcal{G} \,|\, N = n) \stackrel{d}{=} \sqrt{2\Gamma_{n-\frac{1}{2}}}\, \mathcal{T}_n \quad (7.20)$$

where $\Gamma_{n-\frac{1}{2}}$ is independent of $\mathcal{T}_n :=$ SUBTREE$(2B^{\text{ex}}; \{0, U_1, \ldots, U_n\})$.

Proof. Part (i) is obtained by application of Lemmas 7.3 and 7.7, using the strong Markov property of B at the times T_i, and the consequence of Williams' path decomposition of B at the time $S_1 \in [0, T_1]$ when $B(S_1) = \underline{B}[0, T_1]$, that

$-B(S_1)$ and $B(T_1) - B(S_1)$ are independent exponential(1) variables. Part (ii) is then read from (7.5). As for (iii), it is clear by Brownian scaling that

$$(\mathcal{G} \mid D_{T_1} - G_{T_1} = t, N = n) \stackrel{d}{=} \sqrt{t}\,\mathcal{T}_n.$$

From (7.13), the rate of excursions with length in dt which contain exactly n Poisson points is

$$\frac{1}{\sqrt{2\pi}} t^{-3/2} dt\, e^{-t/2} (t/2)^n / n! = c_n t^{n-\frac{1}{2}-1} e^{-t/2} dt.$$

This implies that $(D_{T_1} - G_{T_1} \mid N = n) \stackrel{d}{=} 2\Gamma_{n-\frac{1}{2}}$ and the conclusion follows. □

Proof of Theorem 7.9 This is a translation in terms of trees of an argument given in [355] in terms of alternating walks: by consideration of moments and the gamma duplication formula,

$$\Gamma_{2n-1} \stackrel{d}{=} \sqrt{2\Gamma_{n-1/2}}\,\sqrt{2\Gamma'_n} \tag{7.21}$$

where Γ'_n is assumed independent of $\Gamma_{n-1/2}$ with the same distribution as Γ_n. Now (7.20) and (7.19) give

$$\sqrt{2\Gamma_{n-\frac{1}{2}}}\,\mathcal{T}_n \stackrel{d}{=} \sqrt{2\Gamma_{n-\frac{1}{2}}}\,\sqrt{2\Gamma'_n}\,\mathcal{B}^{n,1}.$$

It is easily argued that the common factor of $\sqrt{2\Gamma_{n-\frac{1}{2}}}$ can be cancelled [355, Lemma 8] to yield (7.17) with $\stackrel{d}{=}$ instead of $=$. To check that the ε_i defined implicitly by (7.17) are independent standard exponential, independent of $(\mathcal{B}^{n,1}, n \geq 1)$, takes a bit more work. But this is essentially elementary, because the conditional law of $(\mathcal{T}_j, 1 \leq j \leq n)$ given $\mathcal{T}_n$ is determined by a simple process of random deletion of vertices and their incident edges. The same elementary considerations yield (ii). □

Alternative proof of Theorem 7.9(i) (Provided by J.-F. Le Gall.) Let $T_1^\lambda < T_2^\lambda < \ldots$ be the times of points of a Poisson process with rate $\lambda > 0$, independent of B, and

$$\mathcal{G}_\lambda := \text{SUBTREE}(2B; \{G_{T_1^\lambda}, T_1^\lambda, \ldots, T_{N_\lambda}^\lambda\}) \text{ where } N_\lambda := \sum_{i=1}^\infty 1(T_i^\lambda \leq D_{T_1^\lambda}).$$

By trivial extensions of (7.19)–(7.20),

$$(\mathcal{G}_\lambda \mid N = n) \stackrel{d}{=} \sqrt{\frac{1}{\lambda}\Gamma_{n-\frac{1}{2}}}\,\mathcal{T}_n$$

and

$$(\mathcal{G}_\lambda \mid N = n) \stackrel{d}{=} \frac{1}{\sqrt{2\lambda}}\Gamma_{2n-1}\mathcal{B}^{n,1}.$$

Thus, if F is any continuous function on the set of trees,

$$\frac{1}{\Gamma(n-\frac{1}{2})}\int_0^\infty dt\, t^{n-3/2}e^{-t}\mathbb{E}\left[F\left(\sqrt{\frac{t}{\lambda}}\mathcal{T}_n\right)\right]$$
$$=\frac{1}{\Gamma(2n-1)}\int_0^\infty ds\, s^{2n-2}e^{-s}\mathbb{E}\left[F\left(\frac{s}{\sqrt{2\lambda}}\mathcal{B}^{n,1}\right)\right].$$

Make the changes of variables $t=\lambda t'$, $s=\sqrt{2\lambda}s'$ and use the fact that

$$e^{-s\sqrt{2\lambda}}=\int_0^\infty dt\frac{s}{\sqrt{2\pi t^3}}e^{-s^2/2t}e^{-\lambda t}.$$

It follows that

$$\frac{1}{\Gamma(n-\frac{1}{2})}\int_0^\infty dt\, t^{n-3/2}e^{-\lambda t}\mathbb{E}\left[F\left(\sqrt{t}\mathcal{T}_n\right)\right]$$
$$=\frac{2^{n-\frac{1}{2}}}{\Gamma(2n-1)\sqrt{2\pi}}\int_0^\infty dt\, e^{-\lambda t}t^{-3/2}\int_0^\infty ds\, s^{2n-1}e^{-s^2/2t}\mathbb{E}\left[F\left(s\mathcal{B}^{n,1}\right)\right].$$

By injectivity of the Laplace transform

$$\mathbb{E}\left[F\left(\sqrt{t}\mathcal{T}_n\right)\right]=\frac{2^{1-n}}{\Gamma(n)}t^{-n}\int_0^\infty ds\, s^{2n-1}e^{-s^2/2t}\mathbb{E}\left[F\left(s\mathcal{B}^{n,1}\right)\right].$$

Take $t=1$ to conclude.

Exercises

7.4.1. (Identification of paths in trees derived from continuous functions) Fix $X\in C[I]$. Let $\tilde{t}:=\{u\in I: u\sim_X t\}$. Note that if t is the minimal element of $\tilde{t}$, then $\tilde{t}=\{u: X(u)=\underline{X}[t,u]=X(t)\}$. For $s,u\in I$ the unique path from $\tilde{s}$ to $\tilde{u}$ at unit speed is determined by its range in $I/\sim_X$. This subset of $I/\sim_X$ is identified in the obvious way with a subset of $[s,u]$, call it the *segment of* TREE(X) *between* s *and* u. This segment is the union of a *falling segment* $\{v\in[s,u]: X(v)=\underline{X}[s,v]\}$, of length $X(s)-\underline{X}[s,u]$, and a *rising segment* $\{v\in[s,u]: X(v)=\underline{X}[v,u]\}$ of length $X(u)-\underline{X}[s,u]$, these segments intersecting at the set of points $\{v\in[s,u]: X(v)=\underline{X}[s,u]$ where X attains its minimum on $[s,u]$. Each of these segments may also be regarded as a subset of $I/\sim_X$ which is isometrically parameterized by an interval. The falling segment is naturally parameterized by $[\underline{X}[s,u],X(s)]$, the rising segment by $[\underline{X}[s,u],X(t)]$, and the whole segment by $[0,d_X(s,u)]$.

7.4.2. Let $\tilde{t}$ be the equivalence class of $t\in\mathbb{R}_{\geq 0}$ equipped with the Brownian tree metric d_B. Deduce from the strong Markov property of B that

$$\mathbb{P}(\ |\tilde{t}|=1 \text{ or } 2 \text{ or } 3 \text{ for all } t\geq 0)=1.$$

Check that $\{t : \tilde{t} = 3\}$ is a.s. equal to the countable set of all local minima of B besides those in the descending ladder set. See [324] for further analysis of local minima of B. Check that $\{t : \tilde{t} = 2\}$ is an uncountable set of Lebesgue measure 0. This is essentially the *skeleton* of $\text{TREE}(B)$, as discussed in Chapter 10. Both sets are a.s. dense in $\mathbb{R}_{\geq 0}$, both in the usual topology and in the topology of $\text{TREE}(B)$.

7.4.3. (Different processes with the same tree)

$$\text{TREE}(B) = \text{TREE}(B - 2\underline{B}) \stackrel{d}{=} \text{TREE}(R_3) \tag{7.22}$$

where the first equality holds for every path $B \in C[0, \infty)$, and second equality in distribution, read from [345] , assumes B is a standard Brownian motion and R_3 is the 3-dimensional Bessel process. Hence for each fixed s,

$$(d_B(s, s+t), t \geq 0) \stackrel{d}{=} (R_3(t), t \geq 0)$$

but it seems there is no simple description of the joint law of $d_B(u, v)$ as both u and v are allowed to vary. Note the implication of (7.22) that there is *loss of information* in passing from X to $\text{TREE}(X)$. Except if $X(t) \geq X(0)$ for all $t \geq 0$, or if it is assumed that X has some particular distribution, the path of X typically cannot be recovered from $\text{TREE}(X)$. Neither is the distribution of X determined by that of $\text{TREE}(X)$.

7.4.4. (The filtration generated by the tree) Let $\mathcal{B}_t := \sigma(B_s, 0 \leq s \leq t)$ be the Brownian filtration, and let $\mathcal{R}_t$ be the σ-field generated by the restriction of $\text{TREE}(B)$ to $[0, t]$. Check that $\mathcal{R}_t = \sigma(R_s, 0 \leq s \leq t)$ where $R_t = d_B(0, t)$ as in Exercise 7.4.3 , and hence that $\mathcal{B}_t$ is generated by $\mathcal{R}_t$ and $\underline{B}(t)$ where the conditional law of $\underline{B}(t)$ given $\mathcal{R}_t$ is uniform on $[-R(t), 0]$, see [345].

7.4.5. (Processes generating Brownian trees) Show that for each probability distribution F of a random variable M with values in $[-\infty, 0]$ there exists a probability distribution on $C[0, \infty)$ for a process X such that $\text{TREE}(X) \stackrel{d}{=} \text{TREE}(B)$ and $\underline{X}[0, \infty) \stackrel{d}{=} M$. In particular, if $M = -\infty$ then $X \stackrel{d}{=} B$, and if $M = 0$ then $X \stackrel{d}{=} R_3$. Describe the law of X by an explicit path decomposition in the case $M = m$ for some fixed $m \in (-\infty, 0)$, and hence give a construction of X corresponding to an arbitrary distribution of M.

7.4.6. (Trees derived from paths falling below their initial value) See Figure 7.4. Generalize Lemma 7.7 to the case when X may fall below its initial value, when the $2n$-step alternating walk is not an excursion, by considering excursions of the alternating walk above its past minimum process. In general, the subtree of $\text{TREE}(X)$ spanned by $0, T_1, \ldots, T_n$ is determined by the $2n$-step alternating walk derived from the $X(T_i)$ and intermediate $X(S_i)$, but not conversely.

7.4.7. (Trees generated by bridges and meanders) Describe as explicitly as possible the laws of the trees generated by each of the following processes indexed by $[0,1]$: Brownian motion, Brownian bridge from 0 to x, Brownian meander, Brownian meander from 0 to x. (Nasty descriptions in terms of alternating walks were given in [355]: there are much nicer descriptions, either in the style of the Poisson line-breaking construction, or in terms of forests of trees associated with excursions).

7.4.8. (Aldous's proof of Theorem 7.9) Let $\mathcal{U}_N$ have uniform distribution on the set of N^{N-1} rooted trees labeled by $[N]$. For $1 \leq n \leq N$ let $\mathcal{U}_{n,N}$ be the subtree of $\mathcal{U}_N$ spanned by the root and $[n]$, let $\mathcal{T}_{n,N}$ be $\mathcal{U}_{n,N}$ regarded as an ordered plane tree (e.g. by imposing a random order), and let $\mathcal{R}_{n,N}$ be the reduced plane tree derived from $\mathcal{T}_{n,N}$ by first giving each edge of $\mathcal{T}_{n,N}$ length $1/\sqrt{N}$, then deleting all degree 2 vertices. Let $B_{n,N}$ be the indicator of the event that $\mathcal{R}_{n,N}$ is a planted binary tree with n leaves, and let $L_{n,N}(i)$ be the length of the ith branch of $\mathcal{R}_{n,N}$, in order of depth first search. Check by direct enumeration that for each fixed n, as $N \to \infty$,

$$(L_{n,N}(i), 1 \leq i \leq 2n-1)1(B_{n,N}) \xrightarrow{d} (X_i, 1 \leq i \leq 2n-1)$$

where the right side is the sequence of lengths of branches of $\mathcal{T}_n$ in Theorem 7.9. Now deduce Theorem 7.9 from Theorem 6.4 . This method has proved effective in characterizing other kinds of continuum random trees besides $\text{TREE}(B^{\text{ex}})$. See [20, 21].

7.4.9. (Combinatorial view of the line-breaking construction) (Aldous [5, 6, 7]) Take $(\xi_i^{(N)}, i \geq 1)$ independent uniform on $[N]$. Make a tree on $[N]$ by declaring 1 to be the root and, for $i = 2, 3, \ldots, N$, put an edge from i to $\min(i-1, \xi_i^{(N)})$. Apply a uniform random permutation to the vertex-labels $[N]$. The resulting tree $\mathcal{U}_N$ is uniform on all N^{N-1} rooted labeled trees, as in Exercise 7.4.8 . (This fact is not obvious; it can be deduced from the Markov chain tree theorem [292] applied to the Markov chain $(\xi_i^{(N)})$; see also Definition 10.5 and papers cited there). Let $R_1^{(N)}, R_2^{(N)}, \ldots$ be the successive values of i for which $\xi_i^{(N)} \leq i-1$. Show that

$$N^{-1}(R_n^{(N)}, n \geq 1) \xrightarrow{d} (\Theta_n, n \geq 1)$$

where the right side is defined by (7.16). Deduce that for fixed n, if we give length $N^{-1/2}$ to the edges of $\mathcal{U}_N$, then the subtrees comprised of the first $R_n^{(N)}$ vertices converge in distribution to $\mathcal{T}_n$ defined in the Poisson line-breaking construction (Theorem 7.9 (ii)).

7.4.10. (The Poisson line-breaking construction of B^{ex}) Let $(\mathcal{T}_n)_{n=1,2,\ldots}$ be the sequence of random trees defined by the Poisson line-breaking construction, as in Theorem 7.9(ii), with $\Theta_n = \text{LENGTH}(\mathcal{T}_n)$. Then there exists on the same probability space an a.s. unique Brownian excursion B^{ex} and a.s. unique independent uniform U_i for $1 \leq i \leq n$ such that $\mathcal{T}_n$ is the subtree of $\text{TREE}(2B^{\text{ex}})$

spanned by $\{0, U_1, \ldots, U_n\}$. To be explicit, if H_n is the Harris path of $\mathcal{T}_n$, then

$$H_n(2\Theta_n u) \to 2B^{\text{ex}}(u) \quad (0 \leq u \leq 1) \tag{7.23}$$

uniformly almost surely. Moreover, U_n is the limiting fraction of times that new edges are added to the left of the nth edge that was added to the tree. If $(W_{n,k}, 0 \leq k \leq 2n)$ is the alternating excursion associated with $\mathcal{T}_n$, then for $1 \leq m \leq n$

$$W_{n,2m-1} = 2B^{\text{ex}}(U_{n,m}) \quad (1 \leq m \leq n) \tag{7.24}$$

where $U_{n,m}$ is the mth smallest among $U_1, \ldots, U_n$, and for $1 \leq m \leq n-1$

$$W_{n,2m} = 2\underline{B^{\text{ex}}}[U_{n,m}, U_{n,m+1}]. \tag{7.25}$$

7.4.11. (An urn-scheme construction of B^{ex}) Let $(\mathcal{S}_n)_{n=1,2,\ldots}$ be the sequence of shapes of the random trees $(\mathcal{T}_n)_{n=1,2,\ldots}$ defined by the Poisson line-breaking construction, as in Theorem 7.9 (ii). So $\mathcal{S}_n \in \text{BINARY}_n$, the set of n-leaf combinatorial planted binary plane trees. Let $\mathcal{S}_{[n]}$ be $\mathcal{S}_n$ with its n leaves labeled by $[n]$, according to their order of addition in the growth process. So $\mathcal{S}_{[n]} \in \text{BINARY}_{[n]}$, the set of n-leaf binary plane trees with leaves labeled by $[n]$, with $|\text{BINARY}[n]| = n!|\text{BINARY}_n|$. Show that both $(\mathcal{S}_n)_{n=1,2,\ldots}$ and $(\mathcal{S}_{[n]})_{n=1,2,\ldots}$ are Markov chains, and describe their transition probabilities. Check in particular that given $\mathcal{S}_{[n]}$, the next tree $\mathcal{S}_{[n+1]}$ is equally likely to be any of the $2(2n-1)$ trees which can be grown from $\mathcal{S}_{[n]}$ by replacing one of the $2n-1$ edges of $\mathcal{S}_{[n]}$, say the edge $a \to b$, by two edges and a new internal vertex c to form $a \to c \to b$, then attaching the leaf labeled $(n+1)$ by an edge $(n+1) \to c$, on one side or the other of $a \to c \to b$. Note the byproduct of this argument that for $n = 1, 2, \ldots$

$$|\text{BINARY}_{[n+1]}| = 2^n(1 \times 3 \times 5 \times \cdots \times (2n-1)) = \binom{2n}{n} n!, \tag{7.26}$$

and hence:

$$|\text{BINARY}_{n+1}| = \frac{|\text{BINARY}_{[n+1]}|}{(n+1)!} = \frac{1}{n+1}\binom{2n}{n}, \tag{7.27}$$

which is the nth Catalan number C_n [407, Exercise 6.19]. Thus, a sequence of trees distributed like $(\mathcal{S}_{[n]})_{n=1,2,\ldots}$ can be grown by a simple urn scheme construction, similar to the Chinese Restaurant Process described in Chapter 3, and $\mathcal{S}_n$ is derived from $\mathcal{S}_{[n]}$ by ignoring the labelling by $[n]$. Deduce from (7.23) and $\Theta_n \sim \sqrt{2n}$ a.s. that if $H_{\mathcal{S}_n}$ is the Harris path of $\mathcal{S}_n$, then

$$\frac{H_{\mathcal{S}_n}(2(2n-1)u)}{\sqrt{2n}} \to 2B^{\text{ex}}(u) \quad (0 \leq u \leq 1) \tag{7.28}$$

uniformly almost surely, where B^{ex} is the same Brownian excursion as that derived similarly from $(\mathcal{T}_n)_{n=1,2,\ldots}$ by (7.23). As a check on the normalization, the weaker form of (7.28) with $\xrightarrow{d}$ instead of almost sure convergence is the instance of Aldous's result Theorem 6.4 for binary branching, due to Gutjahr-Pflug [181]. See also [261, 376, 418, 419, 247] for various other models for random growth of binary trees.

7.4.12. (Independent self-similar growth of subtrees) The combinatorial tree growth process $(\mathcal{S}_n)_{n=1,2,...}$ embedded in B^{ex} as in the previous exercise has the following remarkable property. Suppose that e is one of the $2m-1$ edges of $\mathcal{S}_m$, for some fixed m. Then for each $n \geq m$ there is an obvious way to identify a subtree of $\mathcal{S}_n$ formed by e and all its offshoots, which will be a combinatorial planted binary plane tree. If this sequence of offshoots grows by addition of leaves at random times $m < N_1(e) < N_2(e) < \ldots$ say, then $N_k(e)$ increases to ∞ almost surely, and the sequence of offshoots of e, watched only at the times $m, N_1(e), N_2(e), \ldots$, and re-indexed by $1, 2, \ldots$, defines a tree growth process, call it the *sub-process generated by* e, which has the same distribution as the original growth process. Moreover, as e varies over the $2m-1$ edges of $\mathcal{T}_m$, these $2m-1$ sub-processes are independent.

7.4.13. (Dirichlet distribution of proportions of subtrees) Continuing the previous exercise, for each fixed m, as n varies over $n \geq m$, the process of allocation of new leaves to the offshoots of edges e of $\mathcal{S}_m$ is equivalent to Pólya's urn scheme with $2m-1$ initial balls of different colors, governed by sampling with double replacement. Hence, by Exercise 2.2.2 , as $n \to \infty$, the proportions of edges in the $2m+1$ subtrees converge almost surely to a random vector with the symmetric Dirichlet distribution with $2m+1$ parameters equal to $\frac{1}{2}$. Note that the same is true of proportions of leaves, since a planted plane binary tree with $2n-1$ edges has n leaves. Both processes may be regarded as developments of the Blackwell-MacQueen urn scheme [68, 350].

7.4.14. (An urn-scheme construction of B^{br}) Recall that $\mathcal{S}_{[n]}$ is $\mathcal{S}_n$ with the leaves labeled by their order of appearance in the urn-scheme construction of B^{ex}. Let $\sigma(t), 0 \leq t \leq 2(2n-1)$ be the usual depth-first search of $\mathcal{S}_n$, and define the *trunk* of $\mathcal{S}_n$ to be the path from its root to the vertex labeled 1 in $\mathcal{S}[n]$. For $0 \leq u \leq 1$ let $R_n(u)$ be the distance from $\sigma(2(2n-1)u)$ to the trunk as measured along edges of $\mathcal{S}_n$, and let $S_n(u)$ be $R_n(u)$ with a sign ± 1 according to whether $\sigma(2(2n-1)u)$ is to the left or the right of the trunk. Then, as companions to (7.28), there is almost sure convergence of $S_n/\sqrt{2n}$ to $2B^{\text{br}}$ and $R_n/\sqrt{2n}$ to $2|B^{\text{br}}|$, where B^{br} can be derived as a measurable function of B^{ex} and the independent uniform variable U_1 which is the limiting fraction of leaves of $\mathcal{S}[n]$ which are to the left of leaf 1 at the top of the trunk. This is a variant of transformations between B^{ex} and B^{br} discussed in [53]. Note how the uniform location of 1 among the leaves implies Lévy's theorem that $\int_0^1 1(B_u^{\text{br}} > 0)du$ has uniform $[0,1]$ distribution. See Knight [257] and Kallenberg [229] for generalizations to bridges with exchangeable increments, and related results.

7.4.15. (Weak convergence of conditioned Galton-Watson trees) Consider a critical Galton-Watson branching process with offspring variance σ^2. Write $\mathcal{T}_N$ for this tree conditioned to have exactly N vertices, with each edge given length $N^{-1/2}$. Aldous [7] gave a sense, implied by Theorem Theorem 6.4, in which $\mathcal{T}_N$ converges in distribution to $\text{TREE}(2\sigma^{-1}B^{\text{ex}})$. Check the scaling constant $2/\sigma$ is consistent with what is known by other methods in the following three special cases.

- (a) For Poisson(1) offspring distribution. Here $\text{SHAPE}(\mathcal{T}_N)$ is obtained by randomly ordering the branches of the uniform tree $\mathcal{U}_N$ of Exercise 7.4.8 .
- (b) For geometric(1/2) distribution. Here $\text{SHAPE}(\mathcal{T}_N)$ is uniform on all combinatorial plane trees, which by the Harris correspondence are bijective with simple walk-excursions of length $2N$.
- (c) For binary offspring distribution, i.e. uniform on 0 and 2. Now $\text{SHAPE}(\mathcal{T}_N)$, for $N = 2M - 1$, is uniform on binary plane trees with M leaves, as in Exercise 7.4.11 .

Notes and comments

Neveu [320, 321] introduced a notion of *marked trees*, used also by Le Gall [273], which is essentially equivalent to the notion of plane trees considered here. The structure of $\text{SUBTREE}(B; \{T_0, T_1, \ldots, T_N\})$ for suitable random $T_0, \ldots, T_N$, was first considered in [323, 324], where it was shown that some T_i defined in terms of successive upcrossings and downcrossings by B induce a critical binary Galton-Watson tree. Aldous ([7] and subsequent papers) calls $\text{TREE}(2B^{\text{ex}})$ the Brownian *continuum random tree* (CRT). See also [8] regarding recursive self-similarity properties of $\text{TREE}(2B^{\text{ex}})$, and random triangulations. Kersting [246] discusses symmetry properties of binary branching trees implicit in Aldous's description of $\text{TREE}(2B^{\text{ex}})$.

7.5. Plane forests with edge-lengths

The idea of this section is to consider the tree

$$\mathcal{F} := \text{SUBTREE}(B; \{0, T_1, T_2, \ldots\})$$

for B a Brownian motion, and suitable increasing sequences of random times T_i, called *sampling times*. To avoid uninteresting complications, it will be assumed throughout that

$$0 = T_0 < T_1 < T_2 < \cdots < T_n \uparrow \infty \tag{7.29}$$

almost surely. The assumption that $T_n \uparrow \infty$ implies $\inf_n \underline{B}(T_n) = -\infty$, which means that $\mathcal{F}$ will contain the entire infinite *floor line* of $\text{TREE}(B)$ corresponding to the descending ladder set of B. Thus $\mathcal{F}$ may be regarded as a random *locally finite plane forest*, that is an infinite sequence of plane trees, each with a finite number of edges of finite length, with the roots of these trees located at some strictly increasing sequence of points

$$0 < L_1 < L_2 < \cdots < L_k \uparrow \infty$$

where L_k is the common value of $-\underline{B}(t)$ for t in the kth excursion interval of $B - \underline{B}$ that contains at least one T_i. Moreover, it is clear that with probability one, each of the tree components of $\mathcal{F}$ will be a planted binary plane tree, and that $\mathcal{F}$ can be identified graphically when convenient as a subset of $\mathbb{R}^2$. The definition of Harris paths is extended to such forests (including also subforests with only a finite number of trees) as follows. If $\mathcal{F}_0$ is the trivial forest with no trees, then $H_{\mathcal{F}_0}$ is the *trivial path*, starting at 0, with constant slope -1. For a non-trivial locally finite plane forest $\mathcal{F}$, let $H_{\mathcal{F}}(\cdot)$ be defined by inserting the Harris paths of the component subtrees of $\mathcal{F}$ into the trivial path, in such a way that if $\mathcal{T}_i$ is the tree component of $\mathcal{F}$ whose root is located at ℓ_i, and

$$T_i := \ell_i + \sum_{j<i} 2\text{LENGTH}(\mathcal{T}_j) = \inf\{t : H_{\mathcal{F}}(t) = -\ell_i\}$$

then

$$H_{\mathcal{F}}(T_i + t) = H_{\mathcal{T}_i}(t) \qquad (0 \le t \le 2\text{LENGTH}(\mathcal{T}_i))$$

and $H_{\mathcal{F}}(\cdot)$ is continuous with slope -1 on each of the intervals

$$(\ell_{i-1} + \sum_{j<i} 2\text{LENGTH}(\mathcal{T}_i), \ell_i + \sum_{j<i} 2\text{LENGTH}(\mathcal{T}_i))$$

where $\ell_0 = 0$ and $\ell_k = \infty$ if $\mathcal{F}$ has fewer than k trees. Refer to Figure 7.6.

Intuitively, for reasonable families of sampling times $(T_i^m, i \ge 1)$ whose intervals become small as $m \to \infty$, we expect the forests

$$\mathcal{F}^m = \text{SUBTREE}(B; \{0, T_1^m, T_2^m, \ldots\})$$

to converge to $\text{TREE}(B)$. It will be shown in Sections 7.6 and 7.7 that there are two particular choices of sampling sequences for which the forest-valued processes $(\mathcal{F}^m)$ have interesting autonomous descriptions, given in Definitions 7.16 and 7.21. To get started, note that various results presented in terms of trees in the previous sections now have analogs in terms of forests. For instance, here are the forest analogs of Proposition 7.5, Lemma 7.7, Definition 7.2, and Lemma 7.3.

Proposition 7.11. *Let $H = H_{\mathcal{F}}$ be the Harris path of some locally finite graphically represented plane forest $\mathcal{F}$. Then the depth-first search of $\mathcal{F}$ at unit speed induces a bijection from $[0,\infty)/\sim_H$ to $\mathcal{F}$ whereby the metric space $[0,\infty)/\sim_H$ is isometric to $(\mathcal{F}, d_{\mathcal{F}})$.*

Lemma 7.12. *For $m = 1, 2, \ldots$ let S_m be a time in $[T_{m-1}, T_m]$ at which B attains its minimum on $[T_{m-1}, T_m]$ so*

$$B(S_m) := \underline{B}[T_{m-1}, T_m].$$

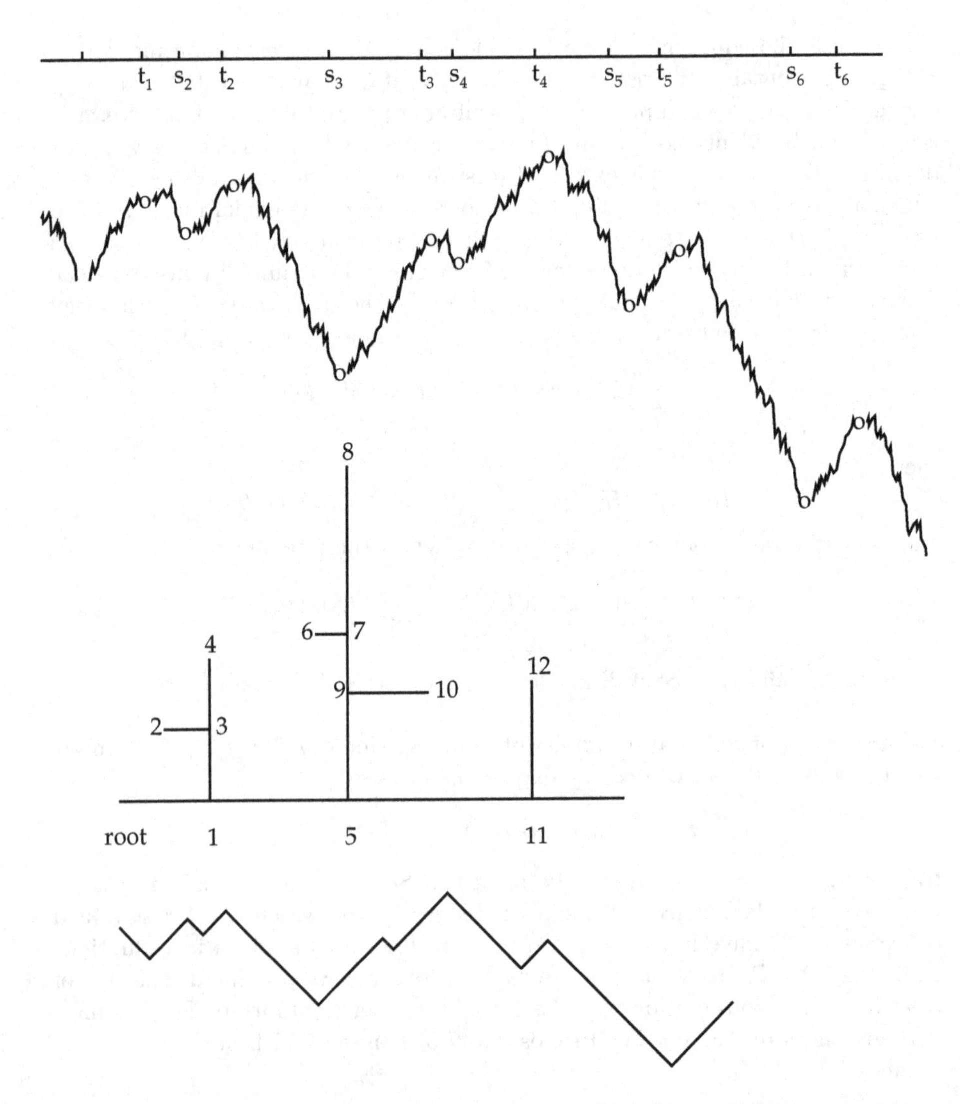

Figure 7.6: A forest SUBTREE$(B; \{0, t_1, t_2, \ldots, t_6\})$ and its Harris path. Note $s_i = \arg\min_{t_{i-1} \leq s \leq t_i} B(s)$. Vertices of the forest are labeled in alternating order, so vertex $2n$ corresponds to time t_n and vertex $2n-1$ to time s_n.

Then SUBTREE$(B; \{0, T_1, T_2, \ldots\})$ *is isometric to the unique locally finite forest defined by a sequence of reduced plane trees attached to points along a floor line, whose Harris path $H_{\mathcal{F}}(\cdot)$ is a time-change with alternating slopes ± 1 of the broken line joining the sequence of points*

$$(0,0),\ (S_1, B(S_1)),\ (T_1, B(T_1),\ (S_2, B(S_2)),\ (T_2, B(T_2)), \ldots.$$

The notation $\mathcal{F}(B; \{0, T_1, T_2, \ldots\})$ may now be used *either* for the locally finite reduced plane forest defined by the Lemma, *or* for the isometric infinite subtree of TREE(B) spanned by the set of points $\{0, T_1, T_2, \ldots\}$.

Definition 7.13. A *binary(λ, μ) forest* is a planted binary plane forest $\mathcal{F}$ with edge-lengths whose trees are a sequence of independent and identically distributed binary(λ, μ) trees, which are rooted at the set of points of an independent Poisson process on $[0, \infty)$ with rate $\mu - \lambda$.

Lemma 7.14. *A locally finite forest $\mathcal{F}$ of reduced plane trees is a binary(λ, μ) forest if and only if the succession of falls and rises of its Harris path $H_{\mathcal{F}}$ is a sequence of independent exponential variables, with exponential$(\mu + \lambda)$ rises and exponential$(\mu - \lambda)$ falls.*

Proof. This follows from Lemma 7.3 and the fact that the exponential rate $\mu - \lambda$ of the spacings between trees in a binary(λ, μ) forest is by definition the same as the rate of the exponential falls of the alternating random walk used to construct the binary(λ, μ) tree. This means that the absolute value of the overshoot of the alternating exponential walk used to make the ith tree of the forest has exactly the correct exponential$(\mu - \lambda)$ distribution to make the spacing on the forest floor between the ith and $(i+1)$th trees. □

Note how the description of the binary(λ, μ) forest is simpler than the corresponding description of the binary(λ, μ) tree by stopping the associated alternating random walk when it first steps negative.

7.6. Sampling at downcrossing times

Throughout this section, let B denote a standard Brownian motion, with $B_0 = 0$. Following [324], consider for each $\theta > 0$ the sequence of *downcrossing times* $\{D^\theta_m\}$ defined inductively as follows. Let $D^\theta_0 = 0$. Given that D^θ_m has been defined, let

$$\begin{aligned} U^\theta_{m+1} &:= \inf\{t : B_t - \underline{B}[D^\theta_m, t] = \theta^{-1}\}, \\ D^\theta_{m+1} &:= \inf\{t : \overline{B}[U^\theta_{m+1}, t] - B_t = \theta^{-1}\}. \end{aligned} \tag{7.30}$$

In words, U^θ_{m+1} is the first time after D^θ_m that B completes an upcrossing of some interval of length θ^{-1}, and D^θ_{m+1} is the next time after U^θ_{m+1} that B completes a down-crossing of some interval of length θ^{-1}. By construction, the D^θ_m are stopping times relative to the filtration of B, and $\underline{B}[D^\theta_m, D^\theta_{m+1}) = B(U^\theta_{m+1}) - \theta^{-1}$. The level θ^{-1} is chosen so that θ is the rate, per unit increment of $-\underline{B}$, of the Poisson process of excursions of $B - \underline{B}$ which reach level θ^{-1}.

Theorem 7.15. [324] *For each fixed* $\theta > 0$,

$$\mathcal{F}_B^\theta := \mathcal{F}(B; \{0, D_1^\theta, D_2^\theta, \ldots\})$$

is a binary$(0, \theta)$ *forest of critical binary trees, in which segments are attached to the floor by a Poisson process with rate* θ*, each segment has either* 0 *or* 2 *children with equal probability, and the lengths of segments are independent exponential*(2θ) *variables.*

Proof. This can be read from the Poisson character of Brownian excursions, as shown in [324]. □

Two alternative viewpoints are useful. Firstly, the forest $\mathcal{F}_B^\theta$ may be regarded as an infinite subtree of TREE(B). From this perspective, the process $(\mathcal{F}_B^\theta)_{\theta \geq 0}$ is a *forest growth process*, meaning that $\mathcal{F}_B^\lambda$ can be identified as a subforest of $\mathcal{F}_B^\mu$ whenever $\lambda < \mu$. Secondly, to regard $(\mathcal{F}_B^\theta)_{\theta \geq 0}$ as a forest-valued process, the space of locally-finite plane forests is identified as a subset of $C[0, \infty)$ via the Harris correspondence, and $C[0, \infty)$ is given the topology of uniform convergence on compacts. The forest growth process $(\mathcal{F}_B^\theta)_{\theta \geq 0}$ has the special property that for each fixed $\lambda > 0$ the entire path $(\mathcal{F}_B^\theta)_{0 < \theta \leq \lambda}$ can be recovered as a measurable function of $\mathcal{F}_B^\lambda$. Indeed, as h increases, the process $(\mathcal{F}_B^{1/h})_{h > 0}$ is the deterministic process of continuous erasure of tips, described by Neveu [321]. The law of the whole forest growth process $(\mathcal{F}_B^\theta)_{\theta \geq 0}$ is determined by this observation and the distribution of $\mathcal{F}_B^\theta$ for fixed θ given by Theorem 7.15. The following definition is suggested by these considerations and the work of Abraham [1, p. 382]:

Definition 7.16. A *tip-growth process* is a time inhomogeneous forest-valued Markov process with continuous paths, $(\mathcal{F}^\theta)_{\theta \geq 0}$, which starts at the empty forest at time $\theta = 0$, and develops as follows:

- at each time θ, along each side of each edge of $\mathcal{F}^\theta$, new tips of length 0 are born according to a Poisson process with unit rate per unit length of side per unit time;
- each tip that is alive at time θ is growing away from the root of its tree at speed θ^{-2}.

In the description of these Poisson rates, each of the subintervals of $[0, \infty)$ between the roots of trees of $\mathcal{F}^\theta$, including the interval before the first root, is regarded as an edge of $\mathcal{F}^\theta$ with only one (upward) side from which new tips can sprout and then grow into larger trees by addition of further tips. All other edges of $\mathcal{F}^\theta$ are regarded as having two sides.

Now there is the following refinement of Theorem 7.15:

Corollary 7.17. *The forest-valued process* $(\mathcal{F}_B^\theta)_{\theta \geq 0}$ *is a tip-growth process such that* $\mathcal{F}_B^\theta$ *is a binary*$(0, \theta)$ *forest of critical binary trees. Conversely, if* $(\mathcal{F}^\theta)_{\theta \geq 0}$ *is such a forest growth process, there exists on the same probability space a unique Brownian motion* B *such that* $(\mathcal{F}^\theta)_{\theta \geq 0} = (\mathcal{F}_B^\theta)_{\theta \geq 0}$ *almost surely. Explicitly,*

$$B(t) = \lim_{\theta \to \infty} H^\theta(\theta t) \qquad (0 \leq t < \infty) \tag{7.31}$$

where H^θ is the Harris path associated with $\mathcal{F}^\theta$, and the convergence holds in $C[0,\infty)$ almost surely.

Proof. The first sentence can be read from the previous discussion, and Itô's excursion theory. See Abraham [1, p. 382], where the tip-growth evolution is described just for a single tree born at time θ which is then destined to have ultimate maximum height θ^{-1}. The proof of the converse is similar to the proof of the converse in Exercise 7.7.4 .
See also Abraham [1, V], where a variant of (7.31) is given for recovery of a single excursion of B from a growing family of subtrees. □

To amplify the Brownian interpretation of the tip-growth process, a tip born at time θ will have grown into a branch of total length $\theta^{-1} - \phi^{-1}$ at a subsequent time ϕ, by which time other tips may have started growing from this branch according to the Poisson process of birth of new tips. Thus each tip born at time θ anticipates a local maximum of B at a level exactly θ^{-1} higher than the level where the tip is born. See also Abraham [1] for further discussion.

7.7. Sampling at Poisson times

Let N be a homogeneous Poisson point process on $(0,\infty)^2$, with unit rate per unit area, assumed independent of the Brownian motion B. Consider the forest-valued process

$$\mathcal{F}^\theta_{N,B} := \mathcal{F}(B; \{0, T^\theta_1, T^\theta_2, \ldots\})$$

generated by sampling B at the times

$$T^\theta_m := \inf\{t : N([0,t] \times [0, \tfrac{1}{2}\theta^2]) = m\}. \tag{7.32}$$

To match the parameterizations of the two forest growth processes described by Corollary 7.17 and Theorem 7.18, the level $\frac{1}{2}\theta^2$ is chosen in (7.32) because θ is the rate per unit local time of Brownian excursions which contain at least one point of N below $\frac{1}{2}\theta^2$. It follows that sample times T^θ_m and D^θ_m have the same mean $2/\theta^2$, but these sample times have different variances. From the definition (7.30), the distribution of D^θ_1 is that of the sum of two independent copies of the first hitting time of θ^{-1} by $|B|$, whereas the distribution of T^θ_1 is exponential($\frac{1}{2}\theta^2$).

Theorem 7.18. [399, §6.2],[200] *For each fixed $\theta > 0$, the infinite subtree $\mathcal{F}^\theta_{N,B}$ of the Brownian tree, obtained by sampling at Poisson times with rate $\frac{1}{2}\theta^2$, is a binary$(0,\theta)$ forest of critical binary trees, hence identical in law to $\mathcal{F}^\theta_B$ obtained by sampling at downcrossings of size θ^{-1}.*

Proof. In view of Lemma 7.3, and the coding of forests by alternating walks discussed in Section 7.1, part (i) follows from the strong Markov property of B at the times T^θ_m, and Williams' decomposition of B at the time $S^{(\theta)}_1$ of its minimum before the exponential($\frac{1}{2}\theta^2$) time T^θ_1, whereby the random variables

$B(S_1^\theta) := \underline{B}(T_1^\theta)$ and $B(T_1^\theta) - B(S_1^\theta)$ are independent exponential(θ) variables. See also [399, §6.2], [355], [200] for variations of this argument. □

The parallel between Theorems 7.15 and 7.18 invites an analysis of the forest growth process $(\mathcal{F}^\theta_{N,B})_{\theta \geq 0}$ analogous to Corollary 7.17. An analog for the Poisson-sampled Brownian forest of Neveu's process of erasure of tips is provided by the next lemma, which follows immediately from standard properties of Poisson processes:

Lemma 7.19. *For $\mathcal{F}^\theta := \mathcal{F}^\theta_{N,B}$, the forest growth process is such that for each $0 < \lambda < \mu$, conditionally given $\mathcal{F}^\mu$, the forest $\mathcal{F}^\lambda$ is derived from $\mathcal{F}^\mu$ by taking the subforest of $\mathcal{F}^\mu$ spanned by a random set of leaves of $\mathcal{F}^\mu$ picked by a process of independent Bernoulli trials so for each i the ith leaf of $\mathcal{F}^\mu$ in order of depth-first search is put in the spanning set with probability λ^2/μ^2.*

This lemma and Theorem 7.18 determine the joint distribution of $\mathcal{F}^\lambda$ and $\mathcal{F}^\mu$ for arbitrary $0 \leq \lambda < \mu$. This in turn determines the distribution of the whole forest-growth process $(\mathcal{F}^\theta)_{\theta \geq 0}$, because it turns out to be Markovian. This Markov property is not obvious, but a consequence of the following theorem.

To formulate the theorem, consider more generally a forest growth process $(\mathcal{F}^\theta)_{\theta \geq 0}$. Then for each choice of λ and μ with $0 \leq \lambda \leq \mu$, there exists a joint graphical representation of $\mathcal{F}^\lambda$ and $\mathcal{F}^\mu$, as unions of line segments in the plane, such that $\mathcal{F}^\lambda$ is contained in $\mathcal{F}^\mu$. The set-theoretic difference $\mathcal{F}^\mu - \mathcal{F}^\lambda$ is then a disjoint union of subtrees of $\mathcal{F}^\mu$, each attached to some point on the sides of $\mathcal{F}^\lambda$. For each of these subtrees, if it is attached to point x on the side of $\mathcal{F}^\lambda$, and x is reached at time $t(x)$ in the depth-first search of $\mathcal{F}^\lambda$ at unit speed, let the subtree be detached from x and then rerooted on an empty forest floor $[0, \infty)$ at distance $t(x)$ from 0.

The result is a new forest $\mathcal{F}^{\lambda \to \mu}$, the forest of *innovations* or *increments* in the forest growth process between times λ and μ. Now $\mathcal{F}^\lambda$ can be understood as derived from the earlier forest $\mathcal{F}^\mu$ and the forest of innovations $\mathcal{F}^{\lambda \to \mu}$ by an operation of *composition* of $\mathcal{F}^\lambda$ and $\mathcal{F}^{\lambda \to \mu}$, whereby the forest $\mathcal{F}^{\lambda \to \mu}$ is wrapped around the sides of $\mathcal{F}^\lambda$ according to the depth-first search of $\mathcal{F}^\lambda$ at unit speed. Refer to Figure 7.7.

In terms of Harris paths, this means that the Harris path of $\mathcal{F}^\mu$ is obtained by inserting excursions defined by the Harris path of $\mathcal{F}^{\lambda \to \mu}$ into the Harris path of $\mathcal{F}^\lambda$ at appropriate times. Two precedents for construction of processes by this kind of insertion of excursions are the approximation of continuous time countable state space Markov chains with instantaneous states [157], and the construction of downwards skip free processes driven by a subordinator [3]. Say that a forest growth process $(\mathcal{F}^\theta)_{\theta \geq 0}$ has *independent growth increments* if for each choice of times $0 = \theta_0 < \theta_1 < \ldots < \theta_n$ the increments $\mathcal{F}^{\theta_{i-1} \to \theta_i}$ for $1 \leq i \leq n$ are independent. Obviously, such a process is Markovian.

Theorem 7.20. [364] *The Poisson-sampled Brownian forest $(\mathcal{F}^\theta)_{\theta \geq 0} := (\mathcal{F}^\theta_{N,B})_{\theta \geq 0}$ has independent growth increments, such that for each $0 \leq \lambda < \mu$ the*

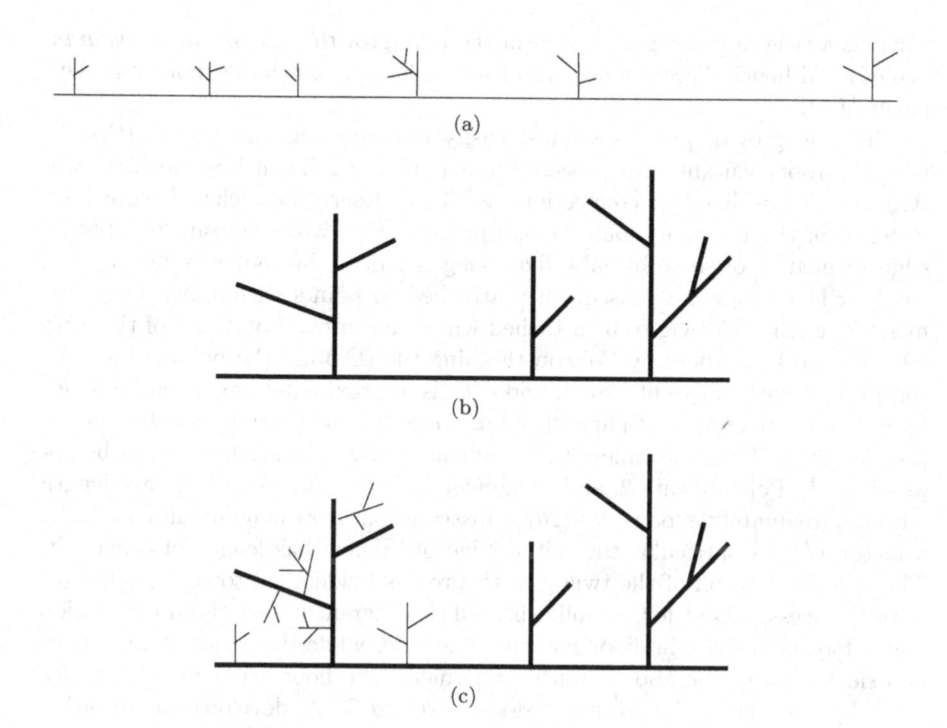

Figure 7.7: Forest growth by wrapping one forest around another.

forest of increments $\mathcal{F}^{\lambda \to \mu}$ is a binary(λ, μ) forest as in Definition 7.13. Moreover, $(\mathcal{F}^\theta)_{\theta \geq 0}$ is a twig-growth process, in the sense of the following definition.

Proof. This is deferred to the next Section 7.8. □

Definition 7.21. cf. [5, p. 2, Process 2] A *twig-growth process* is a forest-valued process $(\mathcal{F}^\theta)_{\theta \geq 0}$ such that, with the conventions discussed in Definition 7.16

- $(\mathcal{F}^\theta)_{\theta \geq 0}$ is a time inhomogeneous forest-valued Markov process with càdlàg paths, which starts as the empty forest at time $\theta = 0$, and develops as follows:
- at each time θ, along each side of each edge of $\mathcal{F}^\theta$, twigs are attached to that side according to a Poisson process with rate 1 per unit length of side per unit time;
- given that a twig is attached to a point x on some side of $\mathcal{F}^\theta$ at time θ, the length of that twig has exponential(2θ) distribution.

Note that the rule for generation of points of attachment of new twigs in the twig-growth process is identical to the rule for generation of new tips in the tip-growth process, according to Definition 7.16. But a tip that appears in the tip-growth process at time θ grows *continuously* to an ultimate length of θ^{-1}, whereas a twig appearing at time θ in the twig-growth process grows *instantaneously* and hence *discontinuously* to its final length, which is exponential with mean $\frac{1}{2}\theta^{-1}$.

The twig growth process is most easily motivated as the "local" (that is, near the root) variant of a process implicit in the Poisson line-breaking construction. To outline this connection, recall the description below Theorem 7.9 in terms of the inhomogenous Poisson process (Θ_i), with intensity tdt. Fix an edge segment I of some initially fixed length ε near the root in some $\mathcal{T}_n$, and consider how twigs are subsequently attached to points within that edge segment. The ith such twig to be attached will be an interval of $[0, \infty)$ of the form $[\hat{\Theta}_i, \hat{\Theta}_i + \eta(\hat{\Theta}_i))$, where by Poisson thinning the $(\hat{\Theta}_i)$ are the points of a Poisson process with constant rate ε, and $\eta(\theta)$ is approximately exponential(θ) for large θ. Now imagine traversing $[0, \infty)$ at speed $2/\varepsilon$, so that at time θ we are at position $2\theta/\varepsilon$, Then the times of attachments of twigs to segment I will be approximately Poisson, rate 2, and a segment attached at time θ will have length with approximately exponential($2\theta/\varepsilon$) distribution. Now magnify all lengths by a factor $1/\varepsilon$; in particular the initial edge of $\mathcal{T}_1$ now has length of order $1/\varepsilon$. The above definition of the twig growth process is identical to the $\varepsilon \to 0$ limit of this process, except for the following subtle difference: Part of our convention about forests is that the floor has only one side, while the edges of trees have two sides. But in the above limiting argument, the floor arises as a limit of a tree edge, so needs to be given two sides. Exercise 7.7.6 develops this variation of the twig growth process, as originally defined by Aldous [5, p. 2, Process 2].

Exercises

7.7.1. Explain why a twig-growth process has independent growth increments.

7.7.2. Show that the first sentence of Theorem 7.20 implies the second one.

7.7.3. (Problem) Theorem 7.20 implies that if $(\mathcal{F}^{\theta}_{\text{twig}})_{\theta \geq 0}$ is a twig-growth process, then

$$\mathcal{F}^{\theta}_{\text{twig}} \text{ is a binary}(0, \theta) \text{ forest of critical binary trees.} \tag{7.33}$$

It must be possible to check this by showing that this family of laws satisfies a system of Kolmogorov forwards equations generated by the jump-hold description of suitable restrictions of the twig-growth process. Formulate the appropriate equations, and show they are uniquely solved by this family of laws. That done, it should be possible to complete a proof of Theorem 7.20 by a repetition of the argument to show that the forest of increments between times λ and μ is a binary(λ, μ) forest.

7.7.4. (Recovery of B and N from a twig-growth process) This corollary of Theorem 7.20 sharpens the results of Aldous [5], who used the twig-growth process to construct a random metric space isometrically equivalent to the self-similar Brownian tree, without reference to Brownian motion. Given a twig-growth process $(\mathcal{F}^{\theta})_{\theta \geq 0}$, there exists on the same probability space a unique

Poisson process N and a unique Brownian motion B such that $(\mathcal{F}^\theta)_{\theta \geq 0} = (\mathcal{F}^\theta_{N,B})_{\theta \geq 0}$ almost surely. Moreover, both N and B are almost surely unique, and they are independent. In particular, $B(\cdot)$ is recovered from $(\mathcal{F}^\theta)_{\theta \geq 0}$ just as in (7.31), and for each fixed m and $\lambda > 0$ the time T^λ_m of the mth point of N in $(0, \infty) \times [0, \frac{1}{2}\lambda^2]$ is recovered as the almost sure limit

$$T^\lambda_m = \lim_{\theta \to \infty} S^{(\lambda,\theta)}_m / \theta \tag{7.34}$$

where $S^{(\lambda,\theta)}_m$ is the a.s. unique time s at which the Harris path of $\mathcal{F}^\theta$ has a local maximum at the height above the forest floor of the mth leaf of $\mathcal{F}^\lambda$.

7.7.5. (Almost sure identity of the two infinite forests) For each fixed $\lambda > 0$ and each fixed $\ell > 0$, there exists an almost surely finite random time $\Theta(\lambda, \ell)$ such that

$$\mathcal{F}^\lambda_B[0, \ell] \subset \mathcal{F}^\theta_{N,B} \text{ and } \mathcal{F}^\lambda_{N,B}[0, \ell] \subset \mathcal{F}^\theta_B \text{ for all } \theta \geq \Theta(\lambda, \ell) \tag{7.35}$$

where $\mathcal{F}[0, \ell]$, the restriction of $\mathcal{F}$ to $[0, \ell]$, is obtained by clear-cutting all trees of $\mathcal{F}$ whose roots fall in (ℓ, ∞).

7.7.6. (The self-similar CRT) Write $B^\pm$ for two copies $(B^+(t), t \geq 0)$ and $(B^-(t), t \geq 0)$ of standard Brownian motion, or equivalently the two sided motion $(B^\pm(t), -\infty, t < \infty)$. Interpret $\text{TREE}(B^\pm)$ as the two-sided forest, in which B^+ defines trees "upwards" from the floor $[0, \infty)$ while B^- defines trees "downwards" from the floor $[0, \infty)$. This is the limit tree suggested by the discussion of Definition 7.21. The scaled tree $\text{TREE}(2B^\pm)$, or equivalently the scaled tree $\text{TREE}(2R^\pm_3)$ for two-sided Bessel(3) $R^\pm_3$ (cf. Exercise 7.4.3) is what Aldous [12] called the *self-similar CRT.*

(a) Extend B^{ex} to all real t by interpreting $B^{\text{ex}}(t) = B^{\text{ex}}(t \bmod 1)$. Observe that as $\varepsilon \downarrow 0$

$$(\varepsilon^{-1/2} B^{\text{ex}}(\varepsilon t), -\infty < t < \infty) \xrightarrow{d} (R^\pm_3(t), -\infty < t < \infty).$$

In this sense, two-sided Bessel(3) is the "local limit" of Brownian excursion near 0.

(b) Combine the discussion of Definition 7.21 with Exercise 7.4.9 to argue informally the following. Let $\mathcal{U}_N$ be uniform on all N^{N-1} rooted labeled trees. Let $k(N) \to 0$, $k(N)/N^{-1/2} \to \infty$. Give length $k(N)$ to each edge of $\mathcal{U}_N$. Then $\mathcal{U}_N$ converges in distribution to the self-similar CRT [12].

7.7.7. (The forest generated by Brownian motion with drift.) For $\delta \in \mathbb{R}$ let $B^\delta(t) := B(t) + \delta t, t \geq 0$. This exercise is a reformulation in terms of trees of David Williams' path decompositions for B^δ, and their explanation in terms of the transformation from B^δ to $B^\delta - 2\underline{B}^\delta$. See [436, 270] for background.

- the distribution of $\text{TREE}(B^\delta)$ depends only on $|\delta|$.
- for $\delta > 0$ the process $\text{TREE}(B^\delta)$ is independent of $-\underline{B}^\delta(\infty)$, which has exponential$(2\delta)$ distribution.

- for $\ell > 0$ let $\mathcal{T}_\ell^\delta$ denote the subtree of $\text{TREE}(B^\delta)$ which is attached to its infinite branch at distance ℓ from 0, let μ_ℓ^δ be the mass of $\mathcal{T}_\ell^\delta$, meaning the length of the corresponding excursion interval of B^δ, and identify $\mathcal{T}_\ell^\delta$ with the isometric tree structure on $[0, \mu_\ell^\delta]$. Then

$$\{(\ell, \mu_\ell^\delta, \mathcal{T}_\ell^\delta) : \mu_\ell^\delta > 0\}$$

is the set of points of a Poisson process with intensity

$$d\ell e^{-\frac{1}{2}\delta^2 t} \frac{dt}{\sqrt{2\pi} t^{3/2}} \mathbb{P}(\text{TREE}(B^{\text{ex},t}) \in d\tau).$$

Notes and comments

Special thanks to David Aldous for help in writing this section, which bridges the gap between descriptions of the Brownian forest given by Neveu-Pitman [323, 324] and Aldous's construction of a self-similar CRT by the twig growth process.

7.8. Path decompositions

Definition 7.22. Given a path B and sampling times $T_1, T_2, \ldots$, the *alternating lower envelope of B derived from $T_1, T_2, \ldots$* is the process with locally bounded variation $A_t := A_t(B, T_1, T_2, \ldots)$ defined as follows:

$$A_t = \begin{cases} \underline{B}[T_m, t] & \text{if } t \in [T_m, S_{m+1}] \\ \underline{B}[t, T_{m+1}] & \text{if } t \in [S_{m+1}, T_{m+1}] \end{cases} \tag{7.36}$$

where $T_0 = 0$ and S_{m+1} is the time in $[T_m, T_{m+1}]$ at which B attains its minimum on that interval. Note that $S_{m+1} = T_m$ or $S_{m+1} = T_{m+1}$ may occur here, causing degeneracy in the sequel. We therefore assume throughout that B and T_m are such that this does not happen, as is the case almost surely in our applications to Poisson-sampled Brownian paths.

Call the reflected process $R_t := B_t - A_t \geq 0$ the *height of B above its alternating lower envelope derived from $T_1, T_2, \ldots$* Define a random sign process σ_t, with $\sigma_t = -1$ if t is in one of the intervals $[T_m, S_{m+1})$ when A is decreasing, and $\sigma_t = +1$ if t is in one of the intervals $[S_{m+1}, T_{m+1})$ when A is increasing, so the process

$$L_t := \int_0^t \sigma_s dA_s$$

is a continuous increasing process, call it the *increasing process derived from B and $T_1, T_2, \ldots$*.

Lemma 7.23. *For an arbitrary continuous path B with $\inf_t B_t = -\infty$, let $\mathcal{F}$ be the forest derived from B by sampling at some finite or infinite increasing sequence of times T_m, and let $\mathcal{F}''$ be the forest derived from B by sampling at some* *sequence of times T''_m, where $\{T''_m\} = \{T_m\} \cup \{T'_m\}$ for some finite or infinite increasing sequence of times T'_m. Let $A_t = A_t(B, T_1, T_2, \ldots)$ be the alternating lower envelope of B induced by the $\{T_m\}$, let $L_t = L_t(B, T_1, T_2, \ldots)$ be the increasing process derived from B and the $\{T_m\}$. Then the forest of innovations grown onto $\mathcal{F}$ to form $\mathcal{F}''$ is identical to the forest derived from $B - A - L$ by sampling at the times $\{T'_m\}$.*

Proof. This is left as an exercise. □

Theorem 7.20 now follows easily from the construction of the forest growth process $(\mathcal{F}^\theta)_{\theta \geq 0}$ by Poisson sampling of B, the previous Lemma, and the following lemma.

Lemma 7.24. *For $\delta \geq 0$ let $\mathbb{P}_{-\delta}$ govern $(B_t, t \geq 0)$ as a Brownian motion with drift $-\delta$, meaning that the $\mathbb{P}_{-\delta}$ distribution of $(B_t, t \geq 0)$ is the $\mathbb{P}_0$ distribution of $(B_t - \delta t, t \geq 0)$. Let $\mathbb{P}_{-\delta}$ also govern T^θ_m for $m = 1, 2, \ldots$ as the points of a Poisson process with rate $\frac{1}{2}\theta^2$ which is independent of B. For $m = 1, 2, \ldots$ let $F^\theta_m := B(T^\theta_{m-1}) - B(S^\theta_m)$ be the mth fall and $H^\theta_m := B(T^\theta_m) - B(S^\theta_m)$ the mth rise of the alternating walk defined by the values $B(T^\theta_m)$ and the intermediate minima $B(S^\theta_m)$. Then under $\mathbb{P}_{-\delta}$ for each $-\delta \leq 0$:*
(i) *the random variables F^θ_m and H^θ_m are independent, with*

$$F^\theta_m \stackrel{d}{=} \text{exponential}(\sqrt{\theta^2 + \delta^2} - \delta) \tag{7.37}$$

$$H^\theta_m \stackrel{d}{=} \text{exponential}(\sqrt{\theta^2 + \delta^2} + \delta). \tag{7.38}$$

(ii) *Let L^θ be the increasing process and A^θ the alternating lower envelope derived from B and the sample times $\{T^\theta_m\}$. Then the process $B - A^\theta - L^\theta$ is a Brownian motion with drift $-\sqrt{\theta^2 + \delta^2}$. Equivalently, $B - A^\theta$ is a Brownian motion with drift $-\sqrt{\theta^2 + \delta^2}$ on $(0, \infty)$ and simple reflection at 0.*
(iii) *The Brownian motion $B - A^\theta - L^\theta$ with drift $-\sqrt{\theta^2 + \delta^2}$ is independent of the bivariate sequence of falls and rises $(F^\theta_m, H^\theta_m)_{m=1,2,\ldots}$, hence also independent of the Poisson-sampled Brownian forest $\mathcal{F}^\theta_{N,B}$ which they encode.*

Proof. The independence assertions in (i), and the exponential form of the distributions of the falls and rises follow from Williams' path decomposition of B at the time S^θ_1 of its minimum on the interval $[0, T^\theta_1]$, and repeated application of the strong Markov property of B at the times T^θ_m. See [436, 178]. Using independence of F^θ_1 and H^θ_1, the parameters of the exponential distributions are easily derived from the the first two moments of $B(T^\theta_1)$.

Or see [73]. Also according to Williams [436], conditionally given $F^{(\theta)}_1 = f$ and $H^{(\theta)}_1 = h$, the fragments of the path of B on the intervals $[0, S^\theta_1]$ and $[S^\theta_1, T^\theta_1]$ are independent, the first fragment is distributed like a Brownian motion with drift $-\sqrt{\theta^2 + \delta^2}$, started at 0 and run until it first hits $-f$, while the second fragment reversed is like a Brownian motion with the same negative drift started at $h - f$ and run until it first hits $-f$. It follows that with the same conditioning, the two reflected path fragments $(B_t - A_t^\theta, 0 \le t \le S_1^\theta)$ and $(B_{S_1^\theta+u} - A^\theta_{S_1^\theta+u}, 0 \le u \le T_1^\theta - S_1^\theta)$ are independent, the first fragment being a reflected BM with drift $-\sqrt{\theta^2+\delta^2}$, run until its local time at 0 reaches f, and the second fragment reversed being a reflected BM with drift $-\sqrt{\theta^2+\delta^2}$, run until its local time at 0 reaches h. But from this description, and the well-known reversibility of a one-dimensional diffusion stopped at an inverse local time, still conditioning on $F_1^\theta = f$ and $H_1^\theta = h$, the process $(B_t - A_t^\theta, 0 \le t \le T_1^\theta)$ is identified as a reflected BM with drift $-\sqrt{\theta^2+\delta^2}$ run until its local time at 0 first reaches $f+h$. Now by repeated use of this argument, the entire process $(B_t - A_t^\theta, t \ge 0)$ conditional on all the rises and falls is a reflected BM with drift $-\sqrt{\theta^2+\delta^2}$ run forever, independent of the given values of the rises and falls, provided they sum to ∞ which they obviously do almost surely. Since $B - A^\theta - L^\theta$ is by construction the Brownian motion driving this reflected process, the conclusions (ii) and (iii) are evident. □

Exercises

7.8.1. (Proof of Lemma 7.23)

7.8.2. For $0 \le \lambda < \mu$ let $\mathcal{F}^{\lambda,\mu}$ be the forest derived from a BM with drift $-\lambda$ by Poisson $(\frac{1}{2}\theta^2)$ sampling with $\theta^2 = \mu^2 - \lambda^2$. Then $\mathcal{F}^{\lambda,\mu}$ is a binary(λ,μ) forest.

7.9. Further developments

Aldous [5, 6, 7] developed a general theory of continuum random trees, special cases of which are the trees TREE(B) and TREE(B^{ex}) discussed here. Aldous, Camarri and Pitman [21, 20, 82] studied a large family of such trees, called *inhomogeneous continuum random trees* which arise as weak limits from a family of combinatorial trees naturally associated with Cayley's multinomial expansion, as discussed in Chapter 10.

Another interesting family of continuum random trees arises from the work of Le Gall, Le Jan and Duquesne [277], [119] on continuous state branching processes. These authors have obtained results on the convergence of trees associated with a sequence of Galton-Watson processes with offspring distribution μ_n under the assumption that these processes, when suitably rescaled, converge in distribution towards a continuous-state branching process with branching mechanism ψ. The corresponding Harris processes also converge in distribution towards a limiting process called the ψ-height process. Informally, this means that whenever a sequence of rescaled Galton-Watson processes converges in distribution, their genealogies also converge to a continuous branching structure coded by the height process. The ψ-height process is constructed in [277, 119] as a local time functional of the Lévy process with Laplace exponent ψ. It can be used to investigate various asymptotic properties of Galton-Watson trees, such as the structure of reduced trees corresponding to ancestors of individuals alive at a fixed (large) time [119]. In the quadratic case $\psi(u) = u^2$ (corresponding to the case where $\mu_n = \mu$ is critical with finite variance), the height process is reflected Brownian motion.

Specializing to the case where $\mu_n = \mu$ is in the domain of attraction of a stable law, Duquesne [118] has shown that the contour process of the μ-Galton-Watson tree conditioned to have exactly n vertices converges after rescaling towards the normalized excursion of the (stable) height process, which should be interpreted as coding a stable CRT. Finite-dimensional distributions of the stable CRT are computed in [119]. The lecture notes [275] provide a survey of these results and some applications to particle systems and models of statistical mechanics. Another reference on spatial branching processes, random snakes and partial differential equations is [274]. This work features Le Gall's Brownian snake construction of Markovian superprocesses, which is a natural development of the branching structure of Brownian trees which has found many applications in the derivation of sample properties of super-Brownian motion and its connections with partial differential equations. Analogously to the Brownian snake approach of quadratic superprocesses, the ψ-height process can be used to construct superprocesses with a general branching mechanism and to study their properties [119, 276]. For some other recent developments, see [94, 295, 217].

For different approaches to the forest growth procedure by sampling at Poisson times and related results see [364]. This approach is extended by Duquesne and Winkel [121] to multiple branching and genealogies of general continuous-state branching processes, including the supercritical case. An appropriate technical setup for tree-valued random processes was provided by Evans, Pitman and Winter [142] who represent trees as (equivalence classes of) certain metric spaces that form a Polish space when equipped with the Gromov-Hausdorff topology. These ideas are applied and further developed in [120, 121, 142, 143].

8

Brownian local times, branching and Bessel processes

It is well known that the random occupation measure induced by the sample path of a Brownian motion $B = (B_t, t \geq 0)$ admits a jointly continuous local time process $(L_t^x(B); x \in \mathbb{R}, t \geq 0)$ such that

$$\int_0^t f(B_s)ds = \int_{-\infty}^{\infty} L_t^x(B) f(x)dx.$$

See [304, 255, 384]. This chapter reviews the basic descriptions of Brownian local times in terms of Bessel processes, due to Ray [381] and Knight [254], and interprets these results in terms of random trees and forests embedded in Brownian paths. Then various developments of these descriptions are discussed.

- **8.1. Stopping at first hitting times** Here the Ray-Knight description of the process $(L_T^x(B); x \in \mathbb{R})$ is recalled for T the first hitting time of a level by either B or by $(L_t^0(B), t \geq 0)$.
- **8.2. Squares of Bessel processes** These are involved in the Ray-Knight theorems and their variants.
- **8.3. Stopping at fixed times** By suitable conditioning, the Ray-Knight theory provides access to the law of $(L_t^x(B), x \in \mathbb{R})$ for fixed times t.
- **8.4. Time-changed local time processes** Making a change of variable in the space-parameter of these local time processes for fixed time t, so that the quadratic variation in x becomes a new time parameter, leads to Jeulin's identity relating Brownian excursion and its local time process.
- **8.5. Branching process approximations** This section offers some further descriptions of local time processes in terms of branching process approximations to squared Bessel processes and their bridges.

8.1. Stopping at an inverse local time

Throughout this section let R denote a reflecting Brownian motion on $[0, \infty)$, which according to Lévy's theorem [384] may be constructed from a standard

Brownian motion either as $R = |B|$, or as $R = B - \underline{B}$, where $\underline{B}_t := \inf_{0 \le s \le t} B_s$. Note that if $R = |B|$ then for $v \ge 0$

$$L_t^v(R) = L_t^v(B) + L_t^{-v}(B) \tag{8.1}$$

and in particular $L_t^0(R) = 2L_t^0(B)$. For $\ell \ge 0$ let

$$\tau_\ell := \inf\{t : L_t^0(R) > \ell\} = \inf\{t : L_t^0(B) > \ell/2\}. \tag{8.2}$$

For $0 \le v < w$ let

$$D(v, w, t) := \text{number of downcrossings of } [v, w] \text{ by } R \text{ before } t$$

From the structure of the Brownian forest described in Section 7.6 there is the following basic description of the process counting downcrossings of intervals up to an inverse local time. See also [426] for more about Brownian downcrossings and their relation to the Ray-Knight theorems.

Corollary 8.1. [323] *The process*

$$(D(v, v + \varepsilon, \tau_\ell), v \ge 0)$$

is a time-homogeneous Markovian birth and death process on $\{0, 1, 2 \ldots\}$, *with state* 0 *absorbing, transition rates*

$$n - 1 \xleftarrow{\frac{n}{\varepsilon}} n \xrightarrow{\frac{n}{\varepsilon}} n + 1$$

for $n = 1, 2, \ldots$, *and initial state* $D(0, \varepsilon, \tau_\ell)$ *which has Poisson*$(\ell/(2\varepsilon))$ *distribution.*

Proof. According to Theorem 7.15 , applied to $R := B - \underline{B}$ run until B first reaches $-\ell/2$, the number $D(v, v+\varepsilon, \tau_\ell)$ is the number of branches at level v in a critical binary $(0, \varepsilon)$ branching process started with a Poisson$(\ell/(2\varepsilon))$ number of initial individuals, so the conclusion is immediate. □

From the Poisson distribution of $D(0, \varepsilon, \tau_\ell)$, and the law of large numbers,

$$\lim_{\varepsilon \downarrow 0} \varepsilon D(0, \varepsilon, \tau_\ell) = \ell \qquad \text{almost surely}$$

and similarly, for each $v > 0$ and $\ell > 0$, by consideration of excursions of R away from level v

$$\lim_{\varepsilon \downarrow 0} 2\varepsilon D(v, v + \varepsilon, \tau_\ell) = L_{\tau_\ell}^v(R) \qquad \text{almost surely.}$$

According to Corollary 8.1, this process $(2\varepsilon D(v, v + \varepsilon, \tau_\ell), v \ge 0)$, which serves as an approximation to $(L_{\tau_\ell}^v(R), v \ge 0)$, is a Markov chain whose state space is the set of integer multiples of 2ε, with transition rates

$$x - 2\varepsilon \xleftarrow{\frac{x}{2\varepsilon^2}} x \xrightarrow{\frac{x}{2\varepsilon^2}} x + 2\varepsilon$$

for $x = 2\varepsilon n > 0$. The generator G_ε of this Markov chain acts on smooth functions f on $(0,\infty)$ according to

$$(G_\varepsilon f)(x) = \frac{x}{2\varepsilon^2} f(x-2\varepsilon) + \frac{x}{2\varepsilon^2} f(x+2\varepsilon) - \frac{x}{\varepsilon^2} f(x)$$

$$= 4x \frac{1}{(2\varepsilon)^2} \left[\tfrac{1}{2} f(x-2\varepsilon) + \tfrac{1}{2} f(x+2\varepsilon) - f(x)\right]$$

$$\to 4x \frac{1}{2} \frac{d^2}{dx^2} f \text{ as } \varepsilon \to 0.$$

Hence, appealing to almost any result on approximation of diffusions by Markov chains [265, 264], we obtain:

Theorem 8.2. (Ray [381], Knight [254]) *For each fixed $\ell > 0$, and $\tau_\ell := \inf\{t : L_t^0(R) > \ell\}$, where $R = |B|$,*

$$(L_{\tau_\ell}^v(R), v \geq 0) \stackrel{d}{=} (Q_{\ell,v}^{(0)}, v \geq 0) \tag{8.3}$$

where the process on the right hand side is the Feller diffusion on $[0,\infty)$ with absorbtion at 0, and generator

$$4x \frac{1}{2} \frac{d^2}{dx^2} \tag{8.4}$$

acting on smooth functions vanishing in a neighbourhood of 0.

Exercises

8.1.1. (Ray-Knight) Let $T_\ell := \tau_{2\ell} := \inf\{t > 0 : L_t^0(B) = \ell\}$. Then the processes $(L_{T_\ell}^v(B), v \geq 0)$ and $(L_{T_\ell}^{-v}(B), v \geq 0)$ are two independent copies of $(Q_{\ell,v}^{(0)}, v \geq 0)$.

Notes and comments

Many proofs and extensions of the basic Ray-Knight theorems can be found in the literature. See for instance [233, 384, 348, 426, 219] and papers cited there.

8.2. Squares of Bessel processes

The Feller diffusion appearing on the right side of (8.3) is the particular case $\delta = 0$ of a *squared Bessel with parameter δ* started at $\ell \geq 0$, denoted $(Q_{\ell,v}^{(\delta)}, v \geq 0)$. This Markov process with state space $[0,\infty)$ can be defined [384, Ch. XI] for each $\delta \geq 0$ as the unique non-negative strong solution of the Itô stochastic differential equation (SDE)

$$Q_0 = \ell; \qquad dQ_v = \delta\, dv + 2\sqrt{Q_v}\, d\beta_v \tag{8.5}$$

where $(\beta_v, v \geq 0)$ is a BM. For $\delta < 0$, such a process can be uniquely defined up to its hitting time of 0 as the solution of this SDE, with either killing or absorbtion at 0. Denote the law of $(Q^{(\delta)}_{\ell,v}, v \geq 0)$ by $BESQ^{(\delta)}_\ell$, and denote by $BESQ^{(\delta)}$ the Markov process with state space $[0,\infty)$ determined by this collection of laws $(BESQ^{(\delta)}_\ell, \ell \geq 0)$. For an account of the basic properties of these processes see [384]. The parameter δ, which simply represents a constant drift coefficient, is often called the *dimension* of the squared Bessel process, due to the following well known consequence of Itô's formula: that if $(B_{i,v}, v \geq 0)$ for $i = 1, 2, \dots$ is a sequence of independent standard Brownian motions, then

$$(Q^{(\delta)}_{\ell,v}, v \geq 0) \stackrel{d}{=} \left((\sqrt{\ell} + B_{1,v})^2 + \sum_{i=2}^{\delta} B^2_{i,v}, v \geq 0 \right) \qquad (\delta = 1, 2, \dots). \tag{8.6}$$

The squared Bessel processes and their bridges, especially for $\delta = 0, 2, 4$, are involved in the description of the local time processes of numerous Brownian path fragments [254, 381, 436, 365]. For instance, if $H_1(X) := \inf\{t : X_t = 1\}$, then according to Ray and Knight

$$(L^v_{H_1(B)}(B), 0 \leq v \leq 1) \stackrel{d}{=} (Q^{(2)}_{0,1-v}, 0 \leq v \leq 1). \tag{8.7}$$

The appearance of $BESQ^{(\delta)}$ processes embedded in the local times of Brownian motion is best understood in terms of the construction of these processes as weak limits of Galton-Watson branching processes with immigration, and their consequent interpretation as continuous state branching processes with immigration [233].

For instance, there is the following expression of the Lévy-Itô representation of squared Bessel processes, and its interpretation in terms of Brownian excursions [365], which are recalled in Exercise 8.2.4 .

Theorem 8.3. Le Gall-Yor [278] *For R a reflecting Brownian motion on $[0,\infty)$, with $R_0 = 0$, let*

$$Y^{(\delta)}_t := R_t + L^0_t(R)/\delta \qquad (t \geq 0).$$

Then for $\delta > 0$ the process of ultimate local times of $Y^{(\delta)}$ is a copy of $BESQ^{(\delta)}_0$:

$$(L^v_\infty(Y^{(\delta)}), v \geq 0) \stackrel{d}{=} (Q^{(\delta)}_{0,v}, v \geq 0). \tag{8.8}$$

This has an immediate interpretation in terms of Brownian trees [2]. If R is constructed as $R = B - \underline{B}$, so $L^0_t(R) = -2\underline{B}(t)$, then

$$\text{TREE}(B) = \text{TREE}(B - 2\underline{B}) = \text{TREE}(Y^{(2)})$$

where $Y^{(2)}$ is a copy of the three-dimensional Bessel process R_3. Thus $L^v_\infty(Y^{(2)})$ is the mass density of vertices in either $\text{TREE}(B)$ or $\text{TREE}(Y^{(2)})$ at distance v from the root 0, where mass refers to Lebesgue measure on $\mathbb{R}_{\geq 0}$. That this process should be regarded as a continuous state branching process with immigration is intuitively obvious from the Poisson structure of the forest of subtrees of TREE(B) attached to its infinite branch Proposition 7.8 . Similarly $L^v_\infty(Y^{(\delta)})$ is the mass density of vertices in TREE$(Y^{(\delta)})$ at distance v from the root 0. By construction of $Y^{(\delta)}$, its tree is derived from TREE$(Y^{(2)})$ by simply stretching distances along the infinite branch of TREE$(Y^{(2)})$ by a factor of $2/\delta$, and leaving the same Poisson forest of subtrees attached to this branch. Thus δ is simply a parameter governing the rate of appearance of Brownian subtrees of various sizes along the infinite branch of $Y^{(\delta)}$, that is to say a rate of immigration, or rate of generation of new Brownian subtrees, in a continous state branching process.

Exercises

8.2.1. (Knight) Use the well known construction of a copy of R from B, by deletion of all negative excursions of B, to show that the process $(L^v_{H_1(R)}(R), 0 \leq v \leq 1)$ has the same distribution as the processes displayed in (8.7). See also Knight [255], McKean [304].

8.2.2. (Williams [436]) Deduce (8.7) from (8.8) for $\delta = 2$, and vice-versa, using Williams' time reversal identity [384] relating the path of B on $[0, H_1(B)]$ to that of R_3 on $[0, K_1]$, where K_1 is the last time that R_3 hits 1.

8.2.3. [384] Deduce from the SDE definition of $BESQ^{(\delta)}_\ell$ the *additivity property* [401] that for all non-negative δ, δ', ℓ and ℓ' the sum of a $BESQ^{(\delta)}_\ell$ process and an independent $BESQ^{(\delta')}_{\ell'}$ process is a $BESQ^{(\delta+\delta')}_{\ell+\ell'}$ process. For positive integer δ, δ' this is obvious from (8.6). Show that the law of $BESQ^{(\delta)}_\ell$ for all $\delta \geq 0$ and $\ell \geq 0$ is uniquely determined by the additivity property and the prescription (8.6) for $\delta = 1$.

8.2.4. [365] As a consequence of the additivity property, the family of $BESQ^{(\delta)}_\ell$ processes can be constructed on a common probability space in such a way that the path-valued process $(Q^{(\delta)}_{\ell,v}, v \geq 0)_{\delta \geq 0, \ell \geq 0}$ is increasing with stationary independent increments. The $C[0,\infty)$-valued processes

$$(Q^{(\delta)}_{0,v}, v \geq 0)_{\delta \geq 0} \text{ and } (Q^{(0)}_{\ell,v}, v \geq 0)_{\ell \geq 0} \tag{8.9}$$

are then two independent increasing processes with stationary independent increments, the jumps of which are two independent Poisson processes of points in $C[0,\infty)$ governed by two σ-finite Lévy measures M and N, which may be described as follows: M is the distribution of the ultimate local time process $(L_{\infty,v}(X), v \geq 0)$ for the co-ordinate process X on $C[0,\infty)$ subject to Itô's σ-finite law ν of Brownian excursions, and $N = \int_0^\infty M_u du$ where M_u is the measure on $C[0,\infty)$ under which $X[0,u] \equiv 0$ and $X[u,\infty)$ has distribution M. Under M the co-ordinate process X is Markovian with the $BESQ^{(0)}$ semigroup, with almost every path starting at 0.

8.2.5. [278] **(Proof of Theorem 8.3)** Deduce (8.8) for general $\delta > 0$ from its special case $\delta = 2$, using the additivity property of $BESQ^{(\delta)}_0$.

Notes and comments

See Le Gall [274] for developments of these ideas in the context of Markovian super-processes. See [365, 348, 384] for descriptions of the infinitely divisible distribution of $\int_0^\infty Q^{(\delta)}_{\ell,v}\mu(dv)$, for suitable measures μ, in terms of the solutions of Sturm-Liouville equations associated with μ, and treatment of the same problem for squared Bessel bridges. Carmona-Petit-Yor [86] found a construction of $BESQ^{(\delta)}_x$ for $\delta < 0 < x$ which is similar to (8.8) See also [340, 343].

8.3. Stopping at fixed times

Features of the distribution of the Brownian local time process $(L^x_t(B), x \in \mathbb{R})$, for a fixed time t, are of interest in a number of contexts. In principle, the family of distributions of these local time processes is determined by the Ray-Knight description of the joint law of $(L^x_T(B), x \in \mathbb{R})$ and B_T for T an exponential time independent of B, as considered in [63]. In practice, it is hard work to invert the implicit Laplace transform to gain anything explicit from this description, though this can certainly be done, as shown by Leuridan [283]. A simpler way to access the structure of the process $(L^x_t(B), x \in \mathbb{R})$ for fixed t is to exploit the branching process approximation to do computations in a combinatorial setting without technical complication, then appeal to some general approximation technique to justify passage to the Brownian limit.

The effectiveness of this approach is well illustrated by the following problem, which arises naturally from the asymptotics of random forests and random mappings. Let $B^{\rm br}_\ell$ denote a standard Brownian bridge $B^{\rm br}$ conditioned on $L^0_1(B^{\rm br}) = \ell$. The problem is to describe the law of the process $(L^v_1(|B^{\rm br}_{\ell/2}|), v \geq 0)$ as explicitly as possible. Note that the conditioning on $L^0_1(B^{\rm br}) = \ell/2$ makes the process $(L^v_1(|B^{\rm br}_{\ell/2}|), v \geq 0)$ start at $L^0_1(|B^{\rm br}_{\ell/2}|) = \ell$. The structure of excursions of $|B^{\rm br}_{\ell/2}|$ away from 0 can be described as follows: the excursion intervals of $|B^{\rm br}_{\ell/2}|$ define an exchangeable interval partition of $[0,1]$, whose law is defined by conditioning the $(\frac{1}{2}, \frac{1}{2})$ partition generated by $B^{\rm br}$ to have $\frac{1}{2}$-diversity equal to $\ell/\sqrt{2}$. By the switching identity (4.71) and the Ray-Knight description (8.3), for each $\ell > 0$

$$(L^v_1(|B^{\rm br}_{\ell/2}|), v \geq 0) \stackrel{d}{=} \left(Q^{(0)}_{\ell,v}, v \geq 0 \,\middle|\, \int_0^\infty Q^{(0)}_{\ell,v} dv = 1 \right) \tag{8.10}$$

Note that for $\ell = 0$ the left side of (8.10) has direct meaning as the local time process of $B^{\rm ex}$, but the right hand side does not because 0 is an absorbing state for $BESQ^{(0)}$. We will return to this point after stating the following theorem.

Theorem 8.4. [358] *For each $t > 0$ and $\ell \geq 0$, a process $Q = (Q_{\ell,t,v}, v \geq 0)$ with continuous paths such that*

$$(L^v_1(|B^{\rm br}_{\ell/2}|), v \geq 0) \stackrel{d}{=} (Q_{\ell,1,v}, v \geq 0), \tag{8.11}$$

can be defined as the unique strong solution of the Itô SDE

$$Q_0 = \ell; \quad dQ_v = \delta_v(Q)dv + 2\sqrt{Q_v}d\beta_v \tag{8.12}$$

where β is a Brownian motion and

$$\delta_v(Q) := 4 - Q_v^2 \left(t - \textstyle\int_0^v Q_u du\right)^{-1} \tag{8.13}$$

with the convention that the equation for Q is to be solved only on $[0, V_t(Q))$ and that $Q_v = 0$ for $v \geq V_t(Q)$ where

$$V_t(Q) := \inf\{v : \textstyle\int_0^v Q_u du = t\}. \tag{8.14}$$

Note that for each $\ell \geq 0$ and $t > 0$ the random time $V_t(Q)$ is strictly positive and finite a.s., and the left limit of Q at time $V_t(Q)$ exists and equals 0 a.s.. A proof of this theorem will be sketched in Section 8.5 , based on a branching process approximation.

It is easily seen by Brownian scaling how to interpret the process $(Q_{\ell,t,v}, v \geq 0)$ for all $t > 0$ and $\ell \geq 0$: if $B^{\mathrm{br},t}$ denotes a Brownian bridge from 0 to 0 of length t, then

$$(L_t^v(|B^{\mathrm{br},t}|), v \geq 0 \,|\, L_t^0(|B^{\mathrm{br},t}|) = \ell) \stackrel{d}{=} (Q_{\ell,t,v}, v \geq 0). \tag{8.15}$$

Provided $\ell > 0$ this process may also be interpreted as a $BESQ_\ell^{(0)}$ process $(Q_{\ell,v}^{(0)}, v \geq 0)$ conditioned on $\int_0^\infty Q_{\ell,v}^{(0)} dv = t$. Call (8.15) the *local time interpretation of Q*. The SDE (8.12) defining Q in (8.15) is a generalization of the $BESQ^{(\delta)}$ SDE (8.5) in which the drift δ_v at level v is a path dependent function of the unknown process Q, namely $\delta_v := 4 - Q_v^2(t - \int_0^v Q_u du)^{-1}$, which may be of either sign. It is well known that for fixed $\delta < 0$ and any initial value $\ell > 0$ the $BESQ^{(\delta)}$ SDE has a pathwise unique solution on a stochastic interval $[0, V_0)$ such that the solution approaches 0 at time V_0. The $BESQ^{(\delta)}$ process for $\delta \leq 0$ is known [365] to arise as a Doob h-process by conditioning a $BESQ^{(4-\delta)}$ process to hit 0. To get some feeling for the effect of the path dependent drift δ_v, keep in mind that the local times process $(L_t^v(|B^{\mathrm{br},t}|), v \geq 0)$ is subject to the constraint

$$\int_0^\infty L_t^u(|B^{\mathrm{br},t}|)du = t$$

so the random variable

$$t - \int_0^v L_t^u(|B^{\mathrm{br},t}|)du = \int_v^\infty L_t^u(|B^{\mathrm{br},t}|)du,$$

representing the time spent by $|B^{\mathrm{br},t}|$ above v, is determined as a function of the local times of $|B^{\mathrm{br},t}|$ at levels below v. According to the theorem, given the local times at all levels below v, the local time process evolves over the next infinitesimal level increment like a squared Bessel process of dimension $\delta_v := 4 - \beta_v$, where β_v is the square of the local time at level v relative to the time above v. Note that if the local time at level v is close to zero, but the time above v is not, then β is close to zero, so the local time process is forced away from 0 in much the same way as a $BESQ_0^{(4)}$ escapes from 0. For $\ell > 0$ it is known [384] that the $BESQ_\ell^{(\delta)}$ process hits zero in finite time only if $\delta < 2$. So the expression $\delta_v = 4 - \beta_v$ implies that the local time process cannot reach zero before a level v such that $\beta_v > 2$, meaning the square of the local time at v exceeds twice the time above v. Much sharper results could be given in the same vein.

As remarked below (8.10), this formula has no direct meaning for $\ell = 0$, and $t > 0$, even though the process $(Q_{0,t,v}, v \geq 0)$ is a well defined process identical in law to the process of local times of $B^{\mathrm{ex},t}$, a Brownian excursion of length t. However, formula (8.10) amounts to the following identity of probability measures on $C[0,\infty)$:

$$BESQ_\ell^{(0)} = \int_0^\infty BESQ_{\ell,t}^{(0)} \, \mathbb{P}(\tau_\ell \in dt) \quad (\ell > 0) \tag{8.16}$$

where $BESQ_\ell^{(0)}$ denotes the law of $C[0,\infty)$ of a $BESQ_\ell^{(0)}$ process $(Q_{\ell,v}^{(0)}, v \geq 0)$,

$$\tau_\ell := \int_0^\infty Q_{\ell,v}^{(0)} dv \stackrel{d}{=} (\ell/2)^2 / B_1^2$$

and $BESQ_{\ell,t}^{(0)}$ is the law of $(Q_{\ell,v}^{(0)}, v \geq 0)$ given $\tau_\ell = t$. Here we exploit the Ray-Knight theorem (8.3) to suppose for convenience that the entire family of $BESQ_\ell^{(0)}$ processes $(Q_{\ell,v}^{(0)}, v \geq 0)$ is defined on the same probability space $(\Omega, \mathcal{F}, \mathbb{P})$ as a basic Brownian motion B, according to the formula $Q_{\ell,v}^{(0)} := L_{\tau_\ell}^v(R)$ where $R = |B|$ and $(\tau_\ell, \ell \geq 0)$ is the inverse local time process of R at 0, as in (8.2). If this form (8.16) of formula (8.10) is divided by ℓ, and the limit taken as $\ell \downarrow 0$, the result is the following formula for the Lévy measure M of the increasing path valued process with stationary independent increments $(Q_{\ell,t,v}, v \geq 0)_{\ell \geq 0}$, as further interpreted in Exercise 8.2.4 :

$$M = \int_0^\infty BESQ_{0,t}^{(0)} \frac{t^{-3/2} dt}{\sqrt{2\pi}} \tag{8.17}$$

where $BESQ_{0,t}^{(0)}$ is the law on $C[0,\infty)$ of

$$(Q_{0,t,v}, v \geq 0) \stackrel{d}{=} (L_t^v(B^{\mathrm{ex},t}), v \geq 0).$$

Thus M is in many respects like a law of excursions of $BESQ^{(0)}$ away from 0. But it is not possible to concatenate such excursions to form a process, because if μ denotes the M distribution of H_0, the return time to 0, then μ fails to satisfy the necessary condition $\int_0^\infty (1 - e^{-x}) \mu(dx) < \infty$ for μ to be the Lévy measure of an inverse local time process.

Notes and comments

This section is based on [358].

8.4. Time-changed local time processes

Keep in mind that the "time" parameter of the local time process $(Q_{\ell,t,v}, v \geq 0)$ is actually a level v, so perhaps we should speak instead of "space-changed" local time processes. Consider some reformulations of Theorem 8.4 which can be made by stochastic calculus. For a non-negative process $Y := (Y_s, 0 \leq s \leq 1)$ admitting a continuous occupation density process $(L_1^v(Y), v \geq 0)$, define a process $\hat{L}(Y) := (\hat{L}^u(Y), 0 \leq u \leq 1)$ by $\hat{L}^u(Y) := L_1^{v(u)}(Y)$ where $v(u) := \sup\{y \geq 0 : \int_y^\infty L_1^x dx > u\}$. So $\hat{L}^u(Y)$ is the local time of Y at a level $v(u)$ above which Y spends time u. This definition is suggested by the remarkable results of Jeulin [220, p. 264] that

$$\hat{L}(B^{\mathrm{ex}}) \stackrel{d}{=} 2B^{\mathrm{ex}} \tag{8.18}$$

and Biane-Yor [61, Th. (5.3)] that

$$\hat{L}(|B^{\mathrm{br}}|) \stackrel{d}{=} 2B^{\mathrm{me}} \tag{8.19}$$

where B^{ex} and B^{me} are standard Brownian bridge and meander, respectively. If we recall that B^{me} conditioned on $B^{\mathrm{me}}(1) = r$ is a BES(3) bridge from 0 to r, say $R_3^{0\to r}$, then (8.19) implies

$$\hat{L}(|B^{\mathrm{br}}_{\ell/2}|) \stackrel{d}{=} 2R_3^{0\to\ell/2}. \tag{8.20}$$

Conversely, (8.19) is recovered from (8.20) and the fact that $L_1^0(B^{\mathrm{br}}) \stackrel{d}{=} B^{\mathrm{me}}(1)$. Moreover, (8.18) is the special case $\ell = 0$ of (8.20). Now either of the descriptions (8.20) and (8.11) can be derived from the other, using the fact that $R_3^{x\to y}$ is the time reversal of $R_3^{y\to x}$, the description of these bridges as solutions of an SDE, and the computation of Lemma 8.5 below, which transforms the SDE solved by $(L_1^v(|B^{\mathrm{br}}_{\ell/2}|), v \geq 0)$ into that solved by $(\hat{L}^u(|B^{\mathrm{br}}_{\ell/2}|), 0 \leq u \leq 1)$.

The transformation between the two SDEs is done by the following Lemma. This corrects [358, Lemma 14], where the process R should obviously be started at $\ell/2$ instead of ℓ.

Lemma 8.5. *For $\ell \geq 0, t > 0$ let R^ℓ be the process derived from $(Q_{\ell,1,v}, v \geq 0)$ via the formula*

$$2R_s^\ell := Q_{\ell,t,v} \text{ for the least } v : \int_0^v Q_{\ell,1,u} du = s, \text{ where } 0 \leq s \leq 1 \tag{8.21}$$

Then $R^\ell \stackrel{d}{=} R_3^{\ell/2\to 0}$ and $(Q_{\ell,1,v}, v \geq 0)$ can be recovered from R^ℓ via the formula

$$Q_{\ell,1,v} = 2R_s^\ell \text{ for the least } s : \int_0^s \frac{dr}{2R_r^\ell} = v. \tag{8.22}$$

Consequently, if $R^\ell \stackrel{d}{=} R_3^{\ell/2\to 0}$, then Q defined by (8.22) has the same distribution as Q defined by the SDE (8.12).

Proof. Starting from three independent standard Brownian bridges $B_i^{\mathrm{br}}, i = 1, 2, 3$, for $x, y \geq 0$ let

$$R_3^{x \to y}(u) := \sqrt{(x + (y - x)u + B_{1,u}^{\mathrm{br}})^2 + (B_{2,u}^{\mathrm{br}})^2 + (B_{3,u}^{\mathrm{br}})^2} \qquad (0 \leq u \leq 1). \tag{8.23}$$

Itô's formula implies that the process $R_3^{x \to y}$ can be characterized for each $x, y \geq 0$ as the solution over $[0, 1]$ of the Itô SDE

$$R_0 = x; \quad dR_s = \left(\frac{1}{R_s} + \frac{(y - R_s)}{(1 - s)} \right) ds + d\gamma_s \tag{8.24}$$

for a Brownian motion γ. The recipe (8.22) for inverting the time change (8.21) is easily checked, so it suffices to show that if $R := (R_s^\ell, 0 \leq s \leq 1)$ solves the SDE (8.24), for $(x, y) = (\ell/2, 0)$ and some Brownian motion γ, then $Q := (Q_{\ell,1,v}, v \geq 0)$ defined by (8.22) solves the SDE (8.12) for some Brownian motion β. But from (8.21) and (8.22)

$$dQ_v = 2\, dR_s \text{ where } s = \int_0^v Q_u du$$

A level increment dv for Q corresponds to a time increment $ds = Q_v dv$ for R, and $R_s = Q_v/2$, so

$$dQ_v = 2 \left(\frac{1}{Q_v/2} - \frac{Q_v/2}{(1 - \int_0^v Q_u du)} \right) Q_v dv + 2\sqrt{Q_v} d\beta_v \tag{8.25}$$

for some other Brownian motion β, where the factor $\sqrt{Q_v}$ appears in the diffusion term due to Brownian scaling, and the equation (8.25) simplifies to (8.12). As a technical point, the definition of β above the level $\int_0^t dr/2R_r^{\ell,0,t}$ when Q hits 0 may require enlargement of the probability space. See [384, Ch. V] for a rigorous discussion of such issues. □

Exercises

8.4.1. Show that

$$\sup_{v \geq 0} L_1^v(|B^{\mathrm{br}}|) \stackrel{d}{=} 2 \sup_{0 \leq u \leq 1} B_u^{\mathrm{me}} \stackrel{d}{=} 4 \sup_{0 \leq u \leq 1} |B_u^{\mathrm{br}}| \tag{8.26}$$

where B^{me} is a Brownian meander of length 1, the second equality is due to Kennedy [238], and the distribution of $\sup_{0 \leq u \leq 1} |B_u^{\mathrm{br}}|$ is given by the well known Kolmogorov-Smirnov formula. Also

$$(L_1^0(|B^{\mathrm{br}}|), \sup_{v \geq 0} L_1^v(|B^{\mathrm{br}}|)) \stackrel{d}{=} 2(B_1^{\mathrm{me}}, \sup_{0 \leq u \leq 1} B_u^{\mathrm{me}}) \tag{8.27}$$

the joint density of which can be read from known results for the Brownian meander [209].

8.4.2. [358, Corollary 18] Conditionally given

$$L_1^0(B^{\text{br}}) = \ell \text{ and } \int_0^1 1(B_u^{\text{br}} > 0)du = a,$$

the processes $(L_1^v(B^{\text{br}}), v \geq 0)$ and $(L_1^{-v}, v \geq 0)$ are independent copies of $(Q_{\ell,a,v}, v \geq 0)$ and $(Q_{\ell,1-a,v}, v \geq 0)$ respectively.

Notes and comments

This section is based on [358]. Perkins [338] showed that for each fixed $t > 0$ the process of local times of B at levels v up to time t is a semi-martingale as v ranges over all real values, and he gave the semi-martingale decomposition of this process. Jeulin [220] gave a version of Perkins results that allows conditioning on B_t. See [167] and [168] for more information regarding the distribution of local times of B^{ex} and $|B^{\text{br}}|$. Knight [258, 259] treats the related problem of describing the distribution of $\int_0^1 |B_\ell^{\text{br}}(u)|du$. See also [342]. Norris-Rogers-Williams [326, Th. 2]showed that the distribution of a local time process derived from another kind of perturbed Brownian motion, with a drift depending on its local time process, can be characterized by a variation of the Bessel square SDE like (8.12), but with a different form of path dependent drift coefficient $\delta_v(Q)$. See also [438] and papers cited there for various other Ray-Knight type descriptions of Brownian local time processes, and further references on this topic.

8.5. Branching process approximations

The following result in the theory of branching processes, first indicated by Feller [149] and further developed by Lamperti [269, 268] and Lindvall [285], generalizes the approximation of $(Q_{\ell,v}^{(0)}, v \geq 0)$ by a continuous time branching process which was made in the previous section. See also [386]. Let $Z_k(h)$ for $h = 0, 1, 2, \ldots$ be the number of individuals in generation h of a Galton-Watson process started with k individuals in which the offspring distribution has mean 1 and finite variance $\sigma^2 > 0$, and let $Z_k(h)$ be defined for all $h \geq 0$ by linear interpolation between integers. Then as both $m \to \infty$ and $k \to \infty$

$$\left(\frac{2}{\sigma m} Z_k(2mv/\sigma), v \geq 0\right) \xrightarrow{d} (Q_{\ell,v}^{(0)}, v \geq 0) \text{ if } \frac{2k}{\sigma m} \to \ell, \tag{8.28}$$

where the limit is $BESQ_\ell^{(0)}$. To check the normalizations in (8.28), observe that if the process on the left side has value x at v such that $2mv/\sigma$ equals an integer h, then $Z_k(h) = x\sigma m/2$. The number $Z_k(h+1)$ in the next generation of the branching process therefore has variance $(x\sigma m/2)\sigma^2$. The increment of the process on the left side over the next v-increment of $\sigma/(2m)$ has this variance multiplied by $(2/\sigma m)^2$. So along the grid of multiples of $\sigma/(2m)$, the variance of increments of the normalized process on the left side per unit v-increment, from one grid point to the next, given the normalized process has value x at the first grid point, is

$$\left(\frac{x\sigma m}{2}\right)\sigma^2\left(\frac{2}{\sigma m}\right)^2\left(\frac{\sigma}{2m}\right)^{-1} = 4x = (2\sqrt{x})^2$$

in accordance with the $BESQ^{(0)}$ SDE. Let $(Z_{k,n}(h), h \geq 0)$ defined by conditioning $(Z_k(h), h \geq 0)$ on the event that its *total progeny* $\sum_{h=0}^{\infty} Z_k(h)$ equals n, so by definition

$$(Z_{k,n}(h), h \geq 0) \stackrel{d}{=} (Z_k(h), h \geq 0 \,|\, \textstyle\sum_{h=0}^{\infty} Z_k(h) = n) \tag{8.29}$$

where it is assumed now that the offspring distribution is aperiodic, so the conditioning event has strictly positive probability for all sufficiently large n. Then it is to be anticipated from (8.28) and (8.29) that as $n \to \infty$

$$\left(\frac{2}{\sigma\sqrt{n}} Z_{k,n}(2\sqrt{n}v/\sigma), v \geq 0\right) \stackrel{d}{\to} (Q_{\ell,1,v}, v \geq 0) \text{ provided } \frac{2k}{\sigma\sqrt{n}} \to \ell \tag{8.30}$$

for some $\ell \geq 0$ where $(Q_{\ell,1,v}, v \geq 0)$ may be identified by conditioning the limit $BESQ_\ell^{(0)}$ process in (8.28) on its integral being equal to 1, as implied by (8.10), and the definition is extended to $\ell = 0$ by weak continuity. This intuitively obvious identification of the limit can be justified by regularity of the solution of the basic SDE (8.12) as a function of its parameters, provided it is shown that the weak convergence (8.30) holds with the limit defined as the pathwise unique solution of the SDE (8.12). So let us now see how to derive the basic SDE (8.12) directly from combinatorial considerations. Note that if $Z_{k,n}(h)$ is interpreted as the number of vertices at level h in a forest with n vertices defined by a collection of k family trees, one for each initial individual in the Galton-Watson process, then for $h = 0, 1, \ldots$ the random variable

$$A_{k,n}(h) := n - \sum_{i=0}^{h} Z_{k,n}(h) \tag{8.31}$$

represents the number of vertices in the forest strictly above level h.

Lemma 8.6. *Fix $1 \leq k < n$. A sequence $(Z(h), h = 0, 1 \ldots)$ has the same distribution as a Galton-Watson process with a Poisson offspring distribution started with k individuals and conditioned on its total progeny being equal to n, if and only if the sequence evolves by the following mechanism: $Z(0) = k$ and for each $h = 0, 1 \ldots$*

$$(Z(h+1) \,|\, Z(i), 0 \leq i < h, Z(h) = z, A(h) = a) \stackrel{d}{=} 1 + \text{binomial}(a-1, z/(a+z)), \tag{8.32}$$

where $A(h) := n - \sum_{i=0}^{h} Z(i)$ and binomial(m,p) *is a binomial(m,p) random variable, with the conventions* binomial$(-1,p) = -1$ *and* binomial$(0,p) = 0$.

See the exercises for a proof and variations of this lemma. Granted the lemma, consider the rescaled process on the left side of (8.30) in the Poisson case, so $\sigma = 1$, in an asymptotic regime with $n \to \infty$ and $2k/\sqrt{n} \to \ell$ for some $\ell \geq 0$. Let $W_{k,n}(h) := (Z_{k,n}(h), A_{k,n}(h))$. From (8.32), in the limit as n, z and a tend to ∞ with $2z/\sqrt{n} \to x$ and $a/n \to p$, for integer h the increment $\Delta_{k,n}(h) := Z_{k,n}(h+1) - Z_{k,n}(h)$ is such that the corresponding normalized increment $\Delta^*_{k,n}(h) := 2\Delta_{k,n}(h)/\sqrt{n}$ has the following conditional mean and variance given a history $(Z_{k,n}(i), 0 \leq i \leq h)$ with $W_{k,n}(h) = (z,a)$:

$$E(\Delta^*_{k,n}(h) \,|\, W_{k,n}(h) = (z,a)) = \frac{2}{\sqrt{n}}\left(1 + \frac{(a-1)z}{a+z} - z\right) \approx \left(4 - \frac{x^2}{p}\right)\frac{1}{2\sqrt{n}}$$

$$Var(\Delta^*_{k,n}(h) \,|\, W_{k,n}(h) = (z,a)) = \frac{4}{n}\frac{(a-1)za}{(a+z)^2} \approx 4x\,\frac{1}{2\sqrt{n}}$$

where the relative errors of approximation are negligible as $n \to \infty$, uniformly in h, provided $x < 1/\epsilon$ and $p > \epsilon$ which can be arranged by a localization argument, stopping the normalized process when either its value exceeds x or its integral exceeds $1 - p$. Since $\Delta^*_{k,n}(h)$ is the increment of the normalized process over a time interval of length $1/(2\sqrt{n})$, and the value of $p \approx A_{k,n}(h)/n$ can be recovered from the path of the normalized process with a negligible error via

$$p \approx \frac{A_{k,n}(h)}{n} = 1 - \frac{1}{n}\sum_{i=0}^{h} Z_{k,n}(h) \approx 1 - \int_0^{h/(2\sqrt{n})} \frac{2}{\sqrt{n}} Z_{k,n}(2\sqrt{n}v) \qquad (8.33)$$

these calculations show that the normalized process is governed asymptotically by the SDE (8.12)-(8.13) presented in Theorem (8.4). It can be shown by standard arguments [384] that the SDE has a unique strong solution. So the preceding argument, combined with known results regarding the approximation of a Markov chain by the solution of an SDE [265] [264, Theorem 5.4], shows that the weak limit of the normalized and conditioned branching process solves the SDE (8.12)–(8.13).

Exercises

8.5.1. (Proof of Lemma 8.6) [358, Lemma 8] Let $X_1, X_2, \ldots$ be a sequence of independent random variables with some distribution p on $\{0,1,2,\ldots\}$, and set $S_j = X_1 + \cdots + X_j$. Fix $1 \leq k < n$ with $P(S_n = n-k) > 0$. Let $(Z_{k,n}(h), h = 0,1,2,\ldots)$ be a Galton-Watson branching process with offspring distribution p started with k individuals and conditioned to have total progeny n, and define $A_{k,n}(h)$ by (8.31), and set $W_{k,n}(h) := (Z_{k,n}(h), A_{k,n}(h))$. Show that $W_{k,n}(h), h = 0,1,2,\ldots$ is a Markov chain, with initial state $(k, n-k)$, and the following transition probabilities which do not depend on n and k: given $W_{k,n}(h) = (z,a)$ the distribution of $Z_{k,n}(h+1)$ is obtained by size-biasing of the distribution of S_z given $S_{z+a} = a$, and $A_{k,n}(h+1) = a - Z_{k,n}(h+1)$. In particular, for a Poisson distribution the law of S_z given $S_{z+a} = a$ is binomial$(a, z/(z+a))$. A size-biased binomial(n, p) variable is 1 plus a binomial$(n-1, p)$ variable, and Lemma 8.6 follows.

8.5.2. (Combinatorial interpretation of Lemma 8.6) Fix k and n with $1 \leq k \leq n$ and for $h = 0, 1, \ldots$ let $L_h := L_h(\mathcal{F}_{k,n})$ be the number of vertices at level h of $\mathcal{F}_{k,n}$, a uniformly distributed random forest of k rooted trees labeled by $[n]$. Then $L_0 = k$ and the distribution of the sequence $L_1, L_2, \ldots$ is determined be the following prescription of conditional laws: for each $h = 0, 1, \ldots$

$$\text{dist}(L_{h+1} \,|\, L_0, \ldots, L_h) = \text{dist}(1 + \text{binomial}(A_h - 1, L_h/(A_h + L_h) \tag{8.34}$$

where $A_h := n - \sum_{i=0}^{h} L_i$ and binomial(n, u) denotes a binomial random variable with parameters n and $u \in (0, 1)$.

8.5.3. (Further combinatorial interpretation of Lemma 8.6) The previous exercise shows that for each (k, n) the bivariate sequence $((L_h, A_h), h = 0, 1, \ldots)$ has the Markov property. The transition probabilities are inhomogeneous, but they have a recursive property, reflecting the fact that for any given h the random variable A_h represents the number of vertices of the forest strictly above level h, while $A_h + L_h$ is the number of vertices at or above level h. The simplifying feature of a uniform forest that given the forest at levels up to and including h leaves room for A_h vertices above h, the number L_{h+1} of these vertices at level $h+1$ are selected as the children of L_h roots in a uniform forest of L_h trees with given roots in a set of size $L_h + A_h$. This basic recursive property, which can be seen by a direct combinatorial argument, can be expressed as follows. Let $L_{n,h}$ denote the number of vertices at level h of a random forest which given $L_{n,0} = k$ is uniformly distributed on all forests of k trees labeled by $[n]$, and set $A_{n,h} := n - \sum_{i=0}^{h} L_{n,i}$ Then for each $h \geq 0$ and $j \geq 0$

$$(L_{n,h+j}, j \geq 0 \,|\, L_{n,i}, 0 \leq i \leq h \text{ with } L_{n,h} = m, A_{n,h} = a)$$

has the same distribution as

$$(L_{a+m,j}, j \geq 0 \,|\, L_{a+m,0} = m).$$

8.5.4. [358] What is the Brownian analog of the property of uniform random forests described in the previous exercise?

8.5.5. (Problem) Does (8.30) require any further conditions on the offspring distribution of the critical branching process besides finite non-zero variance?

Notes and comments

The basic Ray-Knight theorems for Brownian local times have been extended from Brownian motion to real-valued diffusions, using Feller's representation of such diffusions by space and time changes of Brownian motion. These results were extended in [131, 298] to a large class of Markov processes admitting local times. But for these extensions to discontinuous Markov processes, the Markov property of local times in the space variable is lost. See also [282, 348, 130] regarding local times of processes indexed by circles, trees, and graphs.

The properties of the local time processes of Brownian excursion and reflecting Brownian bridge discussed here are closely related to the *interval fragmentation process* $(Y_v, v \geq 0)$ derived from one of the processes $X = B^{\mathrm{ex}}, |B^{\mathrm{br}}|$ or $|B^{\mathrm{br}}_\ell|$ by letting Y_v be the collection of ranked lengths of the random open set $\{t : X_t > v\}$. It is clear from the present discussion that each of these processes $(Y_v, v \geq 0)$ is Markovian, with the same transition mechanism, whereby different interval lengths are fragmented independently according to rescaled copies of the interval fragmentation induced by a single excursion B^{ex}. This process is a basic example of the general class of *self-similar fragmentations* studied by Bertoin [46], which illustrates the phenomenon of "erosion" or loss of mass in the fragmentation process, since the total mass of the process after time v is

$$\int_0^1 1(X_t > v)\, dt = \int_v^\infty L_1^x(X)\, dx.$$

Thus the local time $L_1^v(X)$, which is a measurable function of the sequence of ranked lengths Y_v, is the rate of erosion of mass at time v in the fragmentation process, and the SDE (8.12) may be interpreted as describing the dynamics of erosion in the fragmentation process. One can also think of $L_1^v(X)$ as the density of mass at level v in a forest of Brownian trees of mass $\int_v^\infty L_1^x(X)dx$. As the level v rises, the fragmentation process describes how trees rooted on a valley floor disappear beneath a rising tide as the valley is flooded.

9

Brownian bridge asymptotics for random mappings

This chapter reviews Brownian bridge asymptotics for random mappings, first described in 1994 by Aldous and Pitman. The limit distributions as $n \to \infty$, of various functionals of a uniformly distributed random mapping from an n element set to itself, are those of corresponding functionals of a Brownian bridge. Similar results known to hold for various non-uniform models of random mappings, according to a kind of invariance principle. A mapping $M_n : [n] \to [n]$ can be identified with its digraph $\{i \to M_n(i),\ i \in [n]\}$, as in Figure 1.

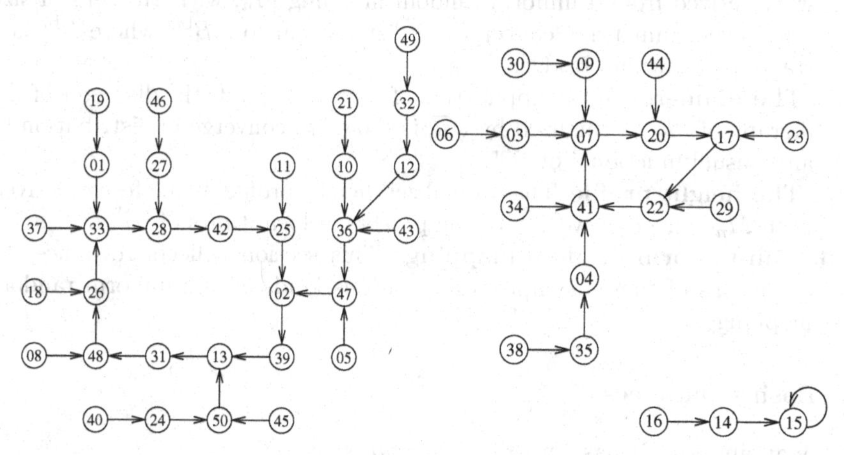

Figure 9.1: Digraph of a mapping $M_{50} : [50] \to [50]$.

Note how the mapping digraph encodes various features of iterates of the mapping. A mapping digraph can be decomposed as a collection of *rooted trees* together with some extra structure (*cycles, basins of attraction*). If each rooted tree is regarded as a plane tree and encoded by its Harris walk, defined by depth-first search following Harris [193], then given some ordering of tree-components, one can concatenate these Harris walks to define a *mapping-walk* which encodes numerous features of M_n.

From now on, we shall be interested in a uniformly distributed random mapping M_n. The connection between random mappings and Brownian bridge, first developed in [17], can be summarized as follows.

- For a uniform random mapping, the induced distribution on tree-components is such that the tree-walks, suitably normalized, converge to Brownian excursion as the tree size increases to infinity. So it is to be expected that the mapping-walks, suitably normalized, should converge to a limit process defined by some concatenation of Brownian excursions.
- With an appropriate choice of ordering of tree components, the weak limit of normalized mapping walks is reflecting Brownian bridge.

The subtle issue is how to order the tree components so that both

a) the mapping-walk encodes structure of cycles and basins of the mapping, and

b) the limit in distribution of the normalized mapping-walk can be explicitly identified.

How this can be done is discussed in some detail in the following sections, which are organized as follows:

9.1. Basins and trees deals with the definitions of basins and trees.

9.2. Mapping walks introduces two variants of the mapping walk.

9.3. Brownian asymptotics contains the main result: the scaled mapping-walk derived from a uniform random mapping M_n, with $2n$ steps of size $\pm 1/\sqrt{n}$ per unit time, converges in distribution to $2|B^{\mathrm{br}}|$ where B^{br} is a standard Brownian bridge.

9.4. The diameter As an application of the main result, the diameter of the digraph of M_n, normalized by $\sqrt{n}$, is shown to converge in distribution to an unusual functional of B^{br}.

9.5. The height profile The normalized height profile of the forest derived from M_n converges weakly to the process of local times of $|B^{\mathrm{br}}|$.

9.6. Non-uniform random mappings This section collects references to extensions of these asymptotics to various kinds of non-uniform random mappings.

9.1. Basins and trees

Fix a mapping M_n. It has a set of *cyclic points*

$$\mathcal{C}_n := \{i \in [n] : M_n^k(i) = i \text{ for some } k \geq 1\},$$

where M_n^k is the kth iterate of M_n. Let $\mathcal{T}_{n,c}$ be the set of vertices of the tree component of the digraph with root $c \in \mathcal{C}_n$. Note that $\mathcal{T}_{n,c}$ might be a trivial tree with just a single root vertex. The tree components are bundled by the disjoint cycles $\mathcal{C}_{n,j} \subseteq \mathcal{C}_n$ to form the *basins of attraction* (connected components) of the mapping digraph, say

$$\mathcal{B}_{n,j} := \bigcup_{c \in \mathcal{C}_{n,j}} \mathcal{T}_{n,c} \supseteq \mathcal{C}_{n,j} \text{ with } \bigcup_j \mathcal{B}_{n,j} = [n] \text{ and } \bigcup_j \mathcal{C}_{n,j} = \mathcal{C}_n \quad (9.1)$$

where all three unions are disjoint unions, and the $\mathcal{B}_{n,j}$ and $\mathcal{C}_{n,j}$ are indexed in some way by $j = 1, \ldots, |\mathcal{C}_n|$. Note that each tree component $\mathcal{T}_{n,c}$ is regarded here just as a subset of $[n]$, which is given the structure of a rooted tree by the action of M_n. The precise meaning of $\mathcal{B}_{n,j}$ and $\mathcal{C}_{n,j}$ now depends on the convention for ordering the cycles, which turns out to be of some importance. Two possible conventions are the *cycles-first ordering*, meaning the $\mathcal{C}_{n,j}$ are put in order of their least elements, and the *basins-first ordering* meaning the $\mathcal{B}_{n,j}$ are put in order of their least elements. Rather than introduce two separate notations for the two orderings, the same notation may be used for either ordering, with an indication of which is meant. Whichever ordering, the definitions of $\mathcal{B}_{n,j}$ and $\mathcal{C}_{n,j}$ are always linked by $\mathcal{B}_{n,j} \supseteq \mathcal{C}_{n,j}$, and (9.1) holds.

The following basic facts are easily deduced from these definitions, and results of Sections Section 2.4 and Section 4.5.

Structure of the basin partition Let Π_n^{basins} be the random partition of $[n]$ whose blocks are the basins of attraction of uniform random mapping M_n. Then Π_n^{basins} is a $\text{Gibbs}_{[n]}(1^\bullet, w_\bullet)$ partition, for w_j the number of mappings of $[j]$ whose digraph is connected. As remarked below Theorem 2.4, that implies the result of Aldous [14] that

$$\Pi_n^{\text{basins}} \xrightarrow{d} \Pi_\infty^{(0,\frac{1}{2})} \tag{9.2}$$

where the limit is a $(0, \frac{1}{2})$ partition of positive integers.

Structure of the tree partition Let Π_n^{trees} be the random partition of $[n]$ whose blocks are the tree components of the uniform random mapping M_n. So Π_n^{trees} is a refinement of Π_n^{basins}, with each basin split into its tree components. Note that the number of components of Π_n^{trees} equals the the number of cyclic points of M_n: $|\Pi_n^{\text{trees}}| = |\mathcal{C}_n|$. From the structure of a mapping digraph, Π_n^{trees} is a $\text{Gibbs}_{[n]}(v_\bullet, w_\bullet)$ partition for $v_k = k!$, the number of different ways that the restriction of M_n can act as a permutation of a given set of k cyclic points, and $w_j = j^{j-1}$ the number of rooted trees labeled by a set of size j. Let $q_j := e^{-j} j^{j-1}/j!$, so $(q_j, j = 0, 1, \ldots)$ is the distribution of total size of a critical Galton-Watson tree with Poisson offspring distribution. Since $q_j \sim (2\pi)^{-1/2} j^{-3/2}$, Theorem 2.5 gives for each $\ell > 0$, as $n \to \infty$

$$(\Pi_n^{\text{trees}} \text{ given } |\Pi_n^{\text{trees}}| = [\ell\sqrt{n}]) \xrightarrow{d} \Pi_\infty^{(\frac{1}{2}|\sqrt{2}\ell)} \tag{9.3}$$

where the limit is the partition of positive integers generated by lengths of excursions of a standard Brownian bridge B^{br} conditioned on $L_1^{\text{br}} = \ell$, where $L_1^{\text{br}} := L_1^0(B^{\text{br}})$. It is well known that L_1^{br} has the *Rayleigh density*

$$\mathbb{P}(L_1^{\text{br}} \in d\ell) = \ell \exp(-\tfrac{1}{2}\ell^2) d\ell \qquad (\ell > 0). \tag{9.4}$$

As a consequence of Cayley's result that kn^{n-k-1} is the number of forests labeled by $[n]$ with a specified set of k roots,

$$\mathbb{P}(|\mathcal{C}_n| = k) = \frac{k}{n} \prod_{i=1}^{k-1} \left(1 - \frac{i}{n}\right) \tag{9.5}$$

and hence that

$$|\mathcal{C}_n|/\sqrt{n} \xrightarrow{d} L_1^{\rm br} \tag{9.6}$$

jointly with

$$\Pi_n^{\rm trees} \xrightarrow{d} \Pi_\infty^{(\frac{1}{2},\frac{1}{2})} \tag{9.7}$$

where $\Pi_\infty^{(\frac{1}{2},\frac{1}{2})}$ is the random partition of positive integers generated by sampling from the interval partition defined by excursions of the standard Brownian bridge $B^{\rm br}$, whose distribution is defined by the $(\frac{1}{2},\frac{1}{2})$ prediction rule. Recall from (4.45) that $L_1^{\rm br}$ is encoded in $\Pi_\infty^{(\frac{1}{2},\frac{1}{2})}$ as the almost sure limit as $n \to \infty$ of $|\Pi_n(\frac{1}{2},\frac{1}{2})|/\sqrt{2n}$, where $|\Pi_n(\frac{1}{2},\frac{1}{2})|$ is the number of distinct excursions of $B^{\rm br}$ discovered by n independent uniform points on $[0,1]$.

Joint distribution of trees and basins As a check on (9.2) and (9.7), and to understand the joint structure of tree and basin partitions generated by a uniform random mapping M_n, it is instructive to compute the joint law of the random variables

$$\#\mathcal{T}_n(1) := \text{ size of the tree containing 1 in the digraph of } M_n \tag{9.8}$$

and

$$\#\mathcal{B}_n(1) := \text{ size of the basin containing 1 in the digraph of } M_n. \tag{9.9}$$

Note that $\#\mathcal{T}_n(1)$ and $\#\mathcal{B}_n(1)$ are size-biased picks from the block-sizes of $\Pi_n^{\rm trees}$ and $\Pi_n^{\rm basins}$ respectively. So their limit distributions as $n \to \infty$, with normalization by n, are the structural distributions of the weak limits of $\Pi_n^{\rm trees}$ and $\Pi_n^{\rm basins}$ respectively.

To expose the combinatorial structure underlying the joint law of $\#\mathcal{T}_n(1)$ and $\#\mathcal{B}_n(1)$, introduce new variables

$$N_{n,1} := \#\mathcal{T}_n(1) - 1; \quad N_{n,2} := \#\mathcal{B}_n(1) - \#\mathcal{T}_n(1); \quad N_{n,3} := n - \#\mathcal{B}_n(1). \tag{9.10}$$

Then for each possible vector of integers

$$(n_1, n_2, n_3) \text{ with } n_i \geq 0 \text{ and } n_1 + n_2 + n_3 = n - 1, \tag{9.11}$$

there is the formula

$$\mathbb{P}(N_{n,i} = n_i, i = 1,2,3) = \binom{n-1}{n_1, n_2, n_3} \frac{(n_1+1)^{n_1} n_2^{n_2} n_3^{n_3}}{n^n}. \tag{9.12}$$

The multinomial coefficient appears here for obvious reasons. For each particular choice $n_1 + 1$ possible elements of the set $\mathcal{T}_n(1)$, the factor $(n_1+1)^{n_1}$ is the number of possible rooted trees induced by the action of M_n on this set, by Cayley's formula (6.24). For each choice n_3 possible elements of $[n]\backslash\mathcal{B}_n(1)$, the factor $n_3^{n_3}$ is the number of possible actions of M_n restricted to this set. This reflects part (i) of the following lemma. Part (ii) of the lemma explains the symmetry of formula (9.12) in (n_2, n_3) for fixed n_1. See also [359] for similar joint distributions derived from random mappings, known as *Abel multinomial distributions*.

Lemma 9.1. *For a uniform random mapping M_n,*

(i) *Conditionally given the restriction of M_n to $\mathcal{B}_n(1)$ with $\mathcal{B}_n(1) = \mathcal{B}$, the restriction of M_n to $[n]-\mathcal{B}$ is a uniform random mapping from $[n]-\mathcal{B}$ to $[n]-\mathcal{B}$.*

(ii) *Conditionally given that $\mathcal{T}_n(1)$ is some subset $\mathcal{T}$ of $[n]$ with $1 \in \mathcal{T}$, the restriction of Π_n^{trees} to $\mathcal{B}_n(1)\backslash\mathcal{T}$ and the restriction of Π_n^{trees} to $[n] - \mathcal{B}_n(1)$ are exchangeable.*

Proof. The first statement is obvious. To clarify statement (ii), given $\mathcal{T}_n(1) = \mathcal{T}$, each restriction of Π_n^{trees} is regarded as a random partition of a random subset of $[n]$, with some notion of a trivial partition if the subset is empty. According to (i), given also $\mathcal{B}_n(1) = \mathcal{B}$, the restriction of Π_n^{trees} to $[n] - \mathcal{B}$ is the tree-partition generated by a uniform random mapping from $\mathcal{B}$ to $\mathcal{B}$. On the other hand, the restriction of Π_n^{trees} to $\mathcal{B} - \mathcal{T}$ is the tree partition generated by a uniformly chosen composite structure on $\mathcal{B} - \mathcal{T}$, whereby $\mathcal{B} - \mathcal{T}$ is partitioned into tree components, and the roots of these components are assigned a linear order. But this is bijectively equivalent to a mapping from $\mathcal{B} - \mathcal{T}$ to $\mathcal{B} - \mathcal{T}$, hence the conclusion. □

By Stirling's formula, the probability in (9.12) is asymptotically equivalent to

$$\frac{1}{n^2}\frac{1}{2\pi}\frac{1}{\sqrt{n_1/n}\sqrt{n_2/n}\sqrt{n_3/n}} \text{ as } n_i \to \infty, i = 1,2,3, \tag{9.13}$$

hence as $n \to \infty$

$$(N_{n,1}, N_{n,2}, N_{n,3})/n \xrightarrow{d} \text{Dirichlet}(\tfrac{1}{2}, \tfrac{1}{2}, \tfrac{1}{2}). \tag{9.14}$$

Recalling the definitions (9.10) of the $N_{n,i}$, this implies

$$\frac{\#\mathcal{B}_n(1)}{n} \xrightarrow{d} \beta_{1,\frac{1}{2}} \text{ and } \frac{\#\mathcal{T}_n(1)}{n} \xrightarrow{d} \beta_{\frac{1}{2},1} \tag{9.15}$$

where $\beta_{a,b}$ has beta(a,b) distribution. As a check, according to Theorem 3.2, $\beta_{1,\frac{1}{2}}$ and $\beta_{\frac{1}{2},1}$ are the structural distributions of $(0, \frac{1}{2})$ and $(\frac{1}{2}, \frac{1}{2})$ partitions respectively. So (9.15) agrees with (9.2) and (9.7). As indicated by Aldous [14], Lemma 9.1 (i) allows recursive application of the second convergence in (9.15) to show that the size-biased frequencies of Π_n^{basins} approach the GEM$(0, \frac{1}{2})$ frequencies (3.8), hence the convergence (9.2) of Π_n^{basins} to a $(0, \frac{1}{2})$ partition.

Exercises

9.1.1. Develop a variation of the above argument to show that the size-biased frequencies of Π_n^{trees} approach the GEM$(\frac{1}{2}, \frac{1}{2})$ frequencies, hence the convergence (9.7) of Π_n^{trees} to a $(\frac{1}{2}, \frac{1}{2})$ partition.

Notes and comments

This section is based on [17, 24]. The theory of random mappings has a long history. See [260, 17, 191] and papers cited there.

9.2. Mapping walks

The construction in [17] encodes the restriction of the digraph of M_n to each tree component $\mathcal{T}_{n,c}$ of size k by the Harris walk of $2k$ steps associated with this tree, which was defined in Section 6.3. This *tree-walk* derived from $\mathcal{T}_{n,c}$, with increments of ± 1 on the non-negative integers, makes an excursion which starts at 0 and returns to 0 for the first time after $2k$ steps, after reaching a maximum level $1 + h_n(c)$, where $h_n(c)$ is the maximal height above c of all vertices of the tree $\mathcal{T}_{n,c}$ with root c, that is

$$h_n(c) = \max\{h : \exists i \in [n] \text{ with } M_n^h(i) = c \text{ and } M_n^j(i) \notin \mathcal{C}_n \text{ for } 0 \leq j < h\}. \tag{9.16}$$

Given that c is a cyclic point such that the set of vertices $\mathcal{T}_{n,c}$ equals K for some subset K of $[n]$ with $c \in K$ and $|K| = k$, the restriction of the digraph of M_n to K has uniform distribution on the set of k^{k-1} trees labeled by K with root c. According to a basic result of Aldous, Theorem 6.4, as $k \to \infty$, the distribution of this tree-walk when scaled to have $2k$ steps of $\pm 1/\sqrt{k}$ per unit time, converges to the distribution of $2B^{\text{ex}}$, for B^{ex} a standard Brownian excursion.

We now define a mapping-walk (to code M_n) as a concatenation of its tree-walks, to make a walk of $2n$ steps starting and ending at 0 with exactly $|\mathcal{C}_n|$ returns to 0, one for each tree component of the mapping digraph. To concatenate the tree-walks, an order of tree-components must be specified. To retain useful information about M_n in the mapping-walk, we want the ordering of tree-walks to respect the cycle and basin structure of the mapping. Here are two orderings that do so.

Definition 9.2. (Cycles-first ordering) Fix a mapping M_n from $[n]$ to $[n]$. First put the cycles in increasing order of their least elements, say $c_{n,1} < c_{n,2} < \cdots < c_{n,|\mathcal{C}_n|}$. Let $\mathcal{C}_{n,j}$ be the cycle containing $c_{n,j}$, and let $\mathcal{B}_{n,j}$ be the basin containing $\mathcal{C}_{n,j}$. Within cycles, list the trees around the cycles, as follows. If the action of M_n takes $c_{n,j} \to c_{n,j,1} \to \cdots \to c_{n,j}$ for each $1 \leq j \leq |\mathcal{C}_n|$, the tree components $\mathcal{T}_{n,c}$ are listed with c in the order

$$(\overbrace{c_{n,1,1}, \ldots, c_{n,1}}^{\mathcal{C}_{n,1}}, \overbrace{c_{n,2,1}, \ldots, c_{n,2}}^{\mathcal{C}_{n,2}}, \ldots\ldots, \overbrace{c_{n,|\mathcal{C}_n|,1}, \ldots, c_{n,|\mathcal{C}_n|}}^{\mathcal{C}_{n,|\mathcal{C}_n|}}). \tag{9.17}$$

The *cycles-first mapping-walk* is obtained by concatenating the tree walks derived from M_n in this order. The *cycles-first search of* $[n]$ is the permutation $\sigma : [n] \to [n]$ where σ_j is the jth vertex of the digraph of M_n which is visited in the corresponding concatenation of tree searches.

Definition 9.3. (Basins-first ordering)[17] First put the basins $\mathcal{B}_{n,j}$ in increasing order of their least elements, say $1 = b_{n,1} < b_{n,2} < \dots b_{n,|\mathcal{C}_n|}$; let $c_{n,j} \in \mathcal{C}_{n,j}$ be the cyclic point at the root of the tree component containing $b_{n,j}$. Now list the trees around the cycles, just as in (9.17), but for the newly defined $c_{n,j}$ and $c_{n,j,i}$. Call the corresponding mapping-walk and search of $[n]$ the *basins-first mapping-walk* and *basins-first search.*

Let us briefly observe some similarities between the two mapping-walks. For each given basin B of M_n with say b elements, the restriction of M_n to B is encoded in a segment of each walk which equals at 0 at some time, and returns again to 0 after $2b$ more steps. If the basin contains exactly c cyclic points, this walk segment of $2b$ steps will be a concatenation of c excursions away from 0. Exactly where this segment of $2b$ steps appears in the mapping-walk depends on the ordering convention, as does the ordering of excursions away from 0 within the segment of $2b$ steps. However, many features of the action of M_n on the basin B are encoded in the same way in the two different stretches of length $2b$ in the two walks, despite the permutation of excursions. One example is the number of elements in the basin whose height above the cycles is h, which is encoded in either walk as the number of upcrossings from h to $h+1$ in the stretch of walk of length $2b$ corresponding to that basin.

9.3. Brownian asymptotics

The idea now is that for either of the mapping walks derived above from a uniform mapping M_n, a suitable rescaling converges weakly in $C[0,1]$ as $n \to \infty$ to the distribution of the reflecting Brownian bridge defined by the absolute value of a *standard Brownian bridge* B^{br} with $B_0^{\mathrm{br}} = B_1^{\mathrm{br}} = 0$ obtained by conditioning a standard Brownian motion B on $B_1 = 0$. Jointly with this convergence, results of [17] imply that for a uniform random mapping, the basin sizes rescaled by n, jointly with corresponding cycle sizes rescaled by $\sqrt{n}$, converge in distribution to a limiting bivariate sequence of random variables $(\lambda_{I_j}, L_{I_j}^{\mathrm{br}})_{j=1,2,\dots}$ where $(I_j)_{j=1,2,\dots}$ is a random interval partition of $[0,1]$, with λ_{I_j} the length of I_j and $L_{I_j}^{\mathrm{br}}$ the increment of local time of B^{br} at 0 over the interval I_j. For the basins-first walk, the limiting interval partition is $(I_j) = (I_j^D)$, according to the following definition. Here $U, U_1, U_2, \dots$ denotes a sequence of independent uniform $(0,1)$ variables, independent of B^{br}, and the local time process of B^{br} at 0 is assumed to be normalized as occupation density relative to Lebesgue measure.

Definition 9.4. (The D-partition [17]) Let $I_j^D := [D_{V_{j-1}}, D_{V_j}]$ where $V_0 = D_{V_0} = 0$ and V_j is defined inductively along with the D_{V_j} for $j \geq 1$ as follows: given that D_{V_i} and V_i have been defined for $0 \leq i < j$, let

$$V_j := D_{V_{j-1}} + U_j(1 - D_{V_{j-1}}),$$

so V_j is uniform on $[D_{V_{j-1}}, 1]$ given B^{br} and (V_i, D_{V_i}) for $0 \leq i < j$, and let

$$D_{V_j} := \inf\{t \geq V_j : B_t^{\mathrm{br}} = 0\}.$$

On the other hand, for the cycles-first walk, the limits involve a different interval partition. This is the partition $(I_j) = (I_j^T)$ defined as follows using the local time process $(L_u^{\mathrm{br}}, 0 \le u \le 1)$ of B^{br} at 0:

Definition 9.5. (The T-partition) Let $I_j^T := [T_{j-1}, T_j]$ where $T_0 := 0$, $\hat{V}_0 :=$ 0, and for $j \ge 1$

$$\hat{V}_j := 1 - \prod_{i=1}^{j}(1 - U_i), \tag{9.18}$$

so $\hat{V}_j$ is uniform on $[\hat{V}_{j-1}, 1]$ given B^{br} and $(\hat{V}_i, T_i)$ for $0 \le i < j$, and

$$T_j := \inf\{u : L_u^{\mathrm{br}}/L_1^{\mathrm{br}} > \hat{V}_j\}.$$

The main result can now be stated as follows:

Theorem 9.6. [24] *The scaled mapping-walk $(M_u^{[n]}, 0 \le u \le 1)$ derived from a uniform random mapping M_n, with $2n$ steps of $\pm 1/\sqrt{n}$ per unit time, for either the cycles-first or the basins-first ordering of excursions corresponding to tree components, converges in distribution to $2|B^{\mathrm{br}}|$ jointly with* (9.6) *and* (9.7), *where $(L_u^{\mathrm{br}}, 0 \le u \le 1)$ is the process of local time at* 0 *of B^{br}, and $\Pi_\infty^{(\frac{1}{2},\frac{1}{2})}$ is the random partition of positive integers generated by sampling from the interval partition defined by excursions B^{br}. Moreover,*
(i) *for the cycles-first ordering, with the cycles $\mathcal{B}_{n,j}$ in order of their least elements, these two limits in distribution hold jointly with*

$$\left(\frac{|\mathcal{B}_{n,j}|}{n}, \frac{|\mathcal{C}_{n,j}|}{\sqrt{n}}\right) \xrightarrow{d} (\lambda_{I_j}, L_{I_j}^{\mathrm{br}}) \tag{9.19}$$

as j varies, where the limits are the lengths and increments of local time of B^{br} at 0 *associated with the interval partition $(I_j) := (I_j^T)$; whereas*
(ii) [17] *for the basins-first ordering, with the basins $\mathcal{B}_{n,j}$ listed in order of their least elements, the same is true, provided the limiting interval partition is defined instead by $(I_j) := (I_j^D)$.*

The result for basins-first ordering is part of [17, Theorem 8]. The variant for cycles-first ordering can be established by a variation of the argument in [17], exploiting the exchangeability of the tree components in the cycles-first ordering. See also [58] and [15] for alternate approaches to the basic result of [17].

The random set of pairs $\{(|\mathcal{B}_{n,j}|/n, |\mathcal{C}_{n,j}|/\sqrt{n}), 1 \le j \le |\mathcal{C}_n|\}$ is the same, no matter what ordering convention is used. So Theorem 9.6 implies that the distribution of the random set of limit points, $\{(\lambda_{I_j}, L_{I_j}^{\mathrm{br}}), j \ge 1\}$, regarded as a point process on $\mathbb{R}^2_{>0}$, is the same for $(I_j) = (I_j^D)$ or $(I_j) = (I_j^T)$. This fact about Brownian bridge is not at all obvious, but can be verified by application of Brownian excursion theory. See [24] for further discussion.

To gain useful information about large random mappings from Theorem 9.6, it is necessary to understand well the joint law of B^{br} and one or other of the limiting interval partitions (I_j) whose definition depends on the path of B^{br}. To be definite, assume from now on that the ordering convention is *basins first.* One feature of natural interest is the maximal height above the cycle of the tallest tree in the basin. Let this maximal height be $H_{n,j}$ for the jth basin. Theorem 9.6 implies

$$\left(\frac{|\mathcal{B}_{n,j}|}{n}, \frac{|\mathcal{C}_{n,j}|}{\sqrt{n}}, \frac{H_{n,j}}{\sqrt{n}}\right) \xrightarrow{d} (\lambda_j, L_j, 2\overline{M}_j)_{j=1,2,\ldots} \tag{9.20}$$

where we abbreviate $\lambda_j := \lambda_{I_j}, L_j := L^{\mathrm{br}}_{I_j}$, and $\overline{M}_j := \overline{|B^{\mathrm{br}}|}(D_{V_{j-1}}, D_{V_j})$ is the maximal value of $|B^{\mathrm{br}}|$ on I_j. It follows easily from Definition 9.4, the strong Markov property of B^{br} at the times D_{V_j}, and Brownian scaling, that

$$\lambda_j = W_j \prod_{i=1}^{j-1}(1 - W_i) \tag{9.21}$$

for a sequence of independent random variables W_j with beta$(1, \frac{1}{2})$ distribution, and that

$$(L_j, \overline{M}_j) = \sqrt{\lambda_j}(\tilde{L}_j, \tilde{M}_j) \tag{9.22}$$

for a sequence of independent and identically distributed random pairs $(\tilde{L}_j, \tilde{M}_j)$, independent of (λ_j). The common distribution of $(\tilde{L}_j, \tilde{M}_j)$ is that of

$$(\tilde{L}_1, \tilde{M}_1) := \left(\frac{L^{\mathrm{br}}_{D_U}}{\sqrt{D_U}}, \frac{M^{\mathrm{br}}_{D_U}}{\sqrt{D_U}}\right) \tag{9.23}$$

where D_U is the time of the first zero of B^{br} after a uniform$[0, 1]$ random time U which is independent of B^{br}, $L^{\mathrm{br}}_t := L^0_t(B^{\mathrm{br}})$, and and $M^{\mathrm{br}}_t := \max_{0 \le u \le t} |B^{\mathrm{br}}_u|$ for $0 \le t \le 1$. It is known [341] that for (λ_j) as in (9.21), assumed independent of B_1, the $B_1^2 \lambda_j$ are the points (in size-biased random order) of a Poisson process on $\mathbb{R}_{>0}$ with intensity measure $\frac{1}{2}t^{-1}e^{-t/2}dt$ which is the Lévy measure of the infinitely divisible gamma$(\frac{1}{2}, \frac{1}{2})$ distribution of B_1^2. Together with standard properties of Poisson processes, this observation and the previous formulae (9.21) to (9.23) yield the following lemma. See also [24] for related results.

Lemma 9.7. *If B_1 is a standard Gaussian variable independent of the sequence of triples $(\lambda_j, L_j, \overline{M}_j)_{j=1,2,\ldots}$ featured in (9.20), then the random vectors*

$$(B_1^2 \lambda_j, |B_1| L_j, |B_1| \overline{M}_j)$$

are the points of a Poisson point process on $\mathbb{R}^3_{>0}$ with intensity measure μ defined by

$$\mu(dt\, d\ell\, dm) = \frac{e^{-t/2}\, dt}{2t} P(\sqrt{t}\tilde{L}_1 \in d\ell, \sqrt{t}\tilde{M}_1 \in dm) \tag{9.24}$$

for $t, \ell, m > 0$, where $(\tilde{L}_1, \tilde{M}_1)$ is the pair of random variables derived from a Brownian bridge by (9.23).

For a process $X := (X_t, t \in J)$ parameterized by an interval J, and $I = [G_I, D_I]$ a subinterval of J with length $\lambda_I := D_I - G_I > 0$, we denote by $X[I]$ or $X[G_I, D_I]$ the *fragment of X on I*, that is the process

$$X[I]_u := X_{G_I + u} \qquad (0 \le u \le \lambda_I). \tag{9.25}$$

Denote by $X_*[I]$ or $X_*[G_I, D_I]$ the *standardized fragment of X on I*, defined by the *Brownian scaling operation*

$$X_*[I]_u := \frac{X_{G_I+u\lambda_I} - X_{G_I}}{\sqrt{\lambda_I}} \qquad (0 \leq u \leq 1). \tag{9.26}$$

The process $\tilde{B}^{\text{br}} := B_*[0, \tau_1]$, where τ_1 is an inverse local time at 0 for the unconditioned Brownian motion B, is known as a *Brownian pseudo-bridge*, and there is the following absolute continuity relation between the laws of $\tilde{B}^{\text{br}}$ and B^{br} found in [59]: for each non-negative measurable function g on $C[0, 1]$,

$$\mathbb{E}[g(\tilde{B}^{\text{br}})] = \sqrt{\frac{2}{\pi}}\mathbb{E}[g(B^{\text{br}})/L_1^{\text{br}}].$$

See Exercise 4.5.2 . It follows from [366, Theorem 1.3] and [17, Proposition 2] that the process $B_*^{\text{br}}[0, D_U]$, obtained by rescaling the path of B^{br} on $[0, D_1]$ to have length 1 by Brownian scaling, has the same distribution as a rearrangement of the path of the pseudo-bridge $\tilde{B}^{\text{br}}$. Neither the maximum nor the local time at 0 are affected by such a rearrangement, so there is the equality in distribution

$$(\tilde{L}_1, \tilde{M}_1) \stackrel{d}{=} (\tilde{L}_1^{\text{br}}, \tilde{M}_1^{\text{br}}) \tag{9.27}$$

where $\tilde{L}_1^{\text{br}} := L_1^0(\tilde{B}^{\text{br}})$ $\tilde{M}_1^{\text{br}} := \max_{0 \leq u \leq 1} |\tilde{B}_u^{\text{br}}|$. So (9.27) yields the formula

$$P(\sqrt{t}\tilde{L}_1 \in d\ell, \sqrt{t}\tilde{M}_1 \leq y) = \sqrt{\frac{2}{\pi}}\frac{\sqrt{t}}{\ell}P(\sqrt{t}L_1^{\text{br}} \in d\ell, \sqrt{t}M_1^{\text{br}} \leq y), \tag{9.28}$$

for $t, \ell, y > 0$. Now the joint law of L_1^{br} and M_1^{br} is characterized by the following identity: for all $\ell > 0$ and $y > 0$

$$\int_0^\infty \frac{e^{-t/2}}{\sqrt{2\pi t}} dt\, P(\sqrt{t}L_1^{\text{br}} \in d\ell, \sqrt{t}M_1^{\text{br}} \leq y) = e^{-\ell} d\ell \exp\left(\frac{-2\ell}{e^{2y} - 1}\right) \tag{9.29}$$

which can be read from [374, Theorem 3, Lemma 4 and (36)], with the following interpretation. Let $(L_t, t \geq 0)$ be the local time process of the Brownian motion B at 0, let T be an exponential random variable with mean 2 independent of B, and let G_T be the time of the last 0 of B before time T. Then (9.29) provides two expressions for

$$P\left(L_T \in d\ell, \sup_{0 \leq u \leq G_T} |B_u| \leq y\right),$$

on the left side by conditioning on G_T, and on the right side by conditioning on L_T. See also [384, Chapter XII, Exercise (4.24)].

Using (9.24), (9.28) and (9.29), we deduce that in the Poisson point process of Lemma 9.7,

$$\mathbb{E}[\text{number of points } (|B_1|L_j, |B_1|\overline{M}_j) \text{ with } |B_1|L_j \in d\ell \text{ and } |B_1|\overline{M}_j \leq y] = \tag{9.30}$$

$$\int_0^\infty \frac{e^{-t/2}\,dt}{2t} P(\sqrt{t}\tilde{L}_1 \in d\ell, \sqrt{t}\tilde{M}_1 \le y) = \ell^{-1} e^{-\ell} d\ell \exp\left(\frac{-2\ell}{e^{2y}-1}\right). \qquad (9.31)$$

A significant check on these calculations can be made as follows. By further integration, the expected number of points j with $|B_1|\overline{M}_j$ greater than y is

$$\eta(y) := \int_0^\infty \ell^{-1} e^{-\ell} d\ell \left[1 - \exp\left(\frac{-2\ell}{e^{2y}-1}\right)\right]. \qquad (9.32)$$

Now the probability of no point greater than y is $e^{-\eta(y)}$, so

$$\mathbb{P}(|B_1| \max_j \overline{M}_j \le y) = e^{-\eta(y)}. \qquad (9.33)$$

But the event $(|B_1| \max_j \overline{M}_j \le y)$ is identical to the event $(M_1^{\mathrm{br}} \le y)$, where $M_1^{\mathrm{br}} := \max_{0 \le u \le 1} |B_u^{\mathrm{br}}|$. And $e^{-\eta(y)} = \frac{1}{1+2/(e^{2y}-1)} = \tanh y$ by application of the Lévy-Khintchine formula for the exponential distribution, that is

$$\frac{1}{1+\lambda} = \exp\left[-\int_0^\infty \ell^{-1} e^{-\ell}(1-e^{-\lambda\ell})d\ell\right],$$

for $\lambda = 2/(e^{2y}-1)$. Thus for B_1 standard Gaussian independent of B^{br} and $y > 0$, there is the remarkable formula

$$\mathbb{P}(|B_1| M_1^{\mathrm{br}} \le y) = \tanh y \qquad (y \ge 0) \qquad (9.34)$$

which is a known equivalent of Kolmogorov's formula

$$\mathbb{P}(M_1^{\mathrm{br}} \le x) = \sum_{n=-\infty}^{\infty} (-1)^n e^{-2n^2x^2} \qquad (x \ge 0) \qquad (9.35)$$

As observed in [60], formula (9.34) allows the Mellin transform of M_1^{br} to be expressed in terms of the Riemann zeta function. See also [339, 371, 373] for closely related Mellin transforms obtained by the technique of multiplication by a suitable independent random factor to introduce Poisson or Markovian structure.

Notes and comments

This section is based on [17] and [24].

9.4. The diameter

The *diameter* of M_n is the random variable

$$\Delta_n := \max_{i \in [n]} T_n(i)$$

where $T_n(i)$ is the number of iterations of M_n starting from i until some value is repeated:

$$T_n(i) := \min\{j \geq 1 : M_n^j(i) = M_n^k(i) \text{ for some } 0 \leq k < j\}$$

where $M_n^0(i) = i$ and $M_n^j(i) := M_n(M_n^{j-1}(i))$ is the image of i under j-fold iteration of M_n for $j \geq 1$. Since by definition $\Delta_n = \max_j(|\mathcal{C}_{n,j}| + H_{n,j})$, it follows from (9.20) that as $n \to \infty$

$$\frac{\Delta_n}{\sqrt{n}} \xrightarrow{d} \Delta := \max_j (L_j + 2\overline{M}_j). \tag{9.36}$$

So we obtain the following corollary of Theorem 9.6:

Corollary 9.8. [22] *Let B_1 be a standard Gaussian variable independent of Δ. Then the distribution of Δ in (9.36) is characterized by*

$$P(|B_1|\Delta \leq v) = e^{-E_1(v) - I(v)} \qquad (v \geq 0) \tag{9.37}$$

where

$$E_1(v) := \int_v^\infty u^{-1} e^{-u} du$$

$$I(v) := \int_0^v u^{-1} e^{-u} \left[1 - \exp\left(\frac{-2u}{e^{v-u} - 1}\right)\right] du.$$

Proof. From (9.36) and Lemma 9.7, the event $|B_1|\Delta \leq v$ is the event that there is no j with $|B_1|L_j + 2|B_1|\overline{M}_j > v$. But from (9.30) - (9.31), $E_1(v)$ is the expected number of j with $|B_1|L_j \geq v$, while $I(v)$ is the expected number of j with $|B_1|L_j < v$ and $|B_1|L_j + 2|B_1|\overline{M}_j > v$. □

Integration of (9.37) gives a formula for $\mathbb{E}(\Delta^p)$ for arbitrary $p > 0$, which is easily shown to be the limit as $n \to \infty$ of $\mathbb{E}((\Delta_n/\sqrt{n})^p)$. This formula was first found for $p = 1$ by Flajolet-Odlyzko [156, Theorem 7] using singularity analysis of generating functions. See also [408, 289, 97] for related asymptotic studies of the diameter of undirected random trees and graphs.

Exercises

9.4.1. (Problem: the diameter of a Brownian tree) Szekeres [408] found an explicit formula for the asymptotic distribution of the diameter of a uniform random tree labeled by $[n]$, with normalization by $\sqrt{n}$. Aldous [6, 3.4] observed that this is the distribution of the diameter of $\mathcal{T}(2B^{\text{ex}})$, and raised the following problem, which is still open: can this distribution be characterized directly in the Brownian world?

Notes and comments

This section is based on [22]. The technique of characterizing the law of some Brownian functional X by first considering the law of $|B_1|X$ for B_1 a standard Gaussian variable independent of X, and the related idea of *random Brownian scaling* have found numerous applications [439, 374, 372].

9.5. The height profile

We continue to suppose that M_n is a uniform random mapping from $[n]$ to $[n]$. For $v \in [n]$ let $h(v, M_n)$ be the least $m \geq 0$ such that $M_n^m(v) \in \mathcal{C}_n$. So $h(v, M_n)$ is the height of v in the forest derived from M_n whose set of roots is the random set $\mathcal{C}_n$ of cyclic points of M_n. For $h = 0, 1, 2, \ldots$ let $Z_{*,n}(h)$ be the number of $v \in [n]$ such that $h(v, M_n) = h$. Call this process $(Z_{*,n}(h), h \geq 0)$ the *height profile of the mapping forest.* Let $(Z_{k,n}(h), h \geq 0)$ be the height profile of the mapping forest conditioned on the event $(Z_{*,n}(0) = k)$ that M_n has exactly k cyclic points. Then $(Z_{k,n}(h), h \geq 0)$ has the same distribution as the height profile generated by a uniform random forest of k rooted trees labeled by $[n]$, to which the limit theorem (8.30) applies, by inspection of (8.34) and (4.9). To review:

Lemma 9.9. *If $(Z_{k,n}(h), h \geq 0)$ is either*
(i) *the height profile of a uniform random forest of k rooted trees labeled by $[n]$, or*
(ii) *the height profile of the forest derived from a random mapping from $[n]$ to $[n]$ conditioned to have k cyclic points,*
then the distribution of the sequence $(Z_{k,n}(h), h \geq 0)$ is that described by Lemma 8.6, and in the limit regime as $n \to \infty$ and $2k/\sqrt{n} \to \ell \geq 0$

$$\left(\frac{2}{\sqrt{n}} Z_{k,n}(2\sqrt{n}v), v \geq 0 \right) \xrightarrow{d} (Q_{\ell,1,v}, v \geq 0) \tag{9.38}$$

where the law of $(Q_{\ell,1,v}, v \geq 0)$ is defined by Theorem 8.4.

The following result is now obtained by mixing the result of the previous lemma with respect to the distribution of the number $Z_{*,n}(0) = |\mathcal{C}_n|$ of cyclic points of M_n. According to (9.6), $|\mathcal{C}_n|/\sqrt{n} \xrightarrow{d} L_1^0(B^{\text{br}})$, so the result is:

Theorem 9.10. Drmota-Gittenberger [116] *The normalized height profile of the forest derived from a uniform random mapping M_n converges weakly to the process of local times of a reflecting Brownian bridge of length 1:*

$$\left(\frac{2}{\sqrt{n}} Z_{*,n}(2\sqrt{n}v), v \geq 0 \right) \xrightarrow{d} (L_1^v(|B^{\text{br}}|, v \geq 0) \tag{9.39}$$

Notes and comments

This section is based on [358]. Presumably the convergence in distribution of height profiles (9.39) holds jointly with all the convergences in distribution described in Theorem 9.6. This must be true, but seems difficult to establish. Corresponding results of joint convergence in distribution of occupation time processes and unconditioned walk paths to their Brownian limits, for simple random walks, can be read from Knight [254]. Presumably corresponding results are known for simple random walks with bridges or excursions as limits, but I do not know a reference.

9.6. Non-uniform random mappings

Definition 9.11. *Let p be a probability distribution on $[n]$. Call M_n* p-mapping *from $[n]$ to $[n]$ if the images $M_n(i)$ of points $i \in [n]$ are independent and identically distributed according to p.*

Combinatorial properties of p-mappings, and some elementary asymptotics are reviewed in [359]. Further asymptotic features of p-mappings were studied in [327]. In [23, 16] it is shown that Brownian bridge asymptotics apply for models of random mappings more general than the uniform model, in particular for *p-mapping* model under suitable conditions. Proofs are simplified by use of Joyal's bijection between mappings and trees, discussed in Exercise 10.1.4 . Another important result on p-mappings is Burtin's formula which is presented in Exercise 10.1.5 . But these results for p-mappings are best considered in connection with *p-trees* and *p-forests*, which are the subject of Chapter 10.

10

Random forests and the additive coalescent

This chapter reviews how various representations of additive coalescent processes, whose state space may be either finite or infinite partitions, can be constructed from random trees and forests. These constructions establish deep connections between the asymptotic behaviour of additive coalescent processes and the theory of Brownian trees and excursions. There are some close parallels with the theory of multiplicative coalescents and the asymptotics of critical random graphs, described in Section 6.4.

10.1. Random p-forests and Cayley's multinomial expansion For each probability distribution p on a set S of n elements, Cayley's multinomial expansion allows the definition of a random p-forest $\mathcal{F}_{n,k}$ of k trees labeled by S. There is a natural way to realize these forests as a forest-valued fragmentation process $(\mathcal{F}_{n,k}, 1 \leq k \leq n)$ where one edge of the forest is lost at each step by uniform random selection from all remaining edges. Time-reversal of this process yields a forest-valued coalescent process. The corresponding sequence of random partitions of S coalesces in such a way that two blocks A_i and A_j merge at each step with probability proportional to $p(A_i) + p(A_j)$.

10.2. The additive coalescent A continuous time variant of this construction yields a partition-valued *additive coalescent* process in which blocks A_i and A_j merge at rate $p(A_i) + p(A_j)$. This is compared with other constructions of additive coalescent processes with various state spaces.

10.3. The standard additive coalescent This continuous time process, parameterized by $\mathbb{R}$, with state space the set $\mathcal{P}_1^{\downarrow}$ of sequences $x = (x_i)_{i \geq 1}$ with $x_1 \geq x_2 \geq \cdots \geq 0$ and $\sum_i x_i = 1$, is obtained as the limit in distribution as $n \to \infty$ of a time-shifted sequence of ranked additive coalescent processes, starting with n equal masses of size $1/n$ at time $-\frac{1}{2} \log n$. It is known that there are many other such "eternal" additive coalescents.

10.4. Poisson cutting of the Brownian tree An explicit construction of the standard additive coalescent is obtained by cutting the branches of a Brownian tree by a Poisson point process of cuts along the skeleton of the tree at rate one per unit time per unit length of skeleton. This yields the *Brownian fragmentation process*, from which the standard additive coalescent is recovered by a non-linear time reversal.

10.1. Random p-forests and Cayley's multinomial expansion

It is hard to overemphasize the importance of Cayley's discovery that in the expansion of $(x_1 + \cdots + x_n)^{n-2}$ the multinomial coefficient of $\prod_i x_i^{n_i}$ is the number of unrooted trees labeled by $[n]$ in which each vertex i has degree $n_i - 1$. Many variations of Cayley's expansion are known. One of the most useful can be presented as follows. For a finite set S and $R \subseteq S$, let $\text{FOR}(S, R)$ be the set of all forests labeled by S, whose set of roots is R. And for $F \in \text{FOR}(S, R)$ and $s \in S$ let F_s denote the set of children of s in F. So $|F_s|$ is the number of children, or *in-degree* of s in F. Recall that edges of F are assumed to be directed towards the roots, and note that for each $F \in \text{FOR}(S, R)$ the sets F_s as s ranges over S are disjoint, possibly empty sets, whose union is $S - R$. Then there is the *forest volume formula*

$$\sum_{F \in \text{FOR}(S,R)} \prod_{s \in S} x_s^{|F_s|} = \left(\sum_{r \in R} x_r \right) \left(\sum_{s \in S} x_s \right)^{|S|-|R|-1} . \tag{10.1}$$

For $|R| = 1$ this amounts to Cayley's expansion of $(\sum_{s \in S} x_s)^{|S|-2}$, and for $x_s \equiv 1$ it yields Cayley's formula

$$|\text{FOR}(S, R)| = |R|\, |S|^{|S|-|R|-1}. \tag{10.2}$$

See [360] for various proofs of the forest volume formula (10.1), and [359] for a number of probabilistic applications. Taking $S = [n]$ and summing (10.1) over all subsets R of $[n]$ with $|R| = k$ gives the cruder identity

$$\sum_{F \in \text{FOR}[n,k]} \prod_{s=1}^{n} x_s^{|F_s|} = \binom{n-1}{k-1} \left(\sum_{s=1}^{n} x_s \right)^{n-k} \tag{10.3}$$

where $\text{FOR}[n, k]$ is the set of all forests of k trees labeled by $[n]$. This was obtained earlier in formula (6.22) as one of several enumerations equivalent to the Lagrange inversion formula Section 6.1.

Take $x_s = p_s$ for a probability distribution $p := (p_s)$ on $[n]$, or any other set S with $|S| = n$, to see that for each $k \in [n]$ the formula

$$\mathbb{P}(\mathcal{F}_{n,k} = F) = \binom{n-1}{k-1}^{-1} \prod_{s \in S} p_s^{|F_s|} \tag{10.4}$$

defines the distribution for a random forest $\mathcal{F}_{n,k}$ of k trees labeled by S with $|S| = n$, call it a *p-forest of k trees labeled by S*. In particular, call $\mathcal{F}_{n,1}$ a *p-tree*. Several natural constructions of p-trees from a sequence of independent and identically distributed random variables with distribution p are recalled in the exercises. The next theorem is fundamental to everything which follows.

Theorem 10.1. [356, Theorem 11] *Let p be a probability distribution on S with $|S| = n$, and let $(\mathcal{F}_{n,k}, 1 \le k \le n)$ be a sequence of random forests labeled by S. The following two descriptions are equivalent, and imply that $\mathcal{F}_{n,k}$ is a p-forest of k trees labeled by S with distribution* (10.4)*:*

- *$\mathcal{F}_{n,1}$ is a p-tree labeled by S, and given $\mathcal{F}_{n,1}$, for each $1 \le k \le n-1$ the forest $\mathcal{F}_{n,k+1}$ is derived from $\mathcal{F}_{n,k}$ by deletion of a single edge e_k, where where $(e_k, 1 \le k \le n-1)$ is a uniform random permutation of the set of $n-1$ edges of $\mathcal{F}_{n,1}$;*
- *$\mathcal{F}_{n,n}$ is the trivial forest with n roots and no edges, and for each $n \ge k \ge 2$, given $\mathcal{F}_{n,j}$ for $n \ge j \ge k$, the forest $\mathcal{F}_{n,k-1}$ is derived from $\mathcal{F}_{n,k}$ by addition of a single directed edge $X_{k-1} \to Y_{k-1}$, where Y_{k-1} has distribution p, and given also Y_{k-1} the vertex X_{k-1} is picked uniformly at random from the set of $k-1$ roots of the tree components of $\mathcal{F}_{n,k}$ other than the one containing Y_{k-1}.*

Proof. Starting from either description of the sequence, the formula (10.4) for the distribution of $\mathcal{F}_{n,k}$ can be established by induction. Then the time-reversed description follows by Bayes rule. □

Corollary 10.2. *For a forest-valued process $(\mathcal{F}_{n,k}, 1 \le k \le n)$ as in* Theorem 10.1*, let $\Pi_{n,k}$ be the partition of S with $|S| = n$ generated by the k tree components of $\mathcal{F}_{n,k}$. Then the sequence $(\Pi_{n,n}, \Pi_{n,n-1}, \ldots, \Pi_{n,1})$ develops according to the following dynamics: given $\Pi_{n,n}, \Pi_{n,n-1}, \ldots, \Pi_{n,k}$ with $\Pi_{n,k} = \{A_1, \ldots, A_k\}$ say, the next partition $\Pi_{n,k-1}$ is obtained from $\{A_1, \ldots, A_k\}$ by merging blocks A_i and A_j with probability $(p(A_i) + p(A_j))/(k-1)$.*

Exercises

10.1.1. (A probabilistic derivation of Cayley's expansion) [356] In Theorem 10.1, starting from Description 2 of a sequence of coalescent random forests, the proof shows that formula (10.4) defines a probability distribution on forests of k trees labeled by S with $|S| = n$, which is equivalent to the form (10.3) of Cayley's multinomial expansion.

10.1.2. (Cayley's expansion over rooted trees.) According to (10.3) for $k = 1$, in the expansion of $(x_1 + \cdots + x_n)^{n-1}$, the multinomial coefficient of $\prod_i x_i^{n_i}$ is the number of rooted trees labeled by $[n]$, with edges directed towards the root, in which vertex i has in-degree n_i for all $i \in [n]$. Deduce (10.3) for general k from this special case of (10.3). Harder [360]: deduce the forest volume formula (10.1) from this case of (10.3).

10.1.3. (Distribution of the root of a p-tree). Show that the root of a p-tree has distribution p.

10.1.4. (Joyal's bijection and p-mappings) [223], [359, §4.1] In the expansion of $(x_1 + \cdots + x_n)^n$, the multinomial coefficient of $\prod_i x_i^{n_i}$ is the number of mappings from $[n]$ to $[n]$ in whose digraph vertex i has in-degree n_i for all $i \in [n]$. Deduce Cayley's expansion over rooted trees from this, by a suitable bijection between mappings m and *marked rooted trees* (t, i), where t is a rooted tree labeled by $[n]$ and $i \in [n]$. To construct the bijection, observe that if the range of the directed path from i to the root of t is a set C of c elements, the path defines a map from $[c]$ to C, which is bijectively equivalent to one from C to C; now rearrange the digraph of t to make a mapping digraph whose set of cyclic points is C. Deduce that if $\mathcal{F}_{n,1}$ is a p-tree, and X is independent of $\mathcal{F}_{n,1}$ with distribution p, then the number of vertices of $\mathcal{F}_{n,1}$ on the path from X to the root of $\mathcal{F}_{n,1}$ has the same distribution as the number of cyclic points of a *p-mapping* $M_n : [n] \to [n]$ where the $M_n(i)$ are independent and identically distributed according to p. See [82, 359] for more in this vein.

10.1.5. (Burtin's formula for p-mappings) [80, 25, 359] Derive the following probabilistic equivalent of the forest volume formula: if M_n is a p-mapping from $[n]$ to $[n]$, as defined in the previous exercise, then for each subset R of $[n]$, the probability that $[n] - R$ contains no cycle of M_n is $p(R)$.

10.1.6. (Restrictions of p-forests) [359, Theorem 19] Let a random forest $\mathcal{F}$ labeled by S be a *p-forest*, meaning that $\mathcal{F}$ is a p-forest of k trees given that $\mathcal{F}$ has k trees. For each non-empty subset B of $\mathcal{F}$, the restriction of $\mathcal{F}$ to B is a $p(\cdot|B)$ forest. Find an explicit formula for the distribution of the number of edges of the restricted forest in terms of $|S|, |B|, p(B)$ and the distribution of the number of edges of $\mathcal{F}$. In particular, if the number of edges of $\mathcal{F}$ has binomial$(|S| - 1, q)$ distribution, then the number of edges of $\mathcal{F}$ contained in B has binomial$(|S| - 1, p(B)q)$ distribution. Note the special case $q = 1$, when $\mathcal{F}$ is a p-tree.

10.1.7. (A symmetry of uniform random forests) [356] Starting with a forest of $m+1$ trees defined by random deletion of m edges from a uniform random tree over $[n]$, given that the tree containing vertex 1 has a particular set V of j vertices, the remaining random forest labelled by $[n] - V$ has the same distribution as if m edges were deleted at random from a uniformly distributed random tree labelled by $[n] - V$.

Notes and comments

This section is based on [356, 359, 360]. See [34, 359] for further properties of p-forests, p-mappings, and connections with other polynomial expansions due to Hurwitz. See [93, 92] for another representation of the additive coalescent related to parking functions.

10.2. The additive coalescent

This section assumes some familiarity with the notion of a coalescent process, as introduced in Section 5.1. Let $\mathcal{P}_1^{\downarrow}$ denote the space of sequences $x = (x_i)_{i \geq 1}$

with $x_1 \geq x_2 \geq \cdots \geq 0$ and $\sum_i x_i = 1$. Think of each $x \in \mathcal{P}_1^{\downarrow}$ as describing the ranked masses in some unlabeled collection of masses. Let $D(\mathbb{R}_{\geq 0}, \mathcal{P}_1^{\downarrow})$ be the Skorokhod space for $\mathcal{P}_1^{\downarrow}$-valued càdlàg processes (with the l_1 metric on $\mathcal{P}_1^{\downarrow}$) For $x \in \mathcal{P}_1^{\downarrow}$ let $\mu_K(x, dx')$ be the σ-finite discrete measure on $\mathcal{P}_1^{\downarrow}$ with a weight of $K(x_i, x_j)$ at $x^{\oplus(i,j)}$, where $x^{\oplus(i,j)}$ is derived from x by first merging the two masses x_i and x_j to form a single mass of size $x_i + x_j$, then re-ranking. Write μ_+ instead of μ_K for the *additive kernel* $K(x,y) = x + y$.

Theorem 10.3. [141] *There exists a unique family* $(\mathbb{P}^x, x \in \mathcal{P}_1^{\downarrow})$ *of distributions on* $D(\mathbb{R}_{\geq 0}, \mathcal{P}_1^{\downarrow})$ *and a unique transition semigroup* $(P_t, t \geq 0)$ *on* $\mathcal{P}_1^{\downarrow}$*, such that if* $\mathbb{P}^x$ *governs* $(X_t^x, t \geq 0)$ *as the Markov process with* $X_0^x = x \in \mathcal{P}_1^{\downarrow}$ *and semigroup* $(P_t, t \geq 0)$*, called the* ranked additive coalescent*, then*

- *If* x *is a finite partition of* 1 *then the process* $(X_t^x, t \geq 0)$ *is the additive coalescent defined as a jump-hold process with transition rates* $\mu_+(x, dx')$.
- X^x *is a binary coalescent, meaning that collisions of more than two clusters at one time do not occur.*
- $\mu_+(x, dx')$ *is a jump kernel for* X^x*, in the usual sense of a Lévy system.*
- X^x *is a càdlàg Hunt process and both* $x \mapsto P_t(x, \cdot)$ *and* $x \mapsto \mathbb{P}^x$ *are weakly continuous mappings from* $\mathcal{P}_1^{\downarrow}$ *to the spaces of probability measures on* $\mathcal{P}_1^{\downarrow}$ *and* $D(\mathbb{R}_{\geq 0}, \mathcal{P}_1^{\downarrow})$ *respectively.*

Intuitively, it is to be expected that a result like this should hold for more general collision kernels than the additive kernel $K(x,y) = x + y$. But this does not seem easy to prove. Aldous [9] gave a variant of this result for the multiplicative kernel $K(xy) = xy$, working in the larger statespace of ranked decreasing square-summable sequences. As discussed in [356], both the additive and multiplicative kernels have natural interpretations in terms of the evolution of random graphs. This means that the existence and uniqueness of both additive and multiplicative coalescents, with very general initial conditions, can be established by direct combinatorial constructions. Such constructions also allow a much deeper analysis of these coalescents than has yet been provided for any other stochastic coalescent processes.

The essentially combinatorial nature of the additive coalescent is exposed by the following continuous time variant of Theorem 10.1 and Corollary 10.2:

Theorem 10.4. [356] *Let* p *be a probability distribution on* S *with* $|S| = n$*. Let* $\varepsilon_i, 1 \leq i \leq n-1$ *be a sequence of independent standard exponential variables. Let* $(\mathcal{F}_n(t), t \geq 0)$ *be the forest-valued process, with state-space the set of all forests labeled by* S*, whose jump times are the* ε_i*, and which may be described in either of the following equivalent ways, according to whether time is run forwards or backwards:*

- $\mathcal{F}_n(0)$ *is the forest of* n *trivial trees with* n *root vertices, and* $\mathcal{F}_n(\varepsilon_i)$ *is derived from and* $\mathcal{F}_n(\varepsilon_i -)$ *for each* $1 \leq i \leq n-1$ *by adding an edge* $R_i \to X_i$*, where* $X_1, X_2, , X_{n-1}$ *are independent random variables with distribution* p *on* $[n]$*, independent of* $\varepsilon_1, \varepsilon_2, , \varepsilon_{n-1}$*, and conditionally given*

$\mathcal{F}_n(t), 0 \le t < \varepsilon_i$ the vertex R_i is picked uniformly at random from the set of root vertices of trees of $\mathcal{F}_n(\varepsilon_i-)$ other than the tree that contains X_i;

- *$\mathcal{F}_n(\infty)$ is a p-tree, and $\mathcal{F}_n(\varepsilon_i-)$ is derived from $\mathcal{F}_n(\varepsilon_i)$ for each i by deleting an edge $R_i \to X_i$, which conditionally given ε_i and $\mathcal{F}_n(t), t \ge \varepsilon_i$, is picked uniformly at random from the edges of $\mathcal{F}_n(\varepsilon_i)$.*

Let $\Pi_n(t)$ be the partition of $[n]$ generated by components of $F_n(t)$. Then the process $(\Pi_n(t), t \ge 0)$ an additive coalescent with mass distribution p, meaning that at each time t, each pair of blocks say A and B of $\Pi_n(t)$ is merging to form a block $A \cup B$ at transition rate $p(A) + p(B)$, as in Definition 5.2

Observe that the forest $\mathcal{F}_n(t)$ constructed in the above theorem contains precisely those edges $R_i \to X_i$ of $\mathcal{F}_n(\infty)$ whose birth times ε_i are such that $\varepsilon_i \le t$. To generalize this construction to yield a ranked additive coalescent whose initial state is some infinite partition (p_i) of 1, the idea is to derive a similar family of coalescing random forests by cutting up a suitable infinite random tree whose distribution is determined by p. The key is to find a definition of p-trees which makes sense for infinitely supported p, and which reduces to the previous definition for finitely supported p. The definition which works is the following:

Definition 10.5. [141, 82] Let $p = (p_j, j \in J)$ be a non-degenerate probability distribution on a finite or infinite set J. Call a random directed graph $\mathcal{T}$ with vertex set J a *p-tree* if the random set of edges of $\mathcal{T}$ has the same distribution as those of the *birthday tree*

$$\mathcal{T}(Y_0, Y_1, \ldots) := \{Y_n \to Y_{n-1} : Y_n \notin \{Y_0, \ldots, Y_{n-1}\} \subseteq J \times J\}$$

derived from a sequence of independent and identically distributed random variables Y_n with distribution p on J, where $i \to j := (i, j) \in J \times J$.

By construction, the root of the birthday tree is Y_0, which has distribution p on J. The term *birthday tree* is suggested by the close relation between $\mathcal{T}(Y_0, Y_1, \ldots)$ and the classical *birthday problem*, concerning the number of repeated values among the first n values of an independent and identically distributed sequence $Y_0, \ldots, Y_{n-1}$. For instance, the construction of the birthday tree starts with a line of $n-1$ edges directed from Y_{n-1} towards the root Y_0 iff there are no repeated values in the first n terms $Y_0, \ldots, Y_{n-1}$. Thinking of J as the set of days of the year, and p_j the probability that someone has day j as their birthday, the first $n-1$ edges of the construction of the birthday tree fall in a line iff no repeat birthday is observed in a sample of n individuals. The fact that $\mathcal{T}(Y_0, Y_1, \ldots)$ is just a p-tree according to the previous definition (10.4) when J is finite is not obvious, but it is a consequence of the *Markov chain tree theorem* [76, Theorem 1],[292] which identifies the distribution of $\mathcal{T}(Y_0, Y_1, \ldots)$ for a stationary ergodic Markov chain (Y_n) with finite state space, up to a normalization constant. For an independent and identically distributed sequence, the fact that the normalization constant is 1 is equivalent to Cayley's multinomial expansion over trees.

A ranked additive coalescent with arbitrary initial state can now be constructed according to the following consequence of Theorem 10.3 and Definition 10.5:

Corollary 10.6. [141, Corollary 20] *Let $x \in \mathcal{P}_1^{\downarrow}$ have $x_1 < 1$, and let $\mathcal{T} = \mathcal{T}(Y_0, Y_1, \ldots)$ be the p-tree derived from independent random variables Y_n with some distribution p on a countable set, such that the sequence of ranked atoms of p is x. Let $(W_i \to X_i, i = 1, 2, \ldots)$ be a list of the directed edges of $\mathcal{T}$, let $(\varepsilon_i)_{i \geq 1}$ be a sequence of independent standard exponential variables independent of $(Y_n)_{n \geq 0}$, let $\mathcal{F}(t)$ be the random forest*

$$\mathcal{F}(t) := \{W_i \to X_i : \varepsilon_i \leq t, i = 1, 2, \ldots\} \tag{10.5}$$

and let $X(t)$ be the ranked p-masses of tree components of $\mathcal{F}(t)$. Then $(X(t), t \geq 0)$ is a realization of the ranked additive coalescent with initial state $X(0) = x$ and càdlàg paths.

Exercises

10.2.1. (Semi-group of the additive coalescent) [356, 141] Let $(\Pi(t), t \geq 0)$ be the $\mathcal{P}_{[n]}$-valued additive coalescent starting with the partition into singletons, with masses $p_1, \ldots, p_n$. Then the distribution of $|\Pi(t)| - 1$ is binomial with parameters $n - 1$ and e^{-t}:

$$\mathbb{P}(|\Pi(t)| = k) = \binom{n-1}{k-1} (1 - e^{-t})^{n-k} e^{-(k-1)t} \quad (1 \leq k \leq n) \tag{10.6}$$

and for any partition $\{A_1, \ldots, A_k\}$ of $[n]$,

$$P(\Pi(t) = \{A_1, \ldots, A_k\}) = e^{-(k-1)t}(1 - e^{-t})^{n-k} \prod_{i=1}^{k} p_{A_i}^{|A_i|-1} \tag{10.7}$$

where $p_A = \sum_{i \in A} p_i$.

10.3. The standard additive coalescent

A process of particular interest is the additive coalescent started with an initial mass distribution that is uniform over n possible values. Recall that $\mathcal{P}_{[n]}$ denotes the set of partitions of $[n] := \{1, \ldots, n\}$. If we consider the $\mathcal{P}_{[n]}$-valued additive coalescent process $(\Pi_n(t), t \geq 0)$, with uniform mass distribution on $[n]$, it is obvious that $\Pi_n(t)$ is an exchangeable random partition of $[n]$. More precisely, using notation from Section 1.5, we read from (10.7) and (10.6) that $\Pi_n(t)$ is a Gibbs $(v_\bullet^{(n,t)}, w_\bullet)$ distribution on partitions of $[n]$, with weight sequences $w_j = j^{j-1}$, the number of ways to assign a rooted tree structure to a set of j elements, and the weight sequence $v_\bullet^{(n,t)}$ is such that the distribution of $|\Pi_n(t)| - 1$ binomial$(n-1, 1-e^{-t})$. It is easily seen that for fixed t, these Gibbs distributions on partitions of $[n]$ are not consistent as n varies. So, unlike the situation for Kingman's coalescent with constant collision kernel, described in Section 5.1, these $\mathcal{P}_{[n]}$-valued additive coalescents processes are not just projections of some process with values in the set of partitions of positive integers.

According to the Gibbs distribution of $\Pi_n(t)$, given $|\Pi_n(t)| = k$ the sizes of the k components of $\Pi_n(t)$, presented in exchangeable random order, are distributed like $(X_1, \dots, X_k)$ given $\sum_{i=1}^k X_i = n$ where the X_i are independent and identically distributed with the *Borel distribution*

$$\mathbb{P}(X_i = m) = \frac{e^{-m} m^{m-1}}{m!}$$

of the total progeny of a critical Poisson branching process. By Stirling's formula

$$\mathbb{P}(X_i = m) \sim \frac{1}{\sqrt{2\pi m^3}} \text{ as } m \to \infty. \tag{10.8}$$

Hence, according to Theorem 2.5, if

$$k \to \infty \text{ and } n \to \infty \text{ with } \frac{k}{\sqrt{n}} \to \lambda \tag{10.9}$$

the ranked sequence derived from

$$(X_i/n, 1 \le i \le k) \text{ given } \sum_{i=1}^k X_i = n$$

converges in distribution to the sequence of ranked jumps of $(T_s, 0 \le s \le \lambda)$ given $T_\lambda = 1$, where $(T_s, s \ge 0)$ is the stable($\frac{1}{2}$) subordinator with $\mathbb{E}(e^{-\xi T_s}) = \exp(-s\sqrt{2\xi})$, so $T_s \stackrel{d}{=} s^2/B_1^2$ for B_1 standard Gaussian. To abbreviate, denote the distribution of this sequence on $\mathcal{P}_1^{\downarrow}$ by $\mathrm{PD}(\frac{1}{2}||\lambda)$, and recall from Section Section 4.5 that this is the law of ranked lengths of excursions of B_λ^{br}, where B_λ^{br} is a standard Brownian bridge B^{br} conditioned on $L_1^0(B^{\mathrm{br}}) = \lambda$, where $L_1^0(B^{\mathrm{br}})$ is the local time of B^{br} at level 0 and time 1, with the usual normalization of occupation density relative to Lebesgue measure. To get a weak limit for the normalized ranked component sizes as $n \to \infty$, the process $(\Pi_n(t), t \ge 0)$ must be run long enough so that $|\Pi_n(t)|$ is of order $\sqrt{n}$. Let

$$h_n := \tfrac{1}{2} \log n.$$

Since $|\Pi_n(t)| - 1$ has binomial$(n-1, e^{-t})$ distribution, if we take $t = h_n + r$ for some $r \in \mathbb{R}$ we find that $\mathbb{E}(|\Pi_n(h_n + r)|) \sim \sqrt{n}e^{-r}$ and the variance is of the same order, so that

$$|\Pi_n(h_n + r)|/\sqrt{n} \stackrel{d}{\to} e^{-r}. \tag{10.10}$$

This brings us to:

Theorem 10.7. [141], [18, Proposition 2] *Let $X_n(t) \in \mathcal{P}_1^{\downarrow}$ be the sequence of masses at time t in a ranked additive coalescent process started with n equal masses of $1/n$ at time $t = 0$. Then as $n \to \infty$*

$$(X_n(h_n + r), -h_n \leq r) \xrightarrow{d} (X_\infty(r), r \in \mathbb{R})$$

in the sense of convergence in distribution on the Skorokhod space of càdlàg paths, with $\mathcal{P}_1^{\downarrow}$ given the l_1 metric, where the limit process is the unique additive coalescent parameterized by $r \in \mathbb{R}$ such that $X_\infty(r)$ has the $PD(\frac{1}{2}||e^{-r})$ distribution of ranked lengths of excursions of B^{br} given given $L_1^0(B^{\mathrm{br}}) = e^{-r}$.

Proof. The convergence in distribution of $X_n(h_n+r)$ to $X_\infty(r)$ for each fixed r was just argued. Convergence in the Skorokhod space then follows immediately from the last regularity property of the ranked additive coalescent process listed in Theorem 10.3. □

The process X_∞ defined by Theorem 10.7 is called the *standard additive coalescent.* Compare with the standard multiplicative coalescent defined by Theorem 6.13. Theorem 10.7 immediately raises the question of whether there exist any other eternal additive coalescents besides time shifts of the standard one. The answer, given in [21], is yes, there are rather a lot of them, but the extreme ones can all be constructed by a natural generalization of the construction of the standard additive coalescent considered in the next section. See also Bertoin [44] for another approach to the solution of this problem, based on processes with exchangeable increments instead of random trees.

10.4. Poisson cutting of the Brownian tree

According to Theorem 10.1, the additive coalescent started with uniform distribution on n masses can be represented in reversed time by successively cutting the edges of a random tree with uniform distribution over the n^{n-1} rooted trees labeled by $[n]$. According to a basic result of Aldous (Theorem 6.4) the structure of this tree, suitably normalized, converges to $\text{TREE}(B^{\mathrm{ex}})$, the tree in a standard Brownian excursion. Moreover, the scaling involved in Theorem 10.7 makes the process of cutting edges of a uniform random tree converge to a Poisson process of cuts along the skeleton of the limiting Brownian tree, with intensity λ per unit length in the limit as $k/\sqrt{n} \to \lambda$, where k is the number of trees in the forest, corresponding to the number of steps backward in time from the terminal state of the discrete coalescent process derived by cutting the uniform tree with n vertices.

This line of reasoning suggests it should be possible to construct the standard additive coalescent by a process of cutting up the branches of a Brownian tree by a Poisson point process along the skeleton of the tree, and keeping track of the ranked masses of tree components so formed. This was shown in [18], along with various other regularity properties of the standard additive coalescent which follow from this construction.

To make this construction, let $\mathcal{T} := \text{TREE}(2B^{\text{ex}})$ be the random tree structure put on $[0,1]$ by a standard Brownian excursion B^{ex}, with all edge-lengths multiplied by 2. As in Theorem 7.9, let $U_1, U_2, \ldots$ be a sequence of independent uniform variables, independent of B^{ex}, and for $n \geq 0$ let

$$\mathcal{T}_n := \text{SUBTREE}(2B^{\text{ex}}; \{0, U_1, \ldots, U_n\}), \tag{10.11}$$

regarded as a subset of $[0,1]$ equipped with the pseudo-metric of $\text{TREE}(2B^{\text{ex}})$. According to Theorem 7.9, the tree $\mathcal{T}_n$ is isometric to a plane tree which can be made by the Poisson line-breaking construction. Define the *skeleton* $\mathcal{T}^\dagger$ of $\mathcal{T}$ to be random subset of $[0,1]$ which is the union over leaves u and v of $\mathcal{T}$ of the open path from u to v in $\mathcal{T}$, meaning the usual closed path from u to v, with endpoints excluded. Note that $\mathcal{T}^\dagger$ is almost surely dense in $[0,1]$, both in the usual topology, and in the tree topology of $\mathcal{T}$. Moreover, $\mathcal{T}^\dagger = \cup_n \mathcal{T}_n^\dagger$ almost surely, where $\mathcal{T}_n^\dagger := \mathcal{T}_n - \{\tilde{0}, \tilde{U}_1, \ldots \tilde{U}_n\}$ where $\tilde{u}$ is the $\mathcal{T}$ equivalence class of u. In fact $\tilde{0} = \{0,1\}$ and $\tilde{U}_i = \{U_i\}$ for all i almost surely. The construction of the sequence of trees $\mathcal{T}_n$ induces a natural *skeleton search map*

$$\sigma^\dagger : \mathbb{R}_{>0} - \{|\mathcal{T}_1|, |\mathcal{T}_2|, \ldots\} \to \mathcal{T}^\dagger$$

whereby for each $n \geq 1$ the open interval $(|\mathcal{T}_n|, |\mathcal{T}_{n+1}|)$ is mapped by depth-first search to the branch in $\mathcal{T}_{n+1}$ which leads from $\mathcal{T}_n$ to U_{n+1}. Here $\mathcal{T}_0 := \{0,1\}$, $|\mathcal{T}_0| = 0$, and $|\mathcal{T}_n|$ for $n \geq 1$ is the total length of the plane tree which is isometric to $\mathcal{T}_n$ regarded as a subset of $\text{TREE}(2B^{\text{ex}})$. The *length measure* on $\mathcal{T}^\dagger$ is Lebesgue measure on $\mathbb{R}_{>0}$ pushed onto $\mathcal{T}^\dagger$ by the skeleton search. There are now two measures on $[0,1]$, which are almost surely mutually singular, and which it is essential to distinguish carefully:

- ordinary Lebesgue measure on $[0,1]$, to be called the *mass measure*, which is concentrated on the random set of leaves of $\mathcal{T}$.
- *length measure on the skeleton* which is an infinite random measure concentrated on the random subset $\mathcal{T}^\dagger$ of $[0,1]$, which is disjoint from the random set of leaves of $\mathcal{T}$.

Now, independent of the lengths $|\mathcal{T}_n|, n \geq 1$ which are the points of a Poisson process on $\mathbb{R}_{>0}$ with intensity $t\,dt$, let N be a Poisson point process in $\mathbb{R}^2_{>0}$. For $\lambda > 0$ let $0 < T_{1,\lambda} < T_{2,\lambda} < \cdots$ be the successive points t such that there is a point (t,u) of N with $u \leq \lambda$, and let $T^\dagger_{1,\lambda}, T^\dagger_{2,\lambda}, \cdots$ be the images of these points via the skeleton search, call them the λ-*cuts* on the skeleton of $\mathcal{T}$. So by construction, conditionally given $\mathcal{T}$, the λ-cuts are the points of a Poisson point process with intensity λ per unit length on the skeleton of $\mathcal{T}$, and as λ increases the set of λ-cuts increases. Now we can formulate:

Theorem 10.8. [18, Theorem 3] *Let $\mathcal{F}_\lambda$ be the random forest whose tree components are the Borel subsets of $[0,1]$ defined to be the equivalence classes for the random equivalence relation $u \sim_\lambda v$ iff the path from u to v in $\mathcal{T}$ does not contain any λ-cut, where the λ-cuts fall on the skeleton of $\mathcal{T}$ according to a Poisson process of rate λ per unit length on the skeleton, which intensifies as λ*

increases. Let $Y(\lambda)$ be the sequence of ranked masses of tree components of $\mathcal{F}_\lambda$. Then $Y(\lambda) \in \mathcal{P}_1^\downarrow$ for all $\lambda > 0$ almost surely, and the process $(Y(e^{-r}), r \geq 0)$ admits a càdlàg modification which is a realization of the ranked additive coalescent.

Some technical points had to be dealt with in [18] to prove this result, but the intuitive idea should be clear: after passage to the limit from a uniform random tree on n vertices, call it u_n-tree,

- the length measure on the skeleton of $\mathcal{T}$ should be regarded as the continuum limit of length measure on the branches of the u_n-tree, with normalization by $\sqrt{n}$;
- the mass measure on the leaves of $\mathcal{T}$ should be regarded as the continuum limit of counting measure on the leaves of the u_n-tree, with normalization by n.

Thus the continuum analog of cutting edges by a process of Bernoulli trials with some probability p is a Poisson process of cuts along the skeleton at some constant rate λ per unit length. Note from Lemma 4.10 and Lemma 3.11 that $Y_m(\lambda)$, the mth term of $Y(\lambda)$, which is the mth largest mass, is such that

$$Y_m(\lambda) \sim \frac{2}{\pi} \frac{\lambda^2}{m^2} \text{ almost surely as } m \to \infty \text{ for each } \lambda > 0 \tag{10.12}$$

so the Poisson intensity parameter λ is encoded almost surely in the state $Y(\lambda)$. In particular, for $\lambda \neq \lambda'$ the laws of $Y(\lambda)$ and $Y(\lambda')$ on $\mathcal{P}_1^\downarrow$ are mutually singular, though the laws of the first m components of $Y(\lambda)$ and $Y(\lambda')$ on $[0,1]^m$ are mutually absolutely continuous for every m.

Definition 10.9. *Call the $\mathcal{P}_1^\downarrow$-valued process $(Y(\lambda), \lambda > 0)$ the* Brownian fragmentation process.

The distribution of $(Y(\lambda), \lambda > 0)$ for each fixed $\lambda > 0$ is the distribution of ranked lengths of excursions of a Brownian bridge B^{br} given $L_1^0(B^{\mathrm{br}}) = \lambda$. So a considerable amount of information about this process can be read from the results of Section 4.5. Following is a selection of such results, with new interpretations.

Moment formulae Let $Y_*(\lambda)$ be a size-biased pick from $Y(\lambda)$, which may be understood as the size of the tree component of the forest $\mathcal{F}_\lambda$ which contains U picked uniformly at random from $[0,1]$ independently of B^{ex}. (One could just as well take the component containing 0, by an obvious symmetry of uniform trees, and passage to the Brownian limit.) Let f_λ denote the density of $Y_*(\lambda)$, which can be read from (4.7), as in [18, (8)]:

$$f_\lambda(y) = (2\pi)^{-1/2}\lambda y^{-1/2}(1-y)^{-3/2}\exp(-\tfrac{1}{2}\lambda^2 y/(1-y)) \quad (0 \leq y < 1). \tag{10.13}$$

Then there is the basic identity (2.23)

$$\mathbb{E}\left(\sum_i g(Y_i(\lambda))\right) = \mathbb{E}\left(\frac{g(Y_*(\lambda))}{Y_*(\lambda)}\right) = \int_0^1 y^{-1} g(y) f_\lambda(y) dy \tag{10.14}$$

which is valid for all $\lambda \geq 0$ and all non-negative measurable functions g. For $q \in \mathbb{R}$ define

$$\mu_q(\lambda) := \mathbb{E}\left(\sum_i Y_i^{q+1}(\lambda)\right) = \mathbb{E}\left[(Y_*(\lambda))^q(\lambda)\right] = \int_0^1 y^q f_\lambda(y)dy. \tag{10.15}$$

For n a positive integer, $\mu_n(\lambda)$ can be interpreted as follows. Let $U_1, U_2, \ldots$ be independent uniform $(0,1)$ variables independent of the Brownian tree $\mathcal{T} :=$ TREE$(2B^{\text{ex}})$. Given the sequence $Y(\lambda)$ of masses of tree components of the forest $\mathcal{F}_\lambda$ derived by cutting the skeleton of $\mathcal{T}$ at rate λ, the event that $0, U_1, \ldots, U_n$ all fall in the same component of $\mathcal{F}_\lambda$ has probability $\sum_i Y_i^{n+1}(\lambda)$. So the unconditional probability of this event is $\mu_n(\lambda)$. On the other hand, this event occurs if and only if the Poisson cut process has no points up to time λ in $\mathcal{T}_n :=$ SUBTREE$(2B^{\text{ex}}; \{0, U_1, \ldots, U_n\})$, as in Theorem 7.9. Given the total length $\Theta_n := |\mathcal{T}_n|$ of this subtree, the event occurs with probability $e^{-\lambda\Theta_n}$. But according to the Poisson line-breaking construction of Theorem 7.9, Θ_n is the time of the nth arrival of an inhomogeneous Poisson process on $(0, \infty)$ with rate t at time t. Thus

$$\mu_n(\lambda) = \mathbb{E}(e^{-\lambda\Theta_n}) \qquad (n = 1, 2, \ldots)$$

where for $t > 0$

$$\mathbb{P}(\Theta_n \in dt) = e^{-\frac{1}{2}t^2} \frac{(\frac{1}{2}t^2)^{n-1}}{(n-1)!} t\, dt \tag{10.16}$$

It follows that for $n = 1, 2 \ldots$

$$\mu_n(\lambda) = \frac{2^{1-n}}{(n-1)!} \int_0^\infty t^{2n-1} e^{-\frac{1}{2}t^2 - \lambda t} dt = \frac{2^{1-n}}{(n-1)!} \Psi_{2n}(\lambda) \tag{10.17}$$

where in terms of a standard Gaussian random variable B_1 with density $\varphi(z) := \frac{1}{\sqrt{2\pi}} e^{-\frac{1}{2}z^2}$, for x real and $p > 0$

$$\Psi_p(x)\varphi(x) = \mathbb{E}[(B_1 - x)^{p-1} 1(B_1 > x)] = \frac{1}{p} \mathbb{E}[B_1 (B_1 - x)^p 1(B_1 > x)] \tag{10.18}$$

where the first equality is read from (10.17) by a change of variable, and the second equality, obtained by integration by parts, is an instance of Stein's identity $\mathbb{E}[f'(B_1)] = \mathbb{E}[B_1 f(B_1)]$ which is valid for all sufficiently smooth f vanishing at $\pm\infty$. It is also known [280] that

$$\Psi_p(z) = \Gamma(p) h_{-p}(z) \qquad (\Re p > 0) \tag{10.19}$$

where h_ν is the Hermite function defined by a different integral representation in (4.59) . The agreement of formulae (10.17) and (4.59) provides a substantial check on the entire circle of results related to the Brownian asymptotics of fragmentation of a uniform random tree by random deletion of edges.

Self-similarity of Brownian fragmentation Brownian scaling yields the following lemma:

Lemma 10.10. *or $0 < t < \infty$ let $\mathcal{T}(t) := \text{TREE}(2B^{\text{ex},t})$ where $B^{\text{ex},t}$ is a Brownian excursion of length t. Then cutting the skeleton of $\mathcal{T}(t)$ according to a Poisson process of rate λ per unit length creates a forest whose component masses are distributed as $tY(t^{1/2}\lambda)$.*

Consider a uniform random tree on $j(n)$ vertices. When we assign each vertex mass $1/j(n)$ and each edge length $1/\sqrt{j(n)}$, then the random tree converges in distribution to $\mathcal{T}$ in the sense of [18, Lemma 9] If instead we assign each vertex mass $1/n$ and each edge length $1/\sqrt{n}$, where $j(n)/n \to t$, then the random tree converges in distribution to $\mathcal{T}(t)$. Now consider the discrete random forest $\mathcal{F}^n(n-m)$ obtained by deleting m random edges from the uniform random tree on n vertices. Conditional on the vertex-sets $(\mathcal{V}_j, j = 1, 2, \ldots)$ of the components of the forest, each component is a uniform random tree on vertex-set $\mathcal{V}_j$, independently as j varies. Because $Y(\mu)$ arises as a limit of relative sizes of the components of $\mathcal{F}^n(n - m(n))$ as $m(n)/n^{1/2} \to \mu$, we deduce:

Lemma 10.11. *Given $Y(\mu) = (t_1, t_2, \ldots)$, the tree components $\mathcal{T}_i$ of $\mathcal{F}_\mu$ can be identified modulo isometry as a sequence of independent copies of $\mathcal{T}(t_i)$, $i = 1, 2, \ldots$.*

Combining with Lemmas 10.10 and 10.11 gives the following statement of a *self-similar Markov branching property* of the Brownian fragmentation process.

Proposition 10.12. *For each $\mu > 0$, the conditional distribution of $Y(\mu + \lambda)$ given $Y(\mu) = y$ is the distribution of the decreasing reordering of*

$$\{y_i Y_j^{(i)}(y_i^{1/2}\lambda); i, j \geq 1\}$$

where $(Y^{(i)}(\cdot), i \geq 1)$ are independent copies of $Y(\cdot)$.

Thus the Brownian fragmentation process is an instance of the general kind of *self-similar fragmentation process* studied by Bertoin [46].

Bertoin's representation of the Brownian fragmentation process Starting from the standard Brownian excursion B^{ex}, for each $\lambda \geq 0$ let $B_\lambda^{\text{ex}} \in C[0, 1]$ be the *excursion dragged down by drift* λ, that is

$$B_\lambda^{\text{ex}}(u) := B^{\text{ex}}(u) - u\lambda \quad (0 \leq u \leq 1),$$

let

$$\underline{B_\lambda^{\text{ex}}}(u) := \inf_{0 \leq t \leq u} B_\lambda^{\text{ex}}(t)$$

and let $Y^{\text{ex}}(\lambda)$ be the sequence of ranked lengths of excursions away from 0 of the process $(B_\lambda^{\text{ex}}(u) - \underline{B_\lambda^{\text{ex}}}(u), 0 \leq u \leq 1)$. Note that these are the ranked masses of subtrees of $\text{TREE}(B_\lambda^{\text{ex}})$, regarded as a forest of subtrees attached to a forest floor of length λ defined by the branch of $\text{TREE}(B_\lambda^{\text{ex}})$ of length λ which joins 0 to 1.

Theorem 10.13. Bertoin [42]. *The process* $(Y^{ex}(\lambda), \lambda \geq 0)$ *is another realization of the Brownian fragmentation process.*

It can be argued that each excursion of B_λ^{ex} above its past minimum process of duration t is simply a Brownian excursion of duration t, and that these excursions are conditionally independent given their lengths. The self-similar fragmentation property of $(Y^{ex}(\lambda), \lambda \geq 0)$ follows by Brownian scaling. So to prove Theorem 10.13, the main thing to check is that $Y^{ex}(\lambda) \stackrel{d}{=} Y(\lambda)$ for each fixed $\lambda > 0$, which is not so obvious. A subtle feature here is the order structure of the sequence of excursion intervals whose ranked lengths is $Y^{ex}(\lambda)$. See [397] and [91] for further discussion.

Exercises

10.4.1. (Evolution of the mass containing 0) [18] Let $Y_*(\lambda)$ be the mass of the tree component of $\mathcal{F}_\lambda$ that contains 0. Then $Y_*(\lambda)$ is a size-biased pick from the components of $Y(\lambda)$, and

$$(Y_*(\lambda), \lambda \geq 0) \stackrel{d}{=} (1/(1+T_\lambda), \lambda \geq 0) \tag{10.20}$$

where $(T_\lambda, \lambda \geq 0)$ is the stable($\frac{1}{2}$) subordinator with $T_\lambda \stackrel{d}{=} \lambda^2/B_1^2$. The same conclusion holds in Bertoin's model, for $Y_*(\lambda)(\lambda)$ the return time to 0 of B_λ^{ex}. See [42].

10.4.2. (Deletion of a size-biased component) [18, Theorem 4 and Lemma 12] Let $Y_*(\lambda)$ be a size-biased pick from the components of $Y(\lambda)$. The conditional distribution of $Y(\lambda)$ given $Y_*(\lambda) = y$ is the unconditional distribution of the decreasing reordering of $\{y\} \cup \{(1-y)Y_i((1-y)^{-1/2}\lambda), i \geq 1\}$.

10.4.3. (A check on the moment formula) Verify that the moment formula (10.17) is consistent with the self-similar Markov branching property of $(Y(\lambda), \lambda \geq 0)$ described in Proposition 10.12.

10.4.4. (Partition probabilities) Fix $n \geq 2$. Let $Y_{[n]}(\lambda) := (Y_{(1)}, \ldots, Y_{(n)}(\lambda))$ where $Y_{(i)}(\lambda)$ is the mass of the tree-component of $\mathcal{F}_\lambda$ containing U_i where the U_i are independent and uniform on $(0,1)$, independent of $\mathcal{F}_\lambda$. Note that the $Y_{(i)}(\lambda)$ are exchangeable random variables, all distributed like $Y_{(1)} = Y_*(\lambda)$ which is a size-biased pick from the components of $Y(\lambda)$. Write $\Pi_n(\lambda)$ for the partition of $[n]$ generated by the values of $Y_{(i)}(\lambda), i \in [n]$ and write $Y_j^*(\lambda), j = 1, 2, \cdots$ for the sequence of distinct values of $Y_{(i)}(\lambda), i \geq 1$. Observe that given $\Pi_n(\lambda) = \{B_1, \ldots, B_k\}$ say, where the B_j for $1 \leq j \leq k$ are arranged in order of their least elements, $Y_{(i)}(\lambda) = Y_j^*(\lambda)$ for all $i \in B_j$. The joint distribution of $(Y_{(1)}(\lambda), \ldots, Y_{(n)}(\lambda))$ for each fixed $\lambda > 0$ is determined by the following formula: for each partition $\{B_1, \ldots, B_k\}$ of $[n]$ such that $\#B_i = n_i$ for $1 \leq i \leq k$, where the n_i are arbitrary positive integers with sum n, and for $y_1, \ldots, y_k$ with $y_j > 0$ and $\sigma := \sum_j y_j < 1$,

$$P(\Pi_n(\lambda) = \{B_1, \ldots, B_k\} \text{ and } Y_j^*(\lambda) \in dy_j, \text{ for all } 1 \le j \le k) \tag{10.21}$$

$$= \frac{\lambda^k}{(2\pi)^{k/2}} \left(\prod_{j=1}^{k} y_j^{n_j - 3/2}\, dy_j \right) (1-\sigma)^{-3/2} \exp\left(-\frac{\lambda^2}{2} \frac{\sigma}{1-\sigma} \right). \tag{10.22}$$

For $n = 1$ this reduces to (10.13). Deduce the previous formula (4.67) for the EPPF of a $(\frac{1}{2}||\lambda)$ partition by integration of (10.21)-(10.22).

10.4.5. (The splitting time) Consider $Y_{[2]}(\lambda) = (Y_{(1)}(\lambda), Y_{(2)}(\lambda))$, the sizes of tree components of $\mathcal{F}_\lambda$ containing independent uniform variables U_1 and U_2. There is a *splitting time* S defined as the smallest λ for which these tree components are distinct. Then the joint density of $(S, Y_{(1)}(S), Y_{(2)}(S))$ is

$$f(s, y_1, y_2) = \frac{s}{2\pi} y_1^{-1/2} y_2^{-1/2} (y_1 + y_2)(1 - y_1 - y_2)^{-3/2} \exp\left(-\frac{s^2}{2} \frac{y_1 + y_2}{1 - y_1 - y_2} \right).$$

Notes and comments

As a generalization of results of this chapter, a class of inhomogeneous continuum random trees arises naturally from asymptotics of p trees for non-uniform p. References, with applications to the entrance boundary of the additive coalescent, are [82], [21] and [20]. See also [44] and [309] for related work.

Bibliography

[1] R. Abraham. Un arbre aléatoire infini associé à l'excursion brownienne. In *Séminaire de Probabilités, XXVI*, pages 374–397. Springer, Berlin, 1992.

[2] R. Abraham and L. Mazliak. Branching properties of Brownian paths and trees. *Exposition. Math.*, 16(1):59–73, 1998.

[3] A. Adhikari. *Skip free processes.* PhD thesis, University of California, Berkeley, 1986. Department of Statistics.

[4] D. Aldous. A random tree model associated with random graphs. *Random Structures Algorithms*, 1:383–402, 1990.

[5] D. Aldous. The continuum random tree I. *Ann. Probab.*, 19:1–28, 1991.

[6] D. Aldous. The continuum random tree II: an overview. In M. Barlow and N. Bingham, editors, *Stochastic Analysis*, pages 23–70. Cambridge University Press, 1991.

[7] D. Aldous. The continuum random tree III. *Ann. Probab.*, 21:248–289, 1993.

[8] D. Aldous. Recursive self-similarity for random trees, random triangulations and Brownian excursion. *Ann. Probab.*, 22:527–545, 1994.

[9] D. Aldous. Brownian excursions, critical random graphs and the multiplicative coalescent. *Ann. Probab.*, 25:812–854, 1997.

[10] D. Aldous. Brownian excursion conditioned on its local time. *Elect. Comm. in Probab.*, 3:79–90, 1998.

[11] D. Aldous. Deterministic and stochastic models for coalescence (aggregation and coagulation): a review of the mean-field theory for probabilists. *Bernoulli*, 5:3–48, 1999.

[12] D. Aldous, B. Fristedt, P. Griffin, and W. Pruitt. The number of extreme points in the convex hull of a random sample. *J. Appl. Probab.*, 28:287–304, 1991.

[13] D. Aldous and V. Limic. The entrance boundary of the multiplicative coalescent. *Electron. J. Probab.*, 3:1–59, 1998.

[14] D. J. Aldous. Exchangeability and related topics. In P. Hennequin, editor, *École d'été de probabilités de Saint-Flour, XIII—1983*, pages 1–198. Springer, Berlin, 1985. Lecture Notes in Mathematics 1117.

[15] D. J. Aldous, G. Miermont, and J. Pitman. Brownian bridge asymptotics for random p-mappings. *Electron. J. Probab.*, 9:37–56, 2004. MR2041828

[16] D. J. Aldous, G. Miermont, and J. Pitman. Weak convergence of random p-mappings and the exploration process of inhomogeneous continuum random trees. *Probab. Th. and Rel. Fields.*, 129:182–218, 2005. arXiv:math.PR/0401115

[17] D. J. Aldous and J. Pitman. Brownian bridge asymptotics for random mappings. *Random Structures and Algorithms*, 5:487–512, 1994. MR1293075

[18] D. J. Aldous and J. Pitman. The standard additive coalescent. *Ann. Probab.*, 26:1703–1726, 1998. MR1675063

[19] D. J. Aldous and J. Pitman. Tree-valued Markov chains derived from Galton-Watson processes. *Ann. Inst. Henri Poincaré*, 34:637–686, 1998. MR1641670

[20] D. J. Aldous and J. Pitman. A family of random trees with random edge lengths. *Random Structures and Algorithms*, 15:176–195, 1999. MR1704343

[21] D. J. Aldous and J. Pitman. Inhomogeneous continuum random trees and the entrance boundary of the additive coalescent. *Probab. Th. Rel. Fields*, 118:455–482, 2000. MR1808372

[22] D. J. Aldous and J. Pitman. The asymptotic distribution of the diameter of a random mapping. *C.R. Acad. Sci. Paris, Ser. I*, 334:1021–1024, 2002. MR1913728

[23] D. J. Aldous and J. Pitman. Invariance principles for non-uniform random mappings and trees. In V. Malyshev and A. M. Vershik, editors, *Asymptotic Combinatorics with Applications in Mathematical Physics*, pages 113–147. Kluwer Academic Publishers, 2002. MR1999358

[24] D. J. Aldous and J. Pitman. Two recursive decompositions of Brownian bridge. Technical Report 595, Dept. Statistics, U.C. Berkeley, 2002. To appear in *Séminaire de Probabilités XXXIX* . arXiv:math.PR/0402399

[25] S. Anoulova, J. Bennies, J. Lenhard, D. Metzler, Y. Sung, and A. Weber. Six ways of looking at Burtin's lemma. *Amer. Math. Monthly*, 106(4):345–351, 1999. MR1682373

[26] C. Antoniak. Mixtures of Dirichlet processes with applications to Bayesian nonparametric problems. *Ann. Statist.*, 2:1152–1174, 1974.

[27] R. Arratia, A. D. Barbour, and S. Tavaré. *Logarithmic combinatorial structures: a probabilistic approach.* EMS Monographs in Mathematics. European Mathematical Society (EMS), Zürich, 2003. MR2032426

[28] J. Baik and E. M. Rains. Symmetrized random permutations. In *Random matrix models and their applications*, volume 40 of *Math. Sci. Res. Inst. Publ.*, pages 1–19. Cambridge Univ. Press, Cambridge, 2001. MR1842780

[29] A.-L. Basdevant. Ruelle's probability cascades seen as a fragmentation process.

[30] M. Bayewitz, J. Yerushalmi, S. Katz, and R. Shinnar. The extent of correlations in a stochastic coalescence process. *J. Atmos. Sci.*, 31:1604–1614, 1974.

[31] E. A. Bender. Central and local limit theorems applied to asymptotic enumeration. *J. Combin. Theory Ser. A*, 15:91–111, 1973.

[32] E. A. Bender. Asymptotic methods in enumeration. *SIAM Rev.*, 16:485–515, 1974.

[33] J. Bennies and G. Kersting. A random walk approach to Galton-Watson trees. *J. Theoret. Probab.*, 13(3):777–803, 2000. MR1785529

[34] J. Bennies and J. Pitman. Asymptotics of the Hurwitz binomial distribution related to mixed Poisson Galton-Watson trees. *Combinatorics, Probability and Computing*, 10:203–211, 2001. MR1841640

[35] J. Berestycki. Ranked fragmentations. *ESAIM Probab. Statist.*, 6:157–175 (electronic), 2002. MR1943145 (2004d:60196)

[36] J. Berestycki. Exchangeable fragmentation-coalescence processes and their equilibrium measures. *Electronic Journal of Probability*, 9:770–824, 2004.

[37] F. Bergeron, G. Labelle, and P. Leroux. *Théorie des espèces et combinatoire des structures arborescentes*, volume 19 of *Publ. du Lab. de Combinatoire et d'Informatique Mathématique.* Univ de Québec, Montréal, 1994.

[38] F. Bergeron, G. Labelle, and P. Leroux. *Combinatorial species and tree-like structures.* Cambridge University Press, Cambridge, 1998. Translated from the 1994 French original by Margaret Readdy, With a foreword by Gian-Carlo Rota. MR1629341

[39] J. Bertoin. An extension of Pitman's theorem for spectrally positive Lévy processes. *Ann. of Probability*, 20:1464 - 1483, 1992.

[40] J. Bertoin. *Lévy processes.* Cambridge University Press, Cambridge, 1996. MR1406564

[41] J. Bertoin. Subordinators: examples and applications. In *Lectures on probability theory and statistics (Saint-Flour, 1997)*, pages 1–91. Springer, Berlin, 1999. 1 746 300

[42] J. Bertoin. A fragmentation process connected to Brownian motion. *Probab. Theory Related Fields*, 117(2):289–301, 2000. MR1771665

[43] J. Bertoin. The asymptotic behaviour of self-similar fragmentations. Prépublication du Laboratoire de Probabilités et modèles aléatoires, Universite Paris VI. Available via `http://www.proba.jussieu.fr`, 2001.

[44] J. Bertoin. Eternal additive coalescents and certain bridges with exchangeable increments. *Ann. Probab.*, 29(1):344–360, 2001. MR1825153

[45] J. Bertoin. Homogeneous fragmentation processes. *Probab. Theory Related Fields*, 121(3):301–318, 2001. MR1867425

[46] J. Bertoin. Self-similar fragmentations. *Ann. Inst. H. Poincaré Probab. Statist.*, 38(3):319–340, 2002. MR1899456

[47] J. Bertoin. The asymptotic behavior of fragmentation processes. *J. Eur. Math. Soc. (JEMS)*, 5(4):395–416, 2003. MR2017852

[48] J. Bertoin, L. Chaumont, and J. Pitman. Path transformations of first passage bridges. *Electronic Comm. Probab.*, 8:155–166, 2003. MR2042754

[49] J. Bertoin and C. Goldschmidt. Dual random fragmentation and coagulation and an application to the genealogy of Yule processes. In *Mathematics and computer science. III*, Trends Math., pages 295–308. Birkhäuser, Basel, 2004. MR2090520

[50] J. Bertoin and J.-F. Le Gall. The Bolthausen-Sznitman coalescent and the genealogy of continuous-state branching processes. *Probab. Theory Related Fields*, 117(2):249–266, 2000. MR1771663

[51] J. Bertoin and J.-F. Le Gall. Stochastic flows associated to coalescent processes. *Probab. Theory Related Fields*, 126(2):261–288, 2003. MR1990057

[52] J. Bertoin and J.-F. Le Gall. Stochastic flows associated to coalescent processes ii : Stochastic differential equations, 2004. MR2086161

[53] J. Bertoin and J. Pitman. Path transformations connecting Brownian bridge, excursion and meander. *Bull. Sci. Math. (2)*, 118:147–166, 1994. MR1268525

[54] J. Bertoin and J. Pitman. Two coalescents derived from the ranges of stable subordinators. *Electron. J. Probab.*, 5:no. 7, 17 pp., 2000. MR1768841

[55] J. Bertoin, J. Pitman, and J. R. de Chavez. Constructions of a brownian path with a given minimum. *Electronic Comm. Probab.*, 4:Paper 5, 1–7, 1999. MR1703609

[56] J. Bertoin and A. Rouault. Additive martingales and probability tilting for homogeneous fragmentations, 2003. Preprint.

[57] P. Biane. Relations entre pont et excursion du mouvement Brownien réel. *Ann. Inst. Henri Poincaré*, 22:1–7, 1986.

[58] P. Biane. Some comments on the paper: "Brownian bridge asymptotics for random mappings" by D. J. Aldous and J. W. Pitman. *Random Structures and Algorithms*, 5:513–516, 1994.

[59] P. Biane, J.-F. L. Gall, and M. Yor. Un processus qui ressemble au pont brownien. In *Séminaire de Probabilités XXI*, pages 270–275. Springer, 1987. Lecture Notes in Math. 1247. 0941990.

[60] P. Biane, J. Pitman, and M. Yor. Probability laws related to the Jacobi theta and Riemann zeta functions, and Brownian excursions. *Bull. Amer. Math. Soc.*, 38:435–465, 2001. arXiv:math.PR/9912170 MR1848256

[61] P. Biane and M. Yor. Valeurs principales associées aux temps locaux Browniens. *Bull. Sci. Math. (2)*, 111:23–101, 1987.

[62] P. Biane and M. Yor. Quelques précisions sur le méandre brownien. *Bull. Sci. Math.*, 112:101–109, 1988.

[63] P. Biane and M. Yor. Sur la loi des temps locaux Browniens pris en un temps exponentiel. In *Séminaire de Probabilités XXII*, pages 454–466. Springer, 1988. Lecture Notes in Math. 1321.

[64] P. Billingsley. *Probability and Measure*. Wiley, New York, 1995. 3rd ed.

[65] P. Billingsley. *Convergence of probability measures*. John Wiley & Sons Inc., New York, second edition, 1999. A Wiley-Interscience Publication. MR1700749

[66] N. H. Bingham, C. M. Goldie, and J. L. Teugels. *Regular variation*, volume 27 of *Encyclopedia of Mathematics and its Applications*. Cambridge University Press, Cambridge, 1989. 90i:26003

[67] M. Birkner, J. Blath, M. Capaldo, A. Etheridge, M. Mhle, J. Schweinsberg, and A. Wakolbinger. Alpha-stable branching and beta-coalescents. *Electronic Journal of Probability*, 10:303–325, 2005.

[68] D. Blackwell and J. MacQueen. Ferguson distributions via Pólya urn schemes. *Ann. Statist.*, 1:353–355, 1973.

[69] R. M. Blumenthal. Weak convergence to Brownian excursion. *Ann. Probab.*, 11:798–800, 1983.

[70] B. Bollobás. *Random Graphs.* Academic Press, London, 1985.

[71] E. Bolthausen. On a functional central limit theorem for random walks conditioned to stay positive. *Ann. Probab.*, 4:480–485, 1976.

[72] E. Bolthausen and A.-S. Sznitman. On Ruelle's probability cascades and an abstract cavity method. *Comm. Math. Phys.*, 197(2):247–276, 1998.

[73] A. N. Borodin and P. Salminen. *Handbook of Brownian motion – facts and formulae, 2nd edition.* Birkhäuser, 2002.

[74] K. Borovkov and Z. Burq. Kendall's identity for the first crossing time revisited. *Electron. Comm. Probab.*, 6:91–94 (electronic), 2001. MR1871697

[75] P. Bougerol and T. Jeulin. Paths in Weyl chambers and random matrices. *Probab. Theory Related Fields*, 124(4):517–543, 2002. MR1942321

[76] A. Broder. Generating random spanning trees. In *Proc. 30'th IEEE Symp. Found. Comp. Sci.*, pages 442–447, 1989.

[77] B. M. Brown. Moments of a stopping rule related to the central limit theorem. *Ann. Math. Statist.*, 40:1236–1249, 1969. MR0243689

[78] W. Brown. Historical note on a recurrent combinatorial problem. *Amer. Math. Monthly*, 72:973–977, 1965.

[79] E. Buffet and J. Pulé. Polymers and random graphs. *J. Statist. Phys.*, 64:87–110, 1991.

[80] Y. D. Burtin. On a simple formula for random mappings and its applications. *J. Appl. Probab.*, 17:403 – 414, 1980.

[81] G. S. C. Banderier, P. Flajolet and M. Soria. Random maps, coalescing saddles, singularity analysis, and Airy phenomena. *Random Structures Algorithms*, 19(3-4):194–246, 2001. Analysis of algorithms (Krynica Morska, 2000). MR1871555

[82] M. Camarri and J. Pitman. Limit distributions and random trees derived from the birthday problem with unequal probabilities. *Electron. J. Probab.*, 5:Paper 2, 1–18, 2000. MR1741774

[83] E. Canfield. Central and local limit theorems for the coefficients of polynomials of binomial type. *J. Comb. Theory A*, 23:275–290, 1977.

[84] C. Cannings. The latent roots of certain Markov chains arising in genetics: a new approach, I. Haploid model. *Adv. Appl. Prob.*, 6:260–290, 1974.

[85] P. Carmona, F. Petit, J. Pitman, and M. Yor. On the laws of homogeneous functionals of the Brownian bridge. *Studia Sci. Math. Hungar.*, 35:445–455, 1999. MR1762255

[86] P. Carmona, F. Petit, and M. Yor. Some extensions of the arc sine law as partial consequences of the scaling property of Brownian motion. *Probab. Th. Rel. Fields*, 100:1–29, 1994.

[87] P. Carmona, F. Petit, and M. Yor. Beta-gamma random variables and intertwining relations between certain Markov processes. *Rev. Mat. Iberoamericana*, 14(2):311–367, 1998.

[88] A. Cayley. A theorem on trees. *Quarterly Journal of Pure and Applied Mathematics*, 23:376–378, 1889. (Also in *The Collected Mathematical Papers of Arthur Cayley. Vol XIII*, 26-28, Cambridge University Press, 1897).

[89] C. A. Charalambides and J. Singh. A review of the Stirling numbers, their generalizations and statistical applications. *Commun. Statist.-Theory Meth.*, 17:2533–2595, 1988.

[90] K. Chase and A. Mekjian. Nuclear fragmentation and its parallels. *Phys. Rev. C*, 49:2164–2176, 1994.

[91] P. Chassaing and S. Janson. A Vervaat-like path transformation for the reflected Brownian bridge conditioned on its local time at 0. *Ann. Probab.*, 29(4):1755–1779, 2001. MR1880241

[92] P. Chassaing and G. Louchard. Phase transition for parking blocks, Brownian excursion and coalescence. *Random Structures Algorithms*, 21:76–119, 2002. MR1913079

[93] P. Chassaing and J.-F. Marckert. Parking functions, empirical processes, and the width of rooted labeled trees. *Electron. J. Combin.*, 8(1):Research Paper 14, 19 pp. (electronic), 2001. MR1814521

[94] P. Chassaing and G. Schaeffer. Random planar lattices and integrated superbrownian excursion. In B. Chauvin, P. Flajolet, D. Gardy, and A. Mokkadem, editors, *Mathematics and Computer Science II*, pages 127–145. Birkhäuser, Basel, 2002. MR1940133

[95] L. Chaumont. Excursion normalisée, méandre et pont pour les processus de Lévy stables. *Bull. Sci. Math.*, 121(5):377–403, 1997. MR1465814

[96] L. Chaumont. An extension of Vervaat's transformation and its consequences. *J. Theoret. Probab.*, 13(1):259–277, 2000. MR1744984

[97] F. Chung and L. Lu. The diameter of sparse random graphs. *Adv. in Appl. Math.*, 26:257–279, 2001.

[98] K. L. Chung. Excursions in Brownian motion. *Arkiv für Matematik*, 14:155–177, 1976.

[99] K. J. Compton. Some methods for computing component distribution probabilities in relational structures. *Discrete Math.*, 66(1-2):59–77, 1987. MR900930

[100] L. Comtet. *Advanced Combinatorics*. D. Reidel Pub. Co., Boston, 1974. (translated from French).

[101] R. Cori. Words and Trees. In M. Lothaire, editor, *Combinatorics on Words*, volume 17 of *Encyclopedia of Mathematics and its Applications*, pages 215–229. Addison-Wesley, Reading, Mass., 1983.

[102] E. Csáki and G. Mohanty. Excursion and meander in random walk. *Canad. J. Statist.*, 9:57–70, 1981.

[103] L. de Lagrange. Nouvelle méthode pour résoudre des équations littérales par le moyen des séries. *Mém. Acad. Roy. Sci. Belles-Lettres de Berlin*, 24, 1770.

[104] A. Dembo, A. Vershik, and O. Zeitouni. Large deviations for integer partitions. *Markov Process. Related Fields*, 6(2):147–179, 2000. MR1778750

[105] B. Derrida. Random-energy model: an exactly solvable model of disordered systems. *Phys. Rev. B (3)*, 24(5):2613–2626, 1981. MR627810

[106] B. Derrida. From random walks to spin glasses. *Phys. D*, 107(2-4):186–198, 1997. Landscape paradigms in physics and biology (Los Alamos, NM, 1996). MR1491962

[107] A. Di Bucchianico. *Probabilistic and analytical aspects of the umbral calculus.* Stichting Mathematisch Centrum, Centrum voor Wiskunde en Informatica, Amsterdam, 1997. MR1431509

[108] P. Diaconis and D. Freedman. Partial exchangeability and sufficiency. In J. K. Ghosh and J. Roy, editors, *Statistics Applications and New Directions; Proceedings of the Indian Statistical Institute Golden Jubilee International Conference; Sankhya A.* Indian Statistical Institute, 205-236, 1984.

[109] P. Diaconis and J. Kemperman. Some new tools for Dirichlet priors. In J. Bernardo, J. Berger, A. Dawid, and A. Smith, editors, *Bayesian Statistics*, pages 95–104. Oxford Univ. Press, 1995.

[110] P. Diaconis, E. Mayer-Wolf, O. Zeitouni, and M. P. W. Zerner. The Poisson-Dirichlet law is the unique invariant distribution for uniform split-merge transformations. *Ann. Probab.*, 32(1B):915–938, 2004. MR2044670

[111] G. Dobiński. Summirung der Reihe $\sum n^m/n!$ für $m = 1, 2, 3, 4, 5, \ldots$. *Grunert Archiv (Arch. für Mat. und Physik)*, 61:333–336, 1877.

[112] G. Doetsch. *Theorie und Anwendung der Laplace-Transformation.* Berlin, 1937.

[113] R. Dong, C. Goldschmidt, and J. B. Martin. Coagulation-fragmentation duality, Poisson-Dirichlet distributions and random recursive trees, 2005. arXiv:math.PR/0507591

[114] P. Donnelly and P. Joyce. Consistent ordered sampling distributions: characterization and convergence. *Adv. Appl. Prob.*, 23:229–258, 1991.

[115] P. Donnelly and S. Tavaré. The ages of alleles and a coalescent. *Adv. Appl. Probab.*, 18:1–19 & 1023, 1986.

[116] M. Drmota and B. Gittenberger. Strata of random mappings—a combinatorial approach. *Stochastic Process. Appl.*, 82(2):157–171, 1999. MR1700003

[117] D. Dufresne. Algebraic properties of beta and gamma distributions, and applications. *Adv. in Appl. Math.*, 20(3):285–299, 1998. MR1618423

[118] T. Duquesne. A limit theorem for the contour process of conditioned Galton-Watson trees. *Ann. Probab.*, 31(2):996–1027, 2003. MR1964956

[119] T. Duquesne and J.-F. Le Gall. Random trees, Lévy processes and spatial branching processes. *Astérisque*, (281):vi+147, 2002. MR1954248

[120] T. Duquesne and J.-F. Le Gall. Probabilistic and fractal aspects of Lévy trees. *To appear in Probab. Theory and Rel. Fields*, 2004.

[121] T. Duquesne and M. Winkel. Growth of Lévy forests, 2005.

[122] R. Durrett. *Probability: theory and examples.* Duxbury Press, Belmont, CA, second edition, 1996.

[123] R. Durrett, B. L. Granovsky, and S. Gueron. The equilibrium behavior of reversible coagulation-fragmentation processes. *J. Theoret. Probab.*, 12(2):447–474, 1999. MR1684753

[124] R. Durrett, D. L. Iglehart, and D. R. Miller. Weak convergence to Brownian meander and Brownian excursion. *Ann. Probab.*, 5:117–129, 1977.

[125] R. Durrett, H. Kesten, and E. Waymire. On weighted heights of random trees. *J. Theoret. Probab.*, 4:223–237, 1991.

[126] R. Durrett and J. Schweinsberg. A coalescent model for the effect of advantageous mutations on the genealogy of a population, 2004. arXiv:math.PR/0411071

[127] M. Dwass. The total progeny in a branching process. *J. Appl. Probab.*, 6:682–686, 1969.

[128] M. Dwass and S. Karlin. Conditioned limit theorems. *Ann. Math. Stat.*, 34:1147–1167, 1963.

[129] E. B. Dynkin. Representation for functionals of superprocesses by multiple stochastic integrals, with applications to self-intersection local times. In *Colloque Paul Lévy sur les Processus Stochastiques*, pages 147–171, 1988. Asterisque 157-158.

[130] N. Eisenbaum and H. Kaspi. On the Markov property of local time for Markov processes on graphs. *Stochastic Process. Appl.*, 64:153–172, 1996.

[131] N. Eisenbaum, H. Kaspi, M. B. Marcus, J. Rosen, and Z. Shi. A Ray-Knight theorem for symmetric Markov processes. *Ann. Probab.*, 28(4):1781–1796, 2000. MR1813843

[132] S. Engen. *Stochastic Abundance Models with Emphasis on Biological Communities and Species Diversity.* Chapman and Hall Ltd., 1978.

[133] A. Erdélyi and I. Etherington. Some problems of non-associative combinations (2). *Edinburgh Math. Notes*, 32:7–12, 1940.

[134] P. Erdős and A. Rényi. On the evolution of random graphs. *Publ. Math. Inst. Hungar. Acad. Sci.*, 5:17–61, 1960.

[135] P. Erdős, R. Guy, and J. Moon. On refining partitions. *J. London Math. Soc. (2)*, 9:565–570, 1970.

[136] A. M. Etheridge. *An introduction to superprocesses.* American Mathematical Society, Providence, RI, 2000. MR1779100

[137] I. Etherington. Some problems of non-associative combinations (1). *Edinburgh Math. Notes*, 32:1–6, 1940.

[138] S. N. Evans. Kingman's coalescent as a random metric space. In *Stochastic models (Ottawa, ON, 1998)*, pages 105–114. Amer. Math. Soc., Providence, RI, 2000. MR1765005

[139] S. N. Evans and E. Perkins. Absolute continuity results for superprocesses with some applications. *Trans. Amer. Math. Soc.*, 325(2):661–681, 1991. MR1012522

[140] S. N. Evans and E. A. Perkins. Collision local times, historical stochastic calculus, and competing superprocesses. *Electron. J. Probab.*, 3:No. 5, 120 pp. (electronic), 1998. MR1615329

[141] S. N. Evans and J. Pitman. Construction of Markovian coalescents. *Ann. Inst. Henri Poincaré*, 34:339–383, 1998. MR1625867

[142] S. N. Evans, J. Pitman, and A. Winter. Rayleigh processes, real trees, and root growth with re-grafting. Technical Report 654, Dept. Statistics, U.C. Berkeley, 2004. To appear in *Probab. Th. and Rel. Fields.* arXiv:math.PR/0402293

[143] S. N. Evans and A. Winter. Subtree prune and re-graft: a reversible real tree valued Markov process. Technical Report 685, Dept. Statistics, U.C. Berkeley. arXiv:math.PR/0502226

[144] W. Ewens. The sampling theory of selectively neutral alleles. *Theor. Popul. Biol.*, 3:87 – 112,1972.

[145] W. J. Ewens and S. Tavaré. The Ewens sampling formula. In N. S. Johnson, S. Kotz, and N. Balakrishnan, editors, *Multivariate Discrete Distributions.* Wiley, New York, 1995.

[146] P. Feinsilver. *Special Functions, Probability Semigroups, and Hamiltonian Flows*, volume 696 of *Lecture Notes in Math.* Springer, New York, 1978.

[147] P. Feinsilver and R. Schott. *Algebraic structures and operator calculus. Vol. I.* Kluwer Academic Publishers Group, Dordrecht, 1993. Representations and probability theory. MR1227095

[148] W. Feller. The fundamental limit theorems in probability. *Bull. Amer. Math. Soc.*, 51:800–832, 1945.

[149] W. Feller. The asymptotic distribution of the range of sums of independent random variables. *Ann. Math. Stat.*, 22:427–432, 1951.

[150] W. Feller. *An Introduction to Probability Theory and its Applications,* Vol 1, 3rd ed. Wiley, New York, 1968.

[151] W. Feller. *An Introduction to Probability Theory and its Applications,* Vol 2, 2nd ed. Wiley, New York, 1971.

[152] S. Feng and F. M. Hoppe. Large deviation principles for some random combinatorial structures in population genetics and Brownian motion. *Ann. Appl. Probab.*, 8(4):975–994, 1998. MR1661315

[153] T. Ferguson. A Bayesian analysis of some nonparametric problems. *Ann. Statist.*, 1:209–230, 1973.

[154] A. F. Filippov. On the distribution of the sizes of particles which undergo splitting. *Th. Probab. Appl*, 6:275–293, 1961.

[155] P. Fitzsimmons, J. Pitman, and M. Yor. Markovian bridges: construction, palm interpretation, and splicing. In E. Çinlar, K. Chung, and M. Sharpe, editors, *Seminar on Stochastic Processes, 1992*, pages 101–134. Birkhäuser, Boston, 1993. MR1278079

[156] P. Flajolet and A. M. Odlyzko. Random mapping statistics. In *Advances in cryptology—EUROCRYPT '89 (Houthalen, 1989)*, volume 434 of *Lecture Notes in Comput. Sci.*, pages 329–354. Springer, Berlin, 1990. MR1083961

[157] D. Freedman. *Approximating Countable Markov Chains.* Holden-Day, San Francisco, 1972.

[158] D. Freedman. Another note on the Borel-Cantelli lemma and the strong law, with the Poisson approximation as a by-product. *The Annals of Probability*, 1:910 – 925, 1973.

[159] B. Fristedt. The structure of partitions of large integers. *T.A.M.S.*, 337:703–735, 1993.

[160] J. Fulman. Random matrix theory over finite fields. *Bull. Amer. Math. Soc. (N.S.)*, 39(1):51–85 (electronic), 2002. MR1864086

[161] J. Geiger. Contour processes in random trees. In A. Etheridge, editor, *Stochastic Partial Differential Equations*, volume 216 of *London Math. Soc. Lect. Notes*, pages 72–96. Cambridge Univ. Press, Cambridge, 1995.

[162] J. Geiger. Size-biased and conditioned random splitting trees. *Stochastic Process. Appl.*, 65:187–207, 1996.

[163] J. Geiger. Growing conditioned Galton-Watson trees from the top. Technical report, Fac. Math., Univ. Frankfurt, 1997.

[164] J. Geiger. Poisson point process limits in size-biased Galton-Watson trees. *Electron. J. Probab.*, 5:no. 17, 12 pp. (electronic), 2000. MR1800073

[165] J. Geiger and G. Kersting. Depth-first search of random trees, and Poisson point processes. In K. B. Athreya and P. Jagers, editors, *Classical and Modern Branching Processes*, volume 84 of *IMA Volumes in Mathematics and its Applications*, pages 111–126. Springer, Berlin, 1996.

[166] S. Ghirlanda and F. Guerra. General properties of overlap probability distributions in disordered spin systems. Towards Parisi ultrametricity. *J. Phys. A*, 31(46):9149–9155, 1998. MR1662161

[167] B. Gittenberger and G. Louchard. The Brownian excursion multidimensional local time density. *J. Appl. Probab.*, 36:350–373, 1999.

[168] B. Gittenberger and G. Louchard. On the local time density of the reflecting Brownian bridge. *J. Appl. Math. Stochastic Anal.*, 13:125–136, 2000.

[169] A. Gnedin and S. Kerov. A characterization of GEM distributions. *Combin. Probab. Comput.*, 10(3):213–217, 2001. MR1841641

[170] A. Gnedin and J. Pitman. Regenerative composition structures. *Ann. Probab.*, 33(2):445–479, 2005. arXiv:math.PR/0307307 MR2122798

[171] A. Gnedin, J. Pitman, and M. Yor. Asymptotic laws for compositions derived from transformed subordinators, 2004. To appear in *Ann. Probab.* arXiv:math.PR/0403438

[172] A. V. Gnedin. The representation of composition structures. *Ann. Probab.*, 25(3):1437–1450, 1997. MR1457625

[173] A. V. Gnedin. On convergence and extensions of size-biased permutations. *J. Appl. Probab.*, 35(3):642–650, 1998. MR1659532

[174] W. Goh and E. Schmutz. Random set partitions. *SIAM J. Discrete Math.*, 7:419–436, 1994.

[175] C. M. Goldie. Records, permutations and greatest convex minorants. *Math. Proc. Cambridge Philos. Soc.*, 106(1):169–177, 1989. MR994088

[176] C. Goldschmidt and J. B. Martin. Random recursive trees and the Bolthausen-Sznitman coalescent, 2005. To appear in *Electron. J. Probab.* arXiv:math.PR/0502263

[177] V. Gončharov. On the field of combinatory analysis. *Amer. Math. Soc. Transl.*, 19:1–46, 1962.

[178] P. Greenwood and J. Pitman. Fluctuation identities for Lévy processes and splitting at the maximum. *Advances in Applied Probability*, 12:893–902, 1980. MR588409

[179] R. Greenwood. The number of cycles associated with the elements of a permutation group. *Amer. Math. Monthly*, 60:407–409, 1953.

[180] W. Gutjahr. Expectation transfer between branching processes and random trees. *Random Structures and Algorithms*, 4:447–467, 1993.

[181] W. Gutjahr and G. C. Pflug. The asymptotic contour process of a binary tree is Brownian excursion. *Stochastic Process. Appl.*, 41:69–90, 1992.

[182] B. Haas. Equilibrium for fragmentation with immigration. *to appear in Ann. App. Probab.*

[183] B. Haas. Loss of mass in deterministic and random fragmentations. *Stochastic Process. Appl.*, 106(2):245–277, 2003. MR1989629

[184] B. Haas. Regularity of formation of dust in self-similar fragmentations. *Ann. Inst. H. Poincaré Probab. Statist.*, 40(4):411–438, 2004. MR2070333

[185] B. Haas and G. Miermont. The genealogy of self-similar fragmentations with negative index as a continuum random tree. *Electron. J. Probab.*, 9:no. 4, 57–97 (electronic), 2004. 2 041 829

[186] A. Hald. The early history of the cumulants and the Gram-Charlier series. *International Statistical Review*, 68:137–153, 2000.

[187] W. J. Hall. On Wald's equations in continuous time. *J. Appl. Probability*, 7:59–68, 1970. MR0258115

[188] B. M. Hambly, J. B. Martin, and N. O'Connell. Pitman's $2M-X$ theorem for skip-free random walks with Markovian increments. *Electron. Comm. Probab.*, 6:73–77 (electronic), 2001. MR1855343

[189] J. Hansen. Order statistics for decomposable combinatorial structures. *Rand. Struct. Alg.*, 5:517–533, 1994.

[190] L. H. Harper. Stirling behavior is asymptotically normal. *Ann. Math. Stat.*, 38:410–414, 1966.

[191] B. Harris. A survey of the early history of the theory of random mappings. In *Probabilistic methods in discrete mathematics (Petrozavodsk, 1992)*, pages 1–22. VSP, Utrecht, 1993. MR1383124

[192] T. Harris. *The Theory of Branching Processes.* Springer-Verlag, New York, 1963.

[193] T. E. Harris. First passage and recurrence distributions. *Trans. Amer. Math. Soc.*, 73:471–486, 1952.

[194] W. Hayman. A generalization of Stirling's formula. *Journal für die reine und angewandte Mathematik*, 196:67–95, 1956.

[195] E. Hendriks, J. Spouge, M. Eibl, and M. Shreckenberg. Exact solutions for random coagulation processes. *Z. Phys. B - Condensed Matter*, 58:219–227, 1985.

[196] F. Hiai and D. Petz. *The semicircle law, free random variables and entropy.* American Mathematical Society, Providence, RI, 2000. MR1746976

[197] P. Hilton and J. Pedersen. Catalan numbers, their generalization, and their uses. *Math. Intelligencer*, 13:64–75, 1991.

[198] U. Hirth and P. Ressel. Random partitions by semigroup methods. *Semigroup Forum*, 59(1):126–140, 1999. MR1847948

[199] U. Hirth and P. Ressel. Exchangeable random orders and almost uniform distributions. *J. Theoret. Probab.*, 13(3):609–634, 2000. MR1785522

[200] D. G. Hobson. Marked excursions and random trees. In *Séminaire de Probabilités, XXXIV*, pages 289–301. Springer, Berlin, 2000. MR1768069

[201] L. Holst. On numbers related to partitions of unlike objects and occupancy problems. *European J. Combin.*, 2(3):231–237, 1981. MR633118

[202] F. M. Hoppe. Size-biased filtering of Poisson-Dirichlet samples with an application to partition structures in genetics. *Journal of Applied Probability*, 23:1008 – 1012, 1986.

[203] L. C. Hsu and P. J.-S. Shiue. A unified approach to generalized Stirling numbers. *Adv. in Appl. Math.*, 20(3):366–384, 1998. MR1618435

[204] I. Ibragimov and Y. V. Linnik. *Independent and Stationary Sequences of Random Variables.* Gröningen, Wolthers-Noordhof, 1971. [Translated from original in Russian (1965), Nauka, Moscow].

[205] D. Iglehart. Functional central limit theorems for random walks conditioned to stay positive. *Ann. Probab.*, 2:608–619, 1974.

[206] T. Ignatov. On a constant arising in the theory of symmetric groups and on Poisson-Dirichlet measures. *Theory Probab. Appl.*, 27:136–147, 1982.

[207] Z. Ignatov. Point processes generated by order statistics and their applications. In *Point processes and queuing problems (Colloq., Keszthely, 1978)*, pages 109–116. North-Holland, Amsterdam-New York, 1981. MR617405

[208] J. P. Imhof. Density factorizations for Brownian motion, meander and the three-dimensional Bessel process, and applications. *J. Appl. Probab.*, 21:500–510, 1984.

[209] J. P. Imhof and P. Kümmerling. Operational derivation of some Brownian motion results. *International Statistical Review*, 54:327–341, 1986.

[210] H. Ishwaran and L. F. James. Generalized weighted Chinese restaurant processes for species sampling mixture models. *Statist. Sinica*, 13(4):1211–1235, 2003. MR2026070

[211] K. Itô and H. P. McKean. *Diffusion Processes and their Sample Paths.* Springer, 1965.

[212] L. F. James. Poisson process partition calculus with applications to exchangeable models and Bayesian nonparametrics. 2002. arXiv:math.PR/0205093

[213] L. F. James. Bayesian calculus for gamma processes with applications to semiparametric intensity models. *Sankhyā*, 65(1):179–206, 2003. MR2016784

[214] S. Janson. *Gaussian Hilbert Spaces.* Number 129 in Cambridge Tracts in Mathematics. Cambridge University Press, 1997.

[215] S. Janson, D. E. Knuth, T. Łuczak, and B. Pittel. The birth of the giant component. *Random Structures and Algorithms*, 4:233–358, 1993. MR1220220

[216] S. Janson, T. Łuczak, and A. Rucinski. *Random graphs.* Wiley-Interscience, New York, 2000. MR1782847

[217] S. Janson and J. F. Marckert. Weak convergence of discrete snakes, 2004. arXiv:math.PR/0403398

[218] I. Jeon. Stochastic fragmentation and some sufficient conditions for shattering transition. *J. Korean Math. Soc.*, 39(4):543–558, 2002. MR1898911

[219] T. Jeulin. Ray-Knight's theorem on Brownian local times and Tanaka's formula. In *Seminar on stochastic processes, 1983 (Gainesville, Fla., 1983)*, volume 7 of *Progr. Probab. Statist.*, pages 131–142. Birkhäuser Boston, Boston, MA, 1984. 88j:60134

[220] T. Jeulin. Temps local et théorie du grossissement: application de la théorie du grossissement à l'étude des temps locaux browniens. In *Grossissements de filtrations: exemples et applications. Séminaire de Calcul Stochastique, Paris 1982/83*, pages 197–304. Springer-Verlag, 1985. Lecture Notes in Math. 1118.

[221] N. L. Johnson and S. Kotz. *Continuous Univariate Distributions, volume 2.* Wiley, 1970.

[222] C. Jordan. *Calculus of Finite Differences.* Rotting and Romwalter, Sorron, Hungary, 1939. Reprod. Chelsea, New York, 1947.

[223] A. Joyal. Une théorie combinatoire des séries formelles. *Adv. in Math.*, 42:1–82, 1981. 0633783.

[224] P. Joyce and S. Tavaré. Cycles, permutations and the structure of the Yule process with immigration. *Stochastic Process. Appl.*, 25:309–314, 1987.

[225] W. D. Kaigh. An invariance principle for random walk conditioned by a late return to zero. *Ann. Probab.*, 4:115 – 121, 1976.

[226] O. Kallenberg. Canonical representations and convergence criteria for processes with interchangeable increments. *Z. Wahrsch. Verw. Gebiete*, 27:23–36, 1973.

[227] O. Kallenberg. The local time intensity of an exchangeable interval partition. In A. Gut and L. Holst, editors, *Probability and Statistics, Essays in Honour of Carl-Gustav Esseen*, pages 85–94. Uppsala University, 1983.

[228] O. Kallenberg. One-dimensional uniqueness and convergence criteria for exchangeable processes. *Stochastic Process. Appl.*, 28(2):159–183, 1988. MR952828

[229] O. Kallenberg. Ballot theorems and sojourn laws for stationary processes. *Ann. Probab.*, 27(4):2011–2019, 1999. MR1742898

[230] O. Kallenberg. *Foundations of modern probability.* Springer-Verlag, New York, second edition, 2002. MR1876169

[231] O. Kallenberg. Paintbox representations of random partitions with general symmetries. To appear as a chapter in the forthcoming monograph *Symmetries in Probability*, 2004.

[232] S. Karlin. Central limit theorems for certain infinite urn schemes. *J. Math. Mech.*, 17:373–401, 1967.

[233] K. Kawazu and S. Watanabe. Branching processes with immigration and related limit theorems. *Theory Probab. Appl.*, 16:36 – 54, 1971.

[234] J. Kemperman. The general one-dimensional random walk with absorbing barriers. Thesis, Excelsior, The Hague, 1950.

[235] J. Kemperman. *The Passage Problem for a Stationary Markov Chain.* University of Chicago Press, 1961.

[236] D. Kendall. Some problems in the theory of queues. *J.R.S.S. B*, 13:151–185, 1951.

[237] D. G. Kendall. Some problems in theory of dams. *J. Roy. Statist. Soc. Ser. B.*, 19:207–212; discussion 212–233, 1957. MR0092290

[238] D. P. Kennedy. The distribution of the maximum Brownian excursion. *J. Appl. Prob.*, 13:371–376, 1976.

[239] S. Kerov. The asymptotics of interlacing sequences and the growth of continual Young diagrams. Technical report, Inst. Electricity and Communications, St Petersburg, 1993.

[240] S. Kerov. Coherent random allocations and the Ewens-Pitman formula. PDMI Preprint, Steklov Math. Institute, St. Petersburg, 1995.

[241] S. Kerov. The boundary of Young lattice and random Young tableaux. In *Formal power series and algebraic combinatorics (New Brunswick, NJ, 1994)*, volume 24 of *DIMACS Ser. Discrete Math. Theoret. Comput. Sci.*, pages 133–158. Amer. Math. Soc., Providence, RI, 1996. MR1363510

[242] S. Kerov, A. Okounkov, and G. Olshanski. The boundary of the Young graph with Jack edge multiplicities. *Internat. Math. Res. Notices*, 4:173–199, 1998. MR1609628

[243] S. Kerov, G. Olshanski, and A. Vershik. Harmonic analysis on the infinite symmetric group. *Invent. Math.*, 158(3):551–642, 2004. arXiv:math.RT/0312270 MR2104794

[244] S. V. Kerov. *Asymptotic representation theory of the symmetric group and its applications in analysis*, volume 219 of *Translations of Mathematical Monographs*. American Mathematical Society, Providence, RI, 2003. Translated from the Russian manuscript by N. V. Tsilevich, With a foreword by A. Vershik and comments by G. Olshanski. MR1984868

[245] S. V. Kerov and A. M. Vershik. Characters of infinite symmetric groups and probability properties of Robinson-Shenstead-Knuth's algorithm. *SIAM J. Algebraic Discrete Methods*, 7:116–124, 1986.

[246] G. Kersting. Symmetry properties of binary branching trees. Preprint, Fachbereich Mathematik, Univ. Frankfurt, 1997.

[247] I. Kessler and M. Sidi. Growing binary trees in a random environment. *IEEE Trans. Inform. Theory*, 39(1):191–194, 1993. MR1211495

[248] H. Kesten. Subdiffusive behavior of random walk on a random cluster. *Ann. Inst. H. Poincaré Probab. Statist.*, 22:425–487, 1987.

[249] J. F. C. Kingman. Random discrete distributions. *J. Roy. Statist. Soc. B*, 37:1–22, 1975.

[250] J. F. C. Kingman. Random partitions in population genetics. *Proc. R. Soc. Lond. A.*, 361:1–20, 1978.

[251] J. F. C. Kingman. The representation of partition structures. *J. London Math. Soc.*, 18:374–380, 1978.

[252] J. F. C. Kingman. *The Mathematics of Genetic Diversity*. SIAM, 1980.

[253] J. F. C. Kingman. The coalescent. *Stochastic Processes and their Applications*, 13:235–248, 1982.

[254] F. B. Knight. Random walks and a sojourn density process of Brownian motion. *Trans. Amer. Math. Soc.*, 107:36–56, 1963.

[255] F. B. Knight. *Essentials of Brownian Motion and Diffusion*. American Math. Soc., 1981. Math. Surveys 18.

[256] F. B. Knight. On the duration of the longest excursion. In E. Cinlar, K. L. Chung, and R. K. Getoor, editors, *Seminar on Stochastic Processes*, pages 117–148. Birkhäuser, 1985.

[257] F. B. Knight. The uniform law for exchangeable and Lévy process bridges. *Astérisque*, (236):171–188, 1996. Hommage à P. A. Meyer et J. Neveu. 97j:60137

[258] F. B. Knight. The moments of the area under reflected Brownian bridge conditional on its local time at zero. *J. Appl. Math. Stochastic Anal.*, 13(2):99–124, 2000. MR1768498

[259] F. B. Knight. On the path of an inert object impinged on one side by a Brownian particle. *Probab. Theory Related Fields*, 121:577–598, 2001. MR1872429

[260] V. F. Kolchin. *Random mappings.* Translation Series in Mathematics and Engineering. Optimization Software Inc. Publications Division, New York, 1986. MR865130

[261] G. Konheim, A and D. J. Newman. A note on growing binary trees. *Discrete Math.*, 4:57–63, 1973. MR0313095

[262] W. König, N. O'Connell, and S. Roch. Non-colliding random walks, tandem queues, and discrete orthogonal polynomial ensembles. *Electron. J. Probab.*, 7:no. 5, 24 pp. (electronic), 2002. MR1887625

[263] R. M. Korwar and M. Hollander. Contributions to the theory of Dirichlet processes. *Ann. Prob.*, 1:705–711, 1973.

[264] T. G. Kurtz and P. Protter. Weak limit theorems for stochastic integrals and stochastic differential equations. *Ann. Probab.*, 19:1035 - 1070, 1991.

[265] H. J. Kushner. On the weak convergence of interpolated Markov chains to a diffusion. *Ann. Probab.*, 2:40 - 50, 1974.

[266] G. Labelle. Une nouvelle démonstration combinatoire des formules d'inversion de Lagrange. *Adv. in Math.*, 42:217–247, 1981.

[267] J. Lamperti. An invariance principle in renewal theory. *Ann. Math. Stat.*, 33:685–696, 1962.

[268] J. Lamperti. Limiting distributions for branching processes. In L. L. Cam and J. Neyman, editors, *Proceedings of the fifth Berkeley Symposium, Vol II, Part 2*, pages 225–241. U. C. Press, Berkeley, 1967.

[269] J. Lamperti. The limit of a sequence of branching processes. *Z. Wahrsch. Verw. Gebiete*, 7:271 - 288, 1967.

[270] J.-F. Le Gall. Une approche élémentaire des théorèmes de décomposition de Williams. In *Séminaire de Probabilités XX*, pages 447–464. Springer, 1986. Lecture Notes in Math. 1204.

[271] J.-F. Le Gall. Marches aléatoires, mouvement Brownien et processus de branchement. In *Séminaire de Probabilités XXIII*, pages 258–274. Springer, 1989. Lecture Notes in Math. 1372.

[272] J.-F. Le Gall. Brownian excursions, trees and measure-valued branching processes. *Ann. Probab.*, 19:1399 - 1439, 1991.

[273] J.-F. Le Gall. The uniform random tree in a Brownian excursion. *Probab. Th. Rel. Fields*, 96:369–383, 1993.

[274] J.-F. Le Gall. *Spatial branching processes, random snakes and partial differential equations.* Birkhäuser Verlag, Basel, 1999. MR1714707

[275] J.-F. Le Gall. *Random Trees and Spatial Branching Processes*, volume 9 of *Maphysto Lecture Notes Series*. University of Aarhus, 2000. Available at `http://www.maphysto.dk/publications/MPS-LN/2000/9.pdf`.

[276] J.-F. Le Gall and Y. Le Jan. Branching processes in Lévy processes: Laplace functionals of snakes and superprocesses. *Ann. Probab.*, 26(4):1407–1432, 1998. MR1675019

[277] J. F. Le Gall and Y. Le Jan. Branching processes in Lévy processes: the exploration process. *Ann. Probab.*, 26:213–252, 1998.

[278] J.-F. Le Gall and M. Yor. Excursions browniennes et carrés de processus de Bessel. *C. R. Acad. Sc. Paris, Série I*, 303:73–76, 1986.

[279] M. R. Leadbetter, G. Lindgren, and H. Rootzén. *Extremes and related properties of random sequences and processes.* Springer-Verlag, New York, 1983. MR691492

[280] N. N. Lebedev. *Special Functions and their Applications.* Prentice-Hall, Englewood Cliffs, N.J., 1965.

[281] S. J. Lee and A. Z. Mekjian. Canonical studies of the cluster distribution, dynamical evolution, and critical temperature in nuclear multifragmentation processes. *Phys. Rev. C*, 45:1284–1310, 1992.

[282] C. Leuridan. Les théorèmes de Ray-Knight et la mesure d'Itô pour le mouvement brownien sur le tore $\mathbb{R}/\mathbb{Z}$. *Stochastics and Stochastic Reports*, 53:109–128, 1995.

[283] C. Leuridan. Le théorème de Ray-Knight à temps fixe. In J. Azéma, M. Émery, M. Ledoux, and M. Yor, editors, *Séminaire de Probabilités XXXII*, pages 376–406. Springer, 1998. Lecture Notes in Math. 1686.

[284] P. Lévy. Sur certains processus stochastiques homogènes. *Compositio Math.*, 7:283–339, 1939. 0000919.

[285] T. Lindvall. Convergence of critical Galton-Watson branching processes. *J. Appl. Probab.*, 9:445 – 450, 1972.

[286] P. Littelmann. Paths and root operators in representation theory. *Ann. of Math. (2)*, 142(3):499–525, 1995. MR1356780

[287] A. Lo, L. Brunner, and A. Chan. Weighted Chinese restaurant processes and Bayesian mixture models. Research report. Hong Kong University of Science and Technology, 1996.

[288] L. Lovász. *Combinatorial Problems and Exercises, 2nd ed.* North-Holland, Amsterdam, 1993.

[289] T. Łuckzak. Random trees and random graphs. *Random Structures and Algorithms*, 13:485–500, 1998. 1662797.

[290] E. Lukacs. Applications of Faà di Bruno's formula in mathematical statistics. *Amer. Math. Monthly*, 62:340–348, 1955. MR0069438

[291] A. Lushnikov. Coagulation in finite systems. *J. Colloid and Interface Science*, 65:276–285, 1978.

[292] R. Lyons and Y. Peres. Probability on trees and networks. Cambridge University Press. In preparation. Current version available at http://php.indiana.edu/~rdlyons/, 2003.

[293] J.-F. Marckert and A. Mokkadem. The depth first processes of Galton-Watson trees converge to the same Brownian excursion. *Ann. Probab.*, 31(3):1655–1678, 2003. MR1989446

[294] J.-F. Marckert and A. Mokkadem. Ladder variables, internal structure of Galton-Watson trees and finite branching random walks. *J. Appl. Probab.*, 40(3):671–689, 2003. MR1993260

[295] J.-F. Marckert and A. Mokkadem. States spaces of the snake and its tour—convergence of the discrete snake. *J. Theoret. Probab.*, 16(4):1015–1046 (2004), 2003. MR2033196

[296] A. Marcus. Stochastic coalescence. *Technometrics*, 10:133 – 143, 1968.

[297] M. B. Marcus and J. Rosen. Sample path properties of the local times of strongly symmetric Markov processes via Gaussian processes. *Ann. Probab.*, 20:1603–1684, 1992.

[298] M. B. Marcus and J. Rosen. New perspectives on Ray's theorem for the local times of diffusions. *Ann. Probab.*, 31(2):882–913, 2003. MR1964952

[299] H. Matsumoto and M. Yor. A version of Pitman's $2M - X$ theorem for geometric Brownian motions. *C. R. Acad. Sci. Paris Sér. I Math.*, 328(11):1067–1074, 1999.

[300] R. D. Mauldin, W. D. Sudderth, and S. C. Williams. Pólya trees and random distributions. *Ann. Statist.*, 20(3):1203–1221, 1992. MR1186247

[301] E. Mayer-Wolf, O. Zeitouni, and M. P. W. Zerner. Asymptotics of certain coagulation-fragmentation processes and invariant Poisson-Dirichlet measures. *Electron. J. Probab.*, 7:no. 8, 25 pp. (electronic), 2002. MR1902841

[302] J. W. McCloskey. A model for the distribution of individuals by species in an environment. Ph. D. thesis, Michigan State University, 1965.

[303] H. P. McKean. Excursions of a non-singular diffusion. *Z. Wahrsch. Verw. Gebiete*, 1:230–239, 1963.

[304] H. P. McKean. Brownian local times. *Advances in Mathematics*, 16:91 – 111, 1975.

[305] A. Meir and J. Moon. On the altitude of nodes in random trees. *Canad. J. Math.*, 30:997–1015, 1978.

[306] A. Mekjian and K. Chase. Disordered systems, power laws and random processes. *Phys. Letters A*, 229:340–346, 1997.

[307] A. Z. Mekjian. Cluster distributions in physics and genetic diversity. *Phys. Rev. A*, 44:8361–8374, 1991.

[308] A. Z. Mekjian and S. J. Lee. Models of fragmentation and partitioning phenomena based on the symmetric group S_n and combinatorial analysis. *Phys. Rev. A*, 44:6294–6312, 1991.

[309] G. Miermont. Ordered additive coalescent and fragmentations associated to Levy processes with no positive jumps. *Electron. J. Probab.*, 6:no. 14, 33 pp. (electronic), 2001. MR1844511

[310] G. Miermont. Self-similar fragmentations derived from the stable tree. I. Splitting at heights. *Probab. Theory Related Fields*, 127(3):423–454, 2003. MR2018924

[311] G. Miermont. Self-similar fragmentations derived from the stable tree. II. Splitting at nodes. *Probab. Theory Related Fields*, 131(3):341–375, 2005. MR2123249

[312] G. Miermont and J. Schweinsberg. Self-similar fragmentations and stable subordinators. In *Séminaire de Probabilités XXXVII*, volume 1832 of *Lecture Notes in Math.*, pages 333–359. Springer, Berlin, 2003. MR2053052

[313] M. Möhle and S. Sagitov. A classification of coalescent processes for haploid exchangeable population models. *Ann. Probab.*, 29(4):1547–1562, 2001. MR1880231

[314] S. A. Molchanov and E. Ostrovski. Symmetric stable processes as traces of degenerate diffusion processes. *Theor. Prob. Appl.*, 14, No. 1:128–131, 1969.

[315] J. Moon. Various proofs of Cayley's formula for counting trees. In F. Harary, editor, *A Seminar on Graph Theory*, pages 70–78. Holt, Rinehart and Winston, New York, 1967.

[316] C. N. Morris. Natural exponential families with quadratic variance functions. *Ann. Statist.*, 10(1):65–80, 1982.

[317] L. R. Mutafchiev. Local limit theorems for sums of power series distributed random variables and for the number of components in labelled relational structures. *Random Structures and Algorithms*, 3:403 – 426, 1992.

[318] S. Nacu. Increments of random partitions, 2003. arXiv:math.PR/0310091

[319] J. Neveu. *Martingales à temps discret.* Masson et Cie, éditeurs, Paris, 1972. MR0402914

[320] J. Neveu. Arbres et processus de Galton-Watson. *Ann. Inst. H. Poincaré Probab. Statist.*, 22:199 – 207, 1986.

[321] J. Neveu. Erasing a branching tree. In *Analytic and Geometric Stochastics: Papers in Honour of G. E. H. Reuter (Special supplement to Adv. Appl. Probab.)*, pages 101 – 108. Applied Probability Trust, 1986.

[322] J. Neveu. A continuous state branching process in relation with the GREM model of spin glasses theory. Rapport Interne No. 267, Ecole Polytechnique, Centre de Mathématiques Appliquées, Palaiseau, France, 1989.

[323] J. Neveu and J. Pitman. The branching process in a Brownian excursion. In *Séminaire de Probabilités XXIII*, volume 1372 of *Lecture Notes in Math.*, pages 248–257. Springer, 1989. MR1022915

[324] J. Neveu and J. Pitman. Renewal property of the extrema and tree property of a one-dimensional Brownian motion. In *Séminaire de Probabilités XXIII*, volume 1372 of *Lecture Notes in Math.*, pages 239–247. Springer, 1989. MR1022914

[325] M. Nordborg. Coalescent theory. In D. J. Balding, M. J. Bishop, and C. Cannings, editors, *Handbook of Statistical Genetics*, pages 179–208. John Wiley & Sons, Inc., New York, New York, 2001.

[326] J. R. Norris, L. C. G. Rogers, and D. Williams. Self-avoiding random walk: A Brownian motion model with local time drift. *Probability Theory and Related Fields*, 74:271–287, 1987.

[327] C. A. O'Cinneide and A. V. Pokrovskii. Nonuniform random transformations. *Ann. Appl. Probab.*, 10(4):1151–1181, 2000. MR1810869

[328] N. O'Connell. Conditioned random walks and the RSK correspondence. *J. Phys. A*, 36(12):3049–3066, 2003. Random matrix theory. MR1986407

[329] N. O'Connell. A path-transformation for random walks and the Robinson-Schensted correspondence. *Trans. Amer. Math. Soc.*, 355(9):3669–3697 (electronic), 2003. 2004f:60109

[330] N. O'Connell. Random matrices, non-colliding processes and queues. In *Séminaire de Probabilités, XXXVI*, volume 1801 of *Lecture Notes in Math.*, pages 165–182. Springer, Berlin, 2003. 2004g:15038

[331] N. O'Connell and M. Yor. Brownian analogues of Burke's theorem. *Stochastic Process. Appl.*, 96(2):285–304, 2001. MR1865759

[332] N. O'Connell and M. Yor. A representation for non-colliding random walks. *Electron. Comm. Probab.*, 7:1–12 (electronic), 2002. MR1887169

[333] A. Odlyzko. Asymptotic enumeration methods. In R. Graham, M. Grötschel, and L. Lovász, editors, *Handbook of Combinatorics Vol. II*, pages 1063–1229. Elsevier, New York, 1995.

[334] R. Otter. The multiplicative process. *Ann. Math. Statist.*, 20:206–224, 1949.

[335] Y. L. Pavlov. The asymptotic distribution of maximum tree size in a random forest. *Theory of Probability and its Applications*, 22:509–520, 1977.

[336] Y. L. Pavlov. Limit distributions for the maximum size of a tree in a random forest. *Diskret. Mat.*, 7(3):19–32, 1995. MR1361491

[337] Y. L. Pavlov. *Random Forests.* VSP, Leiden, The Netherlands, 2000.

[338] E. Perkins. Local time is a semimartingale. *Z. Wahrsch. Verw. Gebiete*, 60:79–117, 1982.

[339] M. Perman. Order statistics for jumps of normalized subordinators. *Stoch. Proc. Appl.*, 46:267–281, 1993. 1226412.

[340] M. Perman. An excursion approach to Ray-Knight theorems for perturbed Brownian motion. *Stochastic Process. Appl.*, 63(1):67–74, 1996.

[341] M. Perman, J. Pitman, and M. Yor. Size-biased sampling of Poisson point processes and excursions. *Probab. Th. Rel. Fields*, 92:21–39, 1992. MR1156448

[342] M. Perman and J. A. Wellner. On the distribution of Brownian areas. *Ann. Appl. Probab.*, 6:1091–1111, 1996.

[343] M. Perman and W. Werner. Perturbed Brownian motions. *Probab. Th. Rel. Fields*, 108:357–383, 1997.

[344] J. Pitman. Uniform rates of convergence for Markov chain transition probabilities. *Z. Wahrsch. Verw. Gebiete*, 29:193–227, 1974. MR0373012

[345] J. Pitman. One-dimensional Brownian motion and the three-dimensional Bessel process. *Advances in Applied Probability*, 7:511–526, 1975. MR0375485

[346] J. Pitman. The two-parameter generalization of Ewens' random partition structure. Technical Report 345, Dept. Statistics, U.C. Berkeley, 1992.

[347] J. Pitman. Exchangeable and partially exchangeable random partitions. *Probab. Th. Rel. Fields*, 102:145–158, 1995. MR1337249

[348] J. Pitman. Cyclically stationary brownian local time processes. *Probab. Th. Rel. Fields*, 106:299–329, 1996. MR1418842

[349] J. Pitman. Random discrete distributions invariant under size-biased permutation. *Adv. Appl. Prob.*, 28:525–539, 1996. MR1387889

[350] J. Pitman. Some developments of the Blackwell-MacQueen urn scheme. In T. F. et al., editor, *Statistics, Probability and Game Theory; Papers in honor of David Blackwell*, volume 30 of *Lecture Notes-Monograph Series*, pages 245–267. Institute of Mathematical Statistics, Hayward, California, 1996. MR1481784

[351] J. Pitman. Partition structures derived from Brownian motion and stable subordinators. *Bernoulli*, 3:79–96, 1997. MR1466546

[352] J. Pitman. Probabilistic bounds on the coefficients of polynomials with only real zeros. *J. Comb. Theory A.*, 77:279–303, 1997. MR1429082

[353] J. Pitman. Some probabilistic aspects of set partitions. *Amer. Math. Monthly*, 104:201–209, 1997. MR1436042

[354] J. Pitman. Enumerations of trees and forests related to branching processes and random walks. In D. Aldous and J. Propp, editors, *Microsurveys in Discrete Probability*, number 41 in DIMACS Ser. Discrete Math. Theoret. Comp. Sci, pages 163–180, Providence RI, 1998. Amer. Math. Soc. MR1630413

[355] J. Pitman. Brownian motion, bridge, excursion and meander characterized by sampling at independent uniform times. *Electron. J. Probab.*, 4:Paper 11, 1–33, 1999. MR1690315

[356] J. Pitman. Coalescent random forests. *J. Comb. Theory A.*, 85:165–193, 1999. MR1673928

[357] J. Pitman. Coalescents with multiple collisions. *Ann. Probab.*, 27:1870–1902, 1999. MR1742892

[358] J. Pitman. The SDE solved by local times of a Brownian excursion or bridge derived from the height profile of a random tree or forest. *Ann. Probab.*, 27:261–283, 1999. MR1681110

[359] J. Pitman. Random mappings, forests and subsets associated with Abel-Cayley-Hurwitz multinomial expansions. *Séminaire Lotharingien de Combinatoire*, 46:45 pp., 2001. MR1877634

[360] J. Pitman. Forest volume decompositions and Abel-Cayley-Hurwitz multinomial expansions. *J. Comb. Theory A.*, 98:175–191, 2002. MR1897932

[361] J. Pitman. Poisson-Dirichlet and GEM invariant distributions for split-and-merge transformations of an interval partition. *Combinatorics, Probability and Computing*, 11:501–514, 2002. MR1930355

[362] J. Pitman. Poisson-Kingman partitions. In D. R. Goldstein, editor, *Science and Statistics: A Festschrift for Terry Speed*, volume 30 of *Lecture Notes – Monograph Series*, pages 1–34. Institute of Mathematical Statistics, Hayward, California, 2003. arXiv:math.PR/0210396

[363] J. Pitman and R. Stanley. A polytope related to empirical distributions, plane trees, parking functions and the associahedron. *Discrete and Computational Geometry*, 27:603–634, 2002. arXiv:math.CO/9908029 MR1902680

[364] J. Pitman and M. Winkel. Growth of the Brownian forest, 2004. *Ann. Probab.* 33(6): 2188–2211, 2005. `http://dx.doi.org/10.1214/009117905000000442` arXiv:math.PR/0404199

[365] J. Pitman and M. Yor. A decomposition of Bessel bridges. *Z. Wahrsch. Verw. Gebiete*, 59:425–457, 1982. MR656509

[366] J. Pitman and M. Yor. Arcsine laws and interval partitions derived from a stable subordinator. *Proc. London Math. Soc. (3)*, 65:326–356, 1992. MR1168191

[367] J. Pitman and M. Yor. Decomposition at the maximum for excursions and bridges of one-dimensional diffusions. In N. Ikeda, S. Watanabe, M. Fukushima, and H. Kunita, editors, *Itô's Stochastic Calculus and Probability Theory*, pages 293–310. Springer-Verlag, 1996. MR1439532

[368] J. Pitman and M. Yor. Random discrete distributions derived from self-similar random sets. *Electron. J. Probab.*, 1:Paper 4, 1–28, 1996. MR1386296

[369] J. Pitman and M. Yor. On the lengths of excursions of some Markov processes. In *Séminaire de Probabilités XXXI*, volume 1655 of *Lecture Notes in Math.*, pages 272–286. Springer, 1997. MR1478737

[370] J. Pitman and M. Yor. On the relative lengths of excursions derived from a stable subordinator. In *Séminaire de Probabilités XXXI*, volume 1655 of *Lecture Notes in Math.*, pages 287–305. Springer, 1997. MR1478738

[371] J. Pitman and M. Yor. The two-parameter Poisson-Dirichlet distribution derived from a stable subordinator. *Ann. Probab.*, 25:855–900, 1997. MR1434129

[372] J. Pitman and M. Yor. Random Brownian scaling identities and splicing of Bessel processes. *Ann. Probab.*, 26:1683–1702, 1998. MR1675059

[373] J. Pitman and M. Yor. The law of the maximum of a Bessel bridge. *Electron. J. Probab.*, 4:Paper 15, 1–35, 1999. MR1701890

[374] J. Pitman and M. Yor. On the distribution of ranked heights of excursions of a Brownian bridge. *Ann. Probab.*, 29:361–384, 2001. MR1825154

[375] J. Pitman and M. Yor. Infinitely divisible laws associated with hyperbolic functions. *Canadian Journal of Mathematics*, 53(581):292–330, 2003. MR1969794

[376] B. Pittel. On growing random binary trees. *J. Math. Anal. Appl.*, 103(2):461–480, 1984. MR762569

[377] H. Pollard. The representation of e^{-x^λ} as a Laplace integral. *Bull. Amer. Math. Soc.*, 52:908–910, 1946.

[378] H. Prüfer. Neuer Beweis eines Satzes über Permutationen. *Archiv für Mathematik und Physik*, 27:142–144, 1918.

[379] G. Raney. A formal solution of $\sum_{i=1}^{\infty} a_i e^{B_i X} = x$. *Canad. J. Math.*, 16:755–762, 1964.

[380] G. N. Raney. Functional composition patterns and power series reversion. *Trans. Amer. Math. Soc.*, 94:441–451, 1960.

[381] D. B. Ray. Sojourn times of a diffusion process. *Ill. J. Math.*, 7:615–630, 1963.

[382] A. Rényi. Probabilistic methods in combinatorial mathematics. In R. Bose and T. Dowling, editors, *Combinatorial Mathematics and its Applications*, pages 1–13. Univ. of North Carolina Press, Chapel Hill, 1969.

[383] A. Rényi. On the enumeration of trees. In R. Guy, H. Hanani, N. Sauer, and J. Schonheim, editors, *Combinatorial Structures and their Applications*, pages 355–360. Gordon and Breach, New York, 1970.

[384] D. Revuz and M. Yor. *Continuous martingales and Brownian motion.* Springer, Berlin-Heidelberg, 1999. 3rd edition. 1725357.

[385] J. Riordan. *An Introduction to Combinatorial Analysis.* Wiley, New York, 1958.

[386] L. C. G. Rogers. Brownian local times and branching processes. In *Seminar on probability, XVIII*, volume 1059 of *Lecture Notes in Math.*, pages 42–55. Springer, Berlin, 1984. MR770947

[387] L. C. G. Rogers and J. Pitman. Markov functions. *Annals of Probability*, 9:573–582, 1981. MR624684

[388] J. Rosen. Renormalization and limit theorems for self-intersections of superprocesses. *Ann. Probab.*, 20(3):1341–1368, 1992. MR1175265

[389] M. Rosenblatt. *Random Processes.* Springer-Verlag, New York, 1974.

[390] G.-C. Rota, D. Kahaner, and A. Odlyzko. On the foundation of combinatorial theory VIII: Finite operator calculus. *J. Math. Anal. Appl.*, 42:684–760, 1973.

[391] G.-C. Rota and R. Mullin. On the foundation of combinatorial theory III: Theory of binomial enumeration. In B. Harris, editor, *Graph Theory and its Applications*, pages 167–213. Academic Press, New York, 1970.

[392] A. Rouault. Lois de Zipf et sources markoviennes. *Ann. de l'Inst. Henri Poincaré B*, 14:169 – 188, 1978.

[393] S. Sagitov. The general coalescent with asynchronous mergers of ancestral lines. *J. Appl. Probab.*, 36(4):1116–1125, 1999. MR1742154

[394] P. Salminen. Brownian excursions revisited. In *Seminar on Stochastic Processes 1983*, pages 161–187. Birkhäuser Boston, 1984.

[395] J. Schweinsberg. A necessary and sufficient condition for the Λ-coalescent to come down from infinity. *Electron. Comm. Probab.*, 5:1–11 (electronic), 2000. MR1736720

[396] J. Schweinsberg. Coalescents with simultaneous multiple collisions. *Electron. J. Probab.*, 5:Paper no. 12, 50 pp. (electronic), 2000. MR1781024

[397] J. Schweinsberg. Applications of the continuous-time ballot theorem to brownian motion and related processes. *Stochastic Process. Appl.*, 95:151–176, 2001.

[398] J. Schweinsberg. Coalescent processes obtained from supercritical Galton-Watson processes. *Stochastic Process. Appl.*, 106(1):107–139, 2003. MR1983046

[399] J. W. Shapiro. *Capacity of Brownian trace and level sets.* Ph. D. thesis, University of California, Berkeley, 1995.

[400] L. A. Shepp and S. P. Lloyd. Ordered cycle lengths in a random permutation. *Trans. Amer. Math. Soc.*, 121:340–357, 1966.

[401] T. Shiga and S. Watanabe. Bessel diffusions as a one-parameter family of diffusion processes. *Z. Wahrsch. Verw. Gebiete*, 27:37–46, 1973.

[402] P. Shor. A new proof of Cayley's formula for counting labelled trees. *J. Combinatorial Theory A.*, 71:154–158, 1995.

[403] A. Stam. Cycles of random permutations. *Ars Combinatorica*, 16:43–48, 1983.

[404] A. J. Stam. Polynomials of binomial type and compound Poisson processes. *Journal of Mathematical Analysis and Applications*, 130:151, 1988.

[405] A. J. Stam. Some stochastic processes in a random permutation. *Journal of Statistical Planning and Inference*, 19:229 – 244, 1988.

[406] R. Stanley. *Enumerative Combinatorics, Vol. 1.* Cambridge University Press, 1997.

[407] R. Stanley. *Enumerative Combinatorics, Vol. 2.* Cambridge University Press, 1999.

[408] G. Szekeres. Distribution of labelled trees by diameter. In *Combinatorial mathematics, X (Adelaide, 1982)*, volume 1036 of *Lecture Notes in Math.*, pages 392–397. Springer, Berlin, 1983. MR731595

[409] L. Takács. A generalization of the ballot problem and its application to the theory of queues. *J. Amer. Stat. Assoc.*, 57:154–158, 1962.

[410] L. Takács. Queues, random graphs and branching processes. *J. Applied Mathematics and Simulation*, 1:223–243, 1988.

[411] L. Takács. Counting forests. *Discrete Mathematics*, 84:323–326, 1990.

[412] M. Talagrand. *Spin glasses: a challenge for mathematicians*, volume 46 of *Ergebnisse der Mathematik und ihrer Grenzgebiete. 3. Folge. A Series of Modern Surveys in Mathematics.* Springer-Verlag, Berlin, 2003. MR1993891

[413] S. Tavaré. Line-of-descent and genealogical processes and their applications in population genetics. *Theoret. Population Biol.*, 26:119–164, 1984.

[414] S. Tavaré. The birth process with immigration, and the genealogical structure of large populations. *J. Math. Biol.*, 25:161–171, 1987.

[415] L. Toscano. Numeri di Stirling generalizzati operatori differenziali e polinomi ipergeometrici. *Commentationes Pontificia Academica Scientarum*, 3:721–757, 1939.

[416] N. Tsilevich, A. Vershik, and M. Yor. An infinite-dimensional analogue of the Lebesgue measure and distinguished properties of the gamma process. *J. Funct. Anal.*, 185(1):274–296, 2001. MR1853759

[417] K. van Harn and F. W. Steutel. Infinite divisibility and the waiting-time paradox. *Comm. Statist. Stochastic Models*, 11(3):527–540, 1995. MR1340971

[418] J. van Pelt and R. W. H. Verwer. Growth models (including terminal and segmental branching) for topological binary trees. *Bull. Math. Biol.*, 47(3):323–336, 1985. MR803398

[419] J. van Pelt and R. W. H. Verwer. Topological properties of binary trees grown with order-dependent branching probabilities. *Bull. Math. Biol.*, 48(2):197–211, 1986. MR845637

[420] A. Vershik and S. Kerov. Asymptotics of the Plancherel measure of the symmetric group and the limiting form of Young tables. *Soviet Math. Dokl.*, 18:527–531, 1977. Translation of Dokl. Acad. Nauk. SSSR 233 (1977) 1024-1027.

[421] A. Vershik and S. Kerov. Asymptotic behavior of the maximum and generic dimensions of irreducible representations of the symmetric group. *Functional Anal. Appl.*, 19:21–31, 1985.

[422] A. Vershik and A. Shmidt. Limit measures arising in the theory of groups, I. *Theor. Prob. Appl.*, 22:79–85, 1977.

[423] A. Vershik and A. Shmidt. Limit measures arising in the theory of symmetric groups, II. *Theor. Prob. Appl.*, 23:36–49, 1978.

[424] A. Vershik and Y. Yakubovich. The limit shape and fluctuations of random partitions of naturals with fixed number of summands. *Mosc. Math. J.*, 1(3):457–468, 472, 2001. MR1877604

[425] W. Vervaat. A relation between Brownian bridge and Brownian excursion. *Ann. Probab.*, 7:143–149, 1979.

[426] J. Walsh. Downcrossings and the Markov property of local time. In *Temps Locaux*, volume 52-53 of *Astérisque*, pages 89–116. Soc. Math. de France, 1978.

[427] J. Warren and M. Yor. The Brownian burglar: conditioning Brownian motion by its local time process. In J. Azéma, M. Émery, , M. Ledoux, and M. Yor, editors, *Séminaire de Probabilités XXXII*, pages 328–342. Springer, 1998. Lecture Notes in Math. 1686.

[428] G. A. Watterson. The stationary distribution of the infinitely-many neutral alleles diffusion model. *J. Appl. Probab.*, 13:639–651, 1976.

[429] G. A. Watterson. Lines of descent and the coalescent. *Theoret. Population Biol.*, 10:239–253, 1984.

[430] J. Wendel. Left continuous random walk and the Lagrange expansion. *Amer. Math. Monthly*, 82:494–498, 1975.

[431] J. G. Wendel. Zero-free intervals of semi-stable Markov processes. *Math. Scand.*, 14:21 - 34, 1964.

[432] P. Whittle. The equilibrium statistics of a clustering process in the uncondensed phase. *Proc. Roy. Lond. Soc. A*, 285:501–519, 1965.

[433] P. Whittle. Statistical processes of aggregation and polymerisation. *Proc. Camb. Phil. Soc.*, 61:475–495, 1965.

[434] P. Whittle. *Systems in Stochastic Equilibrium.* Wiley, 1986.

[435] D. Williams. Decomposing the Brownian path. *Bull. Amer. Math. Soc.*, 76:871–873, 1970.

[436] D. Williams. Path decomposition and continuity of local time for one dimensional diffusions I. *Proc. London Math. Soc. (3)*, 28:738–768, 1974.

[437] D. Williams. *Diffusions, Markov Processes, and Martingales, Vol. 1: Foundations.* Wiley, Chichester, New York, 1979.

[438] M. Yor. *Some Aspects of Brownian Motion, Part I: Some Special Functionals.* Lectures in Math., ETH Zürich. Birkhäuser, 1992.

[439] M. Yor. Random Brownian scaling and some absolute continuity relationships. In E. Bolthausen, M. Dozzi, and F. Russo, editors, *Seminar on Stochastic Analysis, Random Fields and Applications. Centro Stefano Franscini, Ascona, 1993*, pages 243–252. Birkhäuser, 1995.

[440] J. E. Young. *Partition-valued stochastic processes with applications.* Ph. D. thesis, University of California, Berkeley, 1995.

[441] S. Zabell. The continuum of inductive methods revisited. In J. Earman and J. D. Norton, editors, *The Cosmos of Science*, Pittsburgh-Konstanz Series in the Philosophy and History of Science, pages 351–385. University of Pittsburgh Press/Universitätsverlag Konstanz, 1997.

Index

List of participants

ABRAHAM Romain	Univ. René Descartes, Paris, F
ALILI Larbi	ETH Zurich, Switzerland
ATTOUCH Mohamed Kadi	Univ. Djillali Liabes, Sidi Bel Abbès, Algérie
BEFFARA Vincent	Univ. Paris-Sud, Orsay, F
BELHADJI Lamia	Univ. Mostaganem, Algérie
BERESTYCKI Julien	Univ. Pierre et Marie Curie, Paris, F
BERTOIN Jean	Univ. Pierre et Marie Curie, Paris, F
BLACHE Fabrice	Univ. Blaise Pascal, Clermont-Ferrand, F
CABALLERO Maria-Emilia	UNAM, Mexico D.F., Mexico
CALKA Pierre	Univ. Claude Bernard, Lyon, F
CAMPANINO Massimo	Univ. Bologna, Italia
CAMPI Luciano	Univ. Pierre et Marie Curie, Paris, F
CAPITAINE Mireille	CNRS, Univ. Paul Sabatier, Toulouse, F
CARMONA Philippe	Univ. Paul Sabatier, Toulouse, F
CHASSAING Philippe	Institut Elie Cartan, Nancy, F
CHAUMONT Loïc	Univ. Pierre et Marie Curie, Paris, F
CHELIOTIS Dimitrios	Stanford Univ., USA
CHERIDITO Patrick	ETH Zurich, Switzerland
COUTIN Laure	Univ. Paul Sabatier, Toulouse, F
DAVIAUD Olivier	Stanford Univ., USA
DHERSIN Jean-Stéphane	Univ. René Descartes, Paris, F
DONATI-MARTIN Catherine	CNRS, Univ. Paul Sabatier, Toulouse, F
DOUMERC Yan	Univ. Paul Sabatier, Toulouse, F
DUBEDAT Julien	Ecole Normale Supérieure, Paris, F
DURRINGER Clément	Univ. Paul Sabatier, Toulouse, F
ENRIQUEZ Nathanaël	Univ. Pierre et Marie Curie, Paris, F
FERRALIS Marc	Univ. Pierre et Marie Curie, Paris, F
FRIEDRICH Roland	Univ. Paris-Sud, Orsay, F
FUSCHINI Serena	Univ. Bologna, Italia
GHERIBALLAH Abdelkader	Univ. Djillali Liabes, Sidi Bel Abbès, Algérie
GOLDSCHMIDT Christina	Univ. Cambridge, UK

GREENWOOD Priscilla	Arizona State Univ., Tempe, USA
GRORUD Axel	Univ. Provence, Marseille, F
HAAS Bénédicte	Univ. Pierre et Marie Curie, Paris, F
HERBIN Erick	Dassault Aviation, Saint-Cloud, F
HOLROYD Alexander	Univ. California, Los Angeles, USA
HU Yueyun	Univ. Pierre et Marie Curie, Paris, F
KASPI Haya	Technion, Israel
KOURKOVA Irina	Univ. Pierre et Marie Curie, Paris, F
KUPPER Michael	ETH Zurich, Switzerland
LE GALL Jean-Frančois	Ecole Normale Supérieure, Paris, F
LE JAN Yves	Univ. Paris-Sud, Orsay, F
LEURIDAN Christophe	Institut Fourier, Grenoble, F
LEVY Thierry	CNRS, IRMA, Strasbourg, F
LORANG Gerard	Centre Universitaire de Luxembourg
MAIDA Mylène	Ecole Normale Supérieure, Lyon, F
MANSUY Roger	Univ. Pierre et Marie Curie, Paris, F
MARCHAL Philippe	CNRS, Ecole Normale Supérieure, Paris, F
MARTIN-LOF Anders	Univ. Stockholm, Sweden
MATHIEU Pierre	Univ. Provence, Marseille, F
MEJANE Olivier	Univ. Paul Sabatier, Toulouse, F
MYTNIK Leonid	Technion, Israel
NIEDERHAUSEN Meike	Purdue Univ., West Lafayette, USA
NIKEGHBALI Ashkan	Univ. Pierre et Marie Curie, Paris, F
NUALART David	Univ. Barcelona, Spain
PARVIAINEN Robert	Uppsala Univ., Sweden
PECCATI Giovanni	Univ. Pierre et Marie Curie, Paris, F
PICARD Jean	Univ. Blaise Pascal, Clermont-Ferrand, F
QUER Lluís	Univ. Barcelona, Spain
RIVERO Victor	Univ. Pierre et Marie Curie, Paris, F
RIVIERE Olivier	Univ. René Descartes, Paris, F
ROMIK Dan	Univ. Pierre et Marie Curie, Paris, F
ROUAULT Alain	Univ. Versailles, F
ROUX Daniel	Univ. Blaise Pascal, Clermont-Ferrand, F
SABOT Christophe	CNRS, Univ. Pierre et Marie Curie, Paris, F
SAINT LOUBERT BIE Erwan	Univ. Blaise Pascal, Clermont-Ferrand, F
SAVONA Catherine	Univ. Blaise Pascal, Clermont-Ferrand, F
SCHMITZ Tom	ETH Zurich, Switzerland
SERLET Laurent	Univ. René Descartes, Paris, F
SKOLIMOWSKA Magdalena	Univ. Wroclaw, Poland
SZEKELY Balazs	Budapest Univ. Technol. and Econ., Hungary
TAKAOKA Koichiro	Hitotsubashi Univ., Tokyo, Japan
VALKO Benedek	Technical Univ. Budapest, Hungary
WINKEL Matthias	Univ. Oxford, UK
YASSAI Sadr	Univ. Pierre et Marie Curie, Paris, F
YOR Marc	Univ. Pierre et Marie Curie, Paris, F

List of short lectures

Romain ABRAHAM	Représentation probabiliste des solutions de $\Delta u = 4u^2$ dans un domaine avec condition de Neumann au bord
Vincent BEFFARA	The dimension of the SLE_k curve
Julien BERESTYCKI	Fast and slow points in a fragmentation
Jean BERTOIN	Sur les petites masses dans un processus de fragmentation
Massimo CAMPANINO	Ornstein-Zernike theory for the finite range Ising models above T_c
Pierre CALKA	The distribution of the number of sides of the typical Poisson-Voronoi cell
Philippe CHASSAING	Random planar maps and Brownian snake
Loïc CHAUMONT	Sur une identité de fluctuation pour les marches aléatoires
Patrick CHERIDITO	Moving average representation of Gaussian processes and the semimartingale property
Yan DOUMERC	Combinatorial representations of eigenvalues of random Gaussian matrices
Nathanaël ENRIQUEZ	Correlated random walks and their continuous time analog
Christina GOLDSCHMIDT	Essential edges in Poisson random hypergraphs
Bénédicte HAAS	Perte de masse dans des systèmes de fragmentation
Erick HERBIN	Mouvements browniens multifractionnaires indexés par $\mathbb{R}_+^N$
Alexander HOLROYD	Bootstrap percolation and $\pi^2/18$
Haya KASPI	Lenses in skew Brownian motion

Irina KOURKOVA	Derrida's generalised random energy model of spin glasses: a rigorous analysis
Christophe LEURIDAN	Filtration d'une marche aléatoire stationnaire sur le cercle
Thierry LEVY	Yang-Mills measure: a random geometry on surfaces
Philippe MARCHAL	The simple random walk and the Chinese restaurant
Olivier MEJANE	Upper bound of a volume exponent for directed polymers in random environment
Leonid MYTNIK	Regularity and irregularity of $(1+\beta)$-stable super-Brownian motion
David NUALART	Stochastic calculus with respect to the fractional Brownian motion
Robert PARVIAINEN	Ordering bond percolation critical probabilities on Archimedean and Laves lattices
Giovanni PECCATI	Multiple integral representation for functionals of Dirichlet processes
Lluís QUER	Absolute continuity of the law of the solution to the three-dimensional stochastic wave equation
Victor RIVERO	Sur des ensembles aléatoires associés aux maxima locaux d'un processus de Poisson ponctuel
Dan ROMIK	The hook walk on continual Young diagrams
Laurent SERLET	Poisson snake and self-similar fragmentation
Koichiro TAKAOKA	On Kamazaki's criterion for continuous exponential martingales
Benedek VALKO	Perturbing the hydrodynamic limit
Matthias WINKEL	Subordination in the wide sense of Lévy processes
Marc YOR	q-calcul, fonctionnelles exponentielles du processus de Poisson, et une solution au problème des moments de la loi log-normale.

Lecture Notes in Mathematics

For information about earlier volumes
please contact your bookseller or Springer
LNM Online archive: springerlink.com

Vol. 1674: G. Klaas, C. R. Leedham-Green, W. Plesken, Linear Pro-p-Groups of Finite Width (1997)
Vol. 1675: J. E. Yukich, Probability Theory of Classical Euclidean Optimization Problems (1998)
Vol. 1676: P. Cembranos, J. Mendoza, Banach Spaces of Vector-Valued Functions (1997)
Vol. 1677: N. Proskurin, Cubic Metaplectic Forms and Theta Functions (1998)
Vol. 1678: O. Krupková, The Geometry of Ordinary Variational Equations (1997)
Vol. 1679: K.-G. Grosse-Erdmann, The Blocking Technique. Weighted Mean Operators and Hardy's Inequality (1998)
Vol. 1680: K.-Z. Li, F. Oort, Moduli of Supersingular Abelian Varieties (1998)
Vol. 1681: G. J. Wirsching, The Dynamical System Generated by the 3n+1 Function (1998)
Vol. 1682: H.-D. Alber, Materials with Memory (1998)
Vol. 1683: A. Pomp, The Boundary-Domain Integral Method for Elliptic Systems (1998)
Vol. 1684: C. A. Berenstein, P. F. Ebenfelt, S. G. Gindikin, S. Helgason, A. E. Tumanov, Integral Geometry, Radon Transforms and Complex Analysis. Firenze, 1996. Editors: E. Casadio Tarabusi, M. A. Picardello, G. Zampieri (1998)
Vol. 1685: S. König, A. Zimmermann, Derived Equivalences for Group Rings (1998)
Vol. 1686: J. Azéma, M. Émery, M. Ledoux, M. Yor (Eds.), Séminaire de Probabilités XXXII (1998)
Vol. 1687: F. Bornemann, Homogenization in Time of Singularly Perturbed Mechanical Systems (1998)
Vol. 1688: S. Assing, W. Schmidt, Continuous Strong Markov Processes in Dimension One (1998)
Vol. 1689: W. Fulton, P. Pragacz, Schubert Varieties and Degeneracy Loci (1998)
Vol. 1690: M. T. Barlow, D. Nualart, Lectures on Probability Theory and Statistics. Editor: P. Bernard (1998)
Vol. 1691: R. Bezrukavnikov, M. Finkelberg, V. Schechtman, Factorizable Sheaves and Quantum Groups (1998)
Vol. 1692: T. M. W. Eyre, Quantum Stochastic Calculus and Representations of Lie Superalgebras (1998)
Vol. 1694: A. Braides, Approximation of Free-Discontinuity Problems (1998)
Vol. 1695: D. J. Hartfiel, Markov Set-Chains (1998)
Vol. 1696: E. Bouscaren (Ed.): Model Theory and Algebraic Geometry (1998)
Vol. 1697: B. Cockburn, C. Johnson, C.-W. Shu, E. Tadmor, Advanced Numerical Approximation of Nonlinear Hyperbolic Equations. Cetraro, Italy, 1997. Editor: A. Quarteroni (1998)
Vol. 1698: M. Bhattacharjee, D. Macpherson, R. G. Möller, P. Neumann, Notes on Infinite Permutation Groups (1998)
Vol. 1699: A. Inoue,Tomita-Takesaki Theory in Algebras of Unbounded Operators (1998)
Vol. 1700: W. A. Woyczyński, Burgers-KPZ Turbulence (1998)
Vol. 1701: Ti-Jun Xiao, J. Liang, The Cauchy Problem of Higher Order Abstract Differential Equations (1998)
Vol. 1702: J. Ma, J. Yong, Forward-Backward Stochastic Differential Equations and Their Applications (1999)
Vol. 1703: R. M. Dudley, R. Norvaiša, Differentiability of Six Operators on Nonsmooth Functions and p-Variation (1999)
Vol. 1704: H. Tamanoi, Elliptic Genera and Vertex Operator Super-Algebras (1999)
Vol. 1705: I. Nikolaev, E. Zhuzhoma, Flows in 2-dimensional Manifolds (1999)
Vol. 1706: S. Yu. Pilyugin, Shadowing in Dynamical Systems (1999)
Vol. 1707: R. Pytlak, Numerical Methods for Optimal Control Problems with State Constraints (1999)
Vol. 1708: K. Zuo, Representations of Fundamental Groups of Algebraic Varieties (1999)
Vol. 1709: J. Azéma, M. Émery, M. Ledoux, M. Yor (Eds.), Séminaire de Probabilités XXXIII (1999)
Vol. 1710: M. Koecher, The Minnesota Notes on Jordan Algebras and Their Applications (1999)
Vol. 1711: W. Ricker, Operator Algebras Generated by Commuting Projećtions: A Vector Measure Approach (1999)
Vol. 1712: N. Schwartz, J. J. Madden, Semi-algebraic Function Rings and Reflectors of Partially Ordered Rings (1999)
Vol. 1713: F. Bethuel, G. Huisken, S. Müller, K. Steffen, Calculus of Variations and Geometric Evolution Problems. Cetraro, 1996. Editors: S. Hildebrandt, M. Struwe (1999)
Vol. 1714: O. Diekmann, R. Durrett, K. P. Hadeler, P. K. Maini, H. L. Smith, Mathematics Inspired by Biology. Martina Franca, 1997. Editors: V. Capasso, O. Diekmann (1999)
Vol. 1715: N. V. Krylov, M. Röckner, J. Zabczyk, Stochastic PDE's and Kolmogorov Equations in Infinite Dimensions. Cetraro, 1998. Editor: G. Da Prato (1999)
Vol. 1716: J. Coates, R. Greenberg, K. A. Ribet, K. Rubin, Arithmetic Theory of Elliptic Curves. Cetraro, 1997. Editor: C. Viola (1999)
Vol. 1717: J. Bertoin, F. Martinelli, Y. Peres, Lectures on Probability Theory and Statistics. Saint-Flour, 1997. Editor: P. Bernard (1999)
Vol. 1718: A. Eberle, Uniqueness and Non-Uniqueness of Semigroups Generated by Singular Diffusion Operators (1999)
Vol. 1719: K. R. Meyer, Periodic Solutions of the N-Body Problem (1999)
Vol. 1720: D. Elworthy, Y. Le Jan, X-M. Li, On the Geometry of Diffusion Operators and Stochastic Flows (1999)
Vol. 1721: A. Iarrobino, V. Kanev, Power Sums, Gorenstein Algebras, and Determinantal Loci (1999)

Vol. 1722: R. McCutcheon, Elemental Methods in Ergodic Ramsey Theory (1999)
Vol. 1723: J. P. Croisille, C. Lebeau, Diffraction by an Immersed Elastic Wedge (1999)
Vol. 1724: V. N. Kolokoltsov, Semiclassical Analysis for Diffusions and Stochastic Processes (2000)
Vol. 1725: D. A. Wolf-Gladrow, Lattice-Gas Cellular Automata and Lattice Boltzmann Models (2000)
Vol. 1726: V. Marić, Regular Variation and Differential Equations (2000)
Vol. 1727: P. Kravanja M. Van Barel, Computing the Zeros of Analytic Functions (2000)
Vol. 1728: K. Gatermann Computer Algebra Methods for Equivariant Dynamical Systems (2000)
Vol. 1729: J. Azéma, M. Émery, M. Ledoux, M. Yor (Eds.) Séminaire de Probabilités XXXIV (2000)
Vol. 1730: S. Graf, H. Luschgy, Foundations of Quantization for Probability Distributions (2000)
Vol. 1731: T. Hsu, Quilts: Central Extensions, Braid Actions, and Finite Groups (2000)
Vol. 1732: K. Keller, Invariant Factors, Julia Equivalences and the (Abstract) Mandelbrot Set (2000)
Vol. 1733: K. Ritter, Average-Case Analysis of Numerical Problems (2000)
Vol. 1734: M. Espedal, A. Fasano, A. Mikelić, Filtration in Porous Media and Industrial Applications. Cetraro 1998. Editor: A. Fasano. 2000.
Vol. 1735: D. Yafaev, Scattering Theory: Some Old and New Problems (2000)
Vol. 1736: B. O. Turesson, Nonlinear Potential Theory and Weighted Sobolev Spaces (2000)
Vol. 1737: S. Wakabayashi, Classical Microlocal Analysis in the Space of Hyperfunctions (2000)
Vol. 1738: M. Émery, A. Nemirovski, D. Voiculescu, Lectures on Probability Theory and Statistics (2000)
Vol. 1739: R. Burkard, P. Deuflhard, A. Jameson, J.-L. Lions, G. Strang, Computational Mathematics Driven by Industrial Problems. Martina Franca, 1999. Editors: V. Capasso, H. Engl, J. Periaux (2000)
Vol. 1740: B. Kawohl, O. Pironneau, L. Tartar, J.-P. Zolesio, Optimal Shape Design. Tróia, Portugal 1999. Editors: A. Cellina, A. Ornelas (2000)
Vol. 1741: E. Lombardi, Oscillatory Integrals and Phenomena Beyond all Algebraic Orders (2000)
Vol. 1742: A. Unterberger, Quantization and Non-holomorphic Modular Forms (2000)
Vol. 1743: L. Habermann, Riemannian Metrics of Constant Mass and Moduli Spaces of Conformal Structures (2000)
Vol. 1744: M. Kunze, Non-Smooth Dynamical Systems (2000)
Vol. 1745: V. D. Milman, G. Schechtman (Eds.), Geometric Aspects of Functional Analysis. Israel Seminar 1999-2000 (2000)
Vol. 1746: A. Degtyarev, I. Itenberg, V. Kharlamov, Real Enriques Surfaces (2000)
Vol. 1747: L. W. Christensen, Gorenstein Dimensions (2000)
Vol. 1748: M. Ruzicka, Electrorheological Fluids: Modeling and Mathematical Theory (2001)
Vol. 1749: M. Fuchs, G. Seregin, Variational Methods for Problems from Plasticity Theory and for Generalized Newtonian Fluids (2001)
Vol. 1750: B. Conrad, Grothendieck Duality and Base Change (2001)
Vol. 1751: N. J. Cutland, Loeb Measures in Practice: Recent Advances (2001)
Vol. 1752: Y. V. Nesterenko, P. Philippon, Introduction to Algebraic Independence Theory (2001)
Vol. 1753: A. I. Bobenko, U. Eitner, Painlevé Equations in the Differential Geometry of Surfaces (2001)
Vol. 1754: W. Bertram, The Geometry of Jordan and Lie Structures (2001)
Vol. 1755: J. Azéma, M. Émery, M. Ledoux, M. Yor (Eds.), Séminaire de Probabilités XXXV (2001)
Vol. 1756: P. E. Zhidkov, Korteweg de Vries and Nonlinear Schrödinger Equations: Qualitative Theory (2001)
Vol. 1757: R. R. Phelps, Lectures on Choquet's Theorem (2001)
Vol. 1758: N. Monod, Continuous Bounded Cohomology of Locally Compact Groups (2001)
Vol. 1759: Y. Abe, K. Kopfermann, Toroidal Groups (2001)
Vol. 1760: D. Filipović, Consistency Problems for Heath-Jarrow-Morton Interest Rate Models (2001)
Vol. 1761: C. Adelmann, The Decomposition of Primes in Torsion Point Fields (2001)
Vol. 1762: S. Cerrai, Second Order PDE's in Finite and Infinite Dimension (2001)
Vol. 1763: J.-L. Loday, A. Frabetti, F. Chapoton, F. Goichot, Dialgebras and Related Operads (2001)
Vol. 1764: A. Cannas da Silva, Lectures on Symplectic Geometry (2001)
Vol. 1765: T. Kerler, V. V. Lyubashenko, Non-Semisimple Topological Quantum Field Theories for 3-Manifolds with Corners (2001)
Vol. 1766: H. Hennion, L. Hervé, Limit Theorems for Markov Chains and Stochastic Properties of Dynamical Systems by Quasi-Compactness (2001)
Vol. 1767: J. Xiao, Holomorphic Q Classes (2001)
Vol. 1768: M.J. Pflaum, Analytic and Geometric Study of Stratified Spaces (2001)
Vol. 1769: M. Alberich-Carramiñana, Geometry of the Plane Cremona Maps (2002)
Vol. 1770: H. Gluesing-Luerssen, Linear Delay-Differential Systems with Commensurate Delays: An Algebraic Approach (2002)
Vol. 1771: M. Émery, M. Yor (Eds.), Séminaire de Probabilités 1967-1980. A Selection in Martingale Theory (2002)
Vol. 1772: F. Burstall, D. Ferus, K. Leschke, F. Pedit, U. Pinkall, Conformal Geometry of Surfaces in S^4 (2002)
Vol. 1773: Z. Arad, M. Muzychuk, Standard Integral Table Algebras Generated by a Non-real Element of Small Degree (2002)
Vol. 1774: V. Runde, Lectures on Amenability (2002)
Vol. 1775: W. H. Meeks, A. Ros, H. Rosenberg, The Global Theory of Minimal Surfaces in Flat Spaces. Martina Franca 1999. Editor: G. P. Pirola (2002)
Vol. 1776: K. Behrend, C. Gomez, V. Tarasov, G. Tian, Quantum Comohology. Cetraro 1997. Editors: P. de Bartolomeis, B. Dubrovin, C. Reina (2002)
Vol. 1777: E. García-Río, D. N. Kupeli, R. Vázquez-Lorenzo, Osserman Manifolds in Semi-Riemannian Geometry (2002)
Vol. 1778: H. Kiechle, Theory of K-Loops (2002)
Vol. 1779: I. Chueshov, Monotone Random Systems (2002)
Vol. 1780: J. H. Bruinier, Borcherds Products on O(2,1) and Chern Classes of Heegner Divisors (2002)
Vol. 1781: E. Bolthausen, E. Perkins, A. van der Vaart, Lectures on Probability Theory and Statistics. Ecole d' Eté de Probabilités de Saint-Flour XXIX-1999. Editor: P. Bernard (2002)

Vol. 1782: C.-H. Chu, A. T.-M. Lau, Harmonic Functions on Groups and Fourier Algebras (2002)
Vol. 1783: L. Grüne, Asymptotic Behavior of Dynamical and Control Systems under Perturbation and Discretization (2002)
Vol. 1784: L.H. Eliasson, S. B. Kuksin, S. Marmi, J.-C. Yoccoz, Dynamical Systems and Small Divisors. Cetraro, Italy 1998. Editors: S. Marmi, J.-C. Yoccoz (2002)
Vol. 1785: J. Arias de Reyna, Pointwise Convergence of Fourier Series (2002)
Vol. 1786: S. D. Cutkosky, Monomialization of Morphisms from 3-Folds to Surfaces (2002)
Vol. 1787: S. Caenepeel, G. Militaru, S. Zhu, Frobenius and Separable Functors for Generalized Module Categories and Nonlinear Equations (2002)
Vol. 1788: A. Vasil'ev, Moduli of Families of Curves for Conformal and Quasiconformal Mappings (2002)
Vol. 1789: Y. Sommerhäuser, Yetter-Drinfel'd Hopf algebras over groups of prime order (2002)
Vol. 1790: X. Zhan, Matrix Inequalities (2002)
Vol. 1791: M. Knebusch, D. Zhang, Manis Valuations and Prüfer Extensions I: A new Chapter in Commutative Algebra (2002)
Vol. 1792: D. D. Ang, R. Gorenflo, V. K. Le, D. D. Trong, Moment Theory and Some Inverse Problems in Potential Theory and Heat Conduction (2002)
Vol. 1793: J. Cortés Monforte, Geometric, Control and Numerical Aspects of Nonholonomic Systems (2002)
Vol. 1794: N. Pytheas Fogg, Substitution in Dynamics, Arithmetics and Combinatorics. Editors: V. Berthé, S. Ferenczi, C. Mauduit, A. Siegel (2002)
Vol. 1795: H. Li, Filtered-Graded Transfer in Using Noncommutative Gröbner Bases (2002)
Vol. 1796: J.M. Melenk, hp-Finite Element Methods for Singular Perturbations (2002)
Vol. 1797: B. Schmidt, Characters and Cyclotomic Fields in Finite Geometry (2002)
Vol. 1798: W.M. Oliva, Geometric Mechanics (2002)
Vol. 1799: H. Pajot, Analytic Capacity, Rectifiability, Menger Curvature and the Cauchy Integral (2002)
Vol. 1800: O. Gabber, L. Ramero, Almost Ring Theory (2003)
Vol. 1801: J. Azéma, M. Émery, M. Ledoux, M. Yor (Eds.), Séminaire de Probabilités XXXVI (2003)
Vol. 1802: V. Capasso, E. Merzbach, B.G. Ivanoff, M. Dozzi, R. Dalang, T. Mountford, Topics in Spatial Stochastic Processes. Martina Franca, Italy 2001. Editor: E. Merzbach (2003)
Vol. 1803: G. Dolzmann, Variational Methods for Crystalline Microstructure – Analysis and Computation (2003)
Vol. 1804: I. Cherednik, Ya. Markov, R. Howe, G. Lusztig, Iwahori-Hecke Algebras and their Representation Theory. Martina Franca, Italy 1999. Editors: V. Baldoni, D. Barbasch (2003)
Vol. 1805: F. Cao, Geometric Curve Evolution and Image Processing (2003)
Vol. 1806: H. Broer, I. Hoveijn. G. Lunther, G. Vegter, Bifurcations in Hamiltonian Systems. Computing Singularities by Gröbner Bases (2003)
Vol. 1807: V. D. Milman, G. Schechtman (Eds.), Geometric Aspects of Functional Analysis. Israel Seminar 2000-2002 (2003)
Vol. 1808: W. Schindler, Measures with Symmetry Properties (2003)
Vol. 1809: O. Steinbach, Stability Estimates for Hybrid Coupled Domain Decomposition Methods (2003)
Vol. 1810: J. Wengenroth, Derived Functors in Functional Analysis (2003)
Vol. 1811: J. Stevens, Deformations of Singularities (2003)
Vol. 1812: L. Ambrosio, K. Deckelnick, G. Dziuk, M. Mimura, V. A. Solonnikov, H. M. Soner, Mathematical Aspects of Evolving Interfaces. Madeira, Funchal, Portugal 2000. Editors: P. Colli, J. F. Rodrigues (2003)
Vol. 1813: L. Ambrosio, L. A. Caffarelli, Y. Brenier, G. Buttazzo, C. Villani, Optimal Transportation and its Applications. Martina Franca, Italy 2001. Editors: L. A. Caffarelli, S. Salsa (2003)
Vol. 1814: P. Bank, F. Baudoin, H. Föllmer, L.C.G. Rogers, M. Soner, N. Touzi, Paris-Princeton Lectures on Mathematical Finance 2002 (2003)
Vol. 1815: A. M. Vershik (Ed.), Asymptotic Combinatorics with Applications to Mathematical Physics. St. Petersburg, Russia 2001 (2003)
Vol. 1816: S. Albeverio, W. Schachermayer, M. Talagrand, Lectures on Probability Theory and Statistics. Ecole d'Eté de Probabilités de Saint-Flour XXX-2000. Editor: P. Bernard (2003)
Vol. 1817: E. Koelink, W. Van Assche(Eds.), Orthogonal Polynomials and Special Functions. Leuven 2002 (2003)
Vol. 1818: M. Bildhauer, Convex Variational Problems with Linear, nearly Linear and/or Anisotropic Growth Conditions (2003)
Vol. 1819: D. Masser, Yu. V. Nesterenko, H. P. Schlickewei, W. M. Schmidt, M. Waldschmidt, Diophantine Approximation. Cetraro, Italy 2000. Editors: F. Amoroso, U. Zannier (2003)
Vol. 1820: F. Hiai, H. Kosaki, Means of Hilbert Space Operators (2003)
Vol. 1821: S. Teufel, Adiabatic Perturbation Theory in Quantum Dynamics (2003)
Vol. 1822: S.-N. Chow, R. Conti, R. Johnson, J. Mallet-Paret, R. Nussbaum, Dynamical Systems. Cetraro, Italy 2000. Editors: J. W. Macki, P. Zecca (2003)
Vol. 1823: A. M. Anile, W. Allegretto, C. Ringhofer, Mathematical Problems in Semiconductor Physics. Cetraro, Italy 1998. Editor: A. M. Anile (2003)
Vol. 1824: J. A. Navarro González, J. B. Sancho de Salas, $\mathscr{C}^\infty$ – Differentiable Spaces (2003)
Vol. 1825: J. H. Bramble, A. Cohen, W. Dahmen, Multiscale Problems and Methods in Numerical Simulations, Martina Franca, Italy 2001. Editor: C. Canuto (2003)
Vol. 1826: K. Dohmen, Improved Bonferroni Inequalities via Abstract Tubes. Inequalities and Identities of Inclusion-Exclusion Type. VIII, 113 p, 2003.
Vol. 1827: K. M. Pilgrim, Combinations of Complex Dynamical Systems. IX, 118 p, 2003.
Vol. 1828: D. J. Green, Gröbner Bases and the Computation of Group Cohomology. XII, 138 p, 2003.
Vol. 1829: E. Altman, B. Gaujal, A. Hordijk, Discrete-Event Control of Stochastic Networks: Multimodularity and Regularity. XIV, 313 p, 2003.
Vol. 1830: M. I. Gil', Operator Functions and Localization of Spectra. XIV, 256 p, 2003.
Vol. 1831: A. Connes, J. Cuntz, E. Guentner, N. Higson, J. E. Kaminker, Noncommutative Geometry, Martina Franca, Italy 2002. Editors: S. Doplicher, L. Longo (2004)
Vol. 1832: J. Azéma, M. Émery, M. Ledoux, M. Yor (Eds.), Séminaire de Probabilités XXXVII (2003)
Vol. 1833: D.-Q. Jiang, M. Qian, M.-P. Qian, Mathematical Theory of Nonequilibrium Steady States. On the Frontier of Probability and Dynamical Systems. IX, 280 p, 2004.

Vol. 1834: Yo. Yomdin, G. Comte, Tame Geometry with Application in Smooth Analysis. VIII, 186 p, 2004.
Vol. 1835: O.T. Izhboldin, B. Kahn, N.A. Karpenko, A. Vishik, Geometric Methods in the Algebraic Theory of Quadratic Forms. Summer School, Lens, 2000. Editor: J.-P. Tignol (2004)
Vol. 1836: C. Năstăsescu, F. Van Oystaeyen, Methods of Graded Rings. XIII, 304 p, 2004.
Vol. 1837: S. Tavaré, O. Zeitouni, Lectures on Probability Theory and Statistics. Ecole d'Eté de Probabilités de Saint-Flour XXXI-2001. Editor: J. Picard (2004)
Vol. 1838: A.J. Ganesh, N.W. O'Connell, D.J. Wischik, Big Queues. XII, 254 p, 2004.
Vol. 1839: R. Gohm, Noncommutative Stationary Processes. VIII, 170 p, 2004.
Vol. 1840: B. Tsirelson, W. Werner, Lectures on Probability Theory and Statistics. Ecole d'Eté de Probabilités de Saint-Flour XXXII-2002. Editor: J. Picard (2004)
Vol. 1841: W. Reichel, Uniqueness Theorems for Variational Problems by the Method of Transformation Groups (2004)
Vol. 1842: T. Johnsen, A.L. Knutsen, K3 Projective Models in Scrolls (2004)
Vol. 1843: B. Jefferies, Spectral Properties of Noncommuting Operators (2004)
Vol. 1844: K.F. Siburg, The Principle of Least Action in Geometry and Dynamics (2004)
Vol. 1845: Min Ho Lee, Mixed Automorphic Forms, Torus Bundles, and Jacobi Forms (2004)
Vol. 1846: H. Ammari, H. Kang, Reconstruction of Small Inhomogeneities from Boundary Measurements (2004)
Vol. 1847: T.R. Bielecki, T. Björk, M. Jeanblanc, M. Rutkowski, J.A. Scheinkman, W. Xiong, Paris-Princeton Lectures on Mathematical Finance 2003 (2004)
Vol. 1848: M. Abate, J. E. Fornaess, X. Huang, J. P. Rosay, A. Tumanov, Real Methods in Complex and CR Geometry, Martina Franca, Italy 2002. Editors: D. Zaitsev, G. Zampieri (2004)
Vol. 1849: Martin L. Brown, Heegner Modules and Elliptic Curves (2004)
Vol. 1850: V. D. Milman, G. Schechtman (Eds.), Geometric Aspects of Functional Analysis. Israel Seminar 2002-2003 (2004)
Vol. 1851: O. Catoni, Statistical Learning Theory and Stochastic Optimization (2004)
Vol. 1852: A.S. Kechris, B.D. Miller, Topics in Orbit Equivalence (2004)
Vol. 1853: Ch. Favre, M. Jonsson, The Valuative Tree (2004)
Vol. 1854: O. Saeki, Topology of Singular Fibers of Differential Maps (2004)
Vol. 1855: G. Da Prato, P.C. Kunstmann, I. Lasiecka, A. Lunardi, R. Schnaubelt, L. Weis, Functional Analytic Methods for Evolution Equations. Editors: M. Iannelli, R. Nagel, S. Piazzera (2004)
Vol. 1856: K. Back, T.R. Bielecki, C. Hipp, S. Peng, W. Schachermayer, Stochastic Methods in Finance, Bressanone/Brixen, Italy, 2003. Editors: M. Fritelli, W. Runggaldier (2004)
Vol. 1857: M. Émery, M. Ledoux, M. Yor (Eds.), Séminaire de Probabilités XXXVIII (2005)
Vol. 1858: A.S. Cherny, H.-J. Engelbert, Singular Stochastic Differential Equations (2005)
Vol. 1859: E. Letellier, Fourier Transforms of Invariant Functions on Finite Reductive Lie Algebras (2005)
Vol. 1860: A. Borisyuk, G.B. Ermentrout, A. Friedman, D. Terman, Tutorials in Mathematical Biosciences I. Mathematical Neurosciences (2005)
Vol. 1861: G. Benettin, J. Henrard, S. Kuksin, Hamiltonian Dynamics – Theory and Applications, Cetraro, Italy, 1999. Editor: A. Giorgilli (2005)
Vol. 1862: B. Helffer, F. Nier, Hypoelliptic Estimates and Spectral Theory for Fokker-Planck Operators and Witten Laplacians (2005)
Vol. 1863: H. Führ, Abstract Harmonic Analysis of Continuous Wavelet Transforms (2005)
Vol. 1864: K. Efstathiou, Metamorphoses of Hamiltonian Systems with Symmetries (2005)
Vol. 1865: D. Applebaum, B.V. R. Bhat, J. Kustermans, J. M. Lindsay, Quantum Independent Increment Processes I. From Classical Probability to Quantum Stochastic Calculus. Editors: M. Schürmann, U. Franz (2005)
Vol. 1866: O.E. Barndorff-Nielsen, U. Franz, R. Gohm, B. Kümmerer, S. Thorbjønsen, Quantum Independent Increment Processes II. Structure of Quantum Lévy Processes, Classical Probability, and Physics. Editors: M. Schürmann, U. Franz, (2005)
Vol. 1867: J. Sneyd (Ed.), Tutorials in Mathematical Biosciences II. Mathematical Modeling of Calcium Dynamics and Signal Transduction. (2005)
Vol. 1868: J. Jorgenson, S. Lang, $Pos_n(R)$ and Eisenstein Sereies. (2005)
Vol. 1869: A. Dembo, T. Funaki, Lectures on Probability Theory and Statistics. Ecole d'Eté de Probabilités de Saint-Flour XXXIII-2003. Editor: J. Picard (2005)
Vol. 1870: V.I. Gurariy, W. Lusky, Geometry of Müntz Spaces and Related Questions. (2005)
Vol. 1871: P. Constantin, G. Gallavotti, A.V. Kazhikhov, Y. Meyer, S. Ukai, Mathematical Foundation of Turbulent Viscous Flows, Martina Franca, Italy, 2003. Editors: M. Cannone, T. Miyakawa (2006)
Vol. 1872: A. Friedman (Ed.), Tutorials in Mathematical Biosciences III. Cell Cycle, Proliferation, and Cancer (2006)
Vol. 1873: R. Mansuy, M. Yor, Random Times and Enlargements of Filtrations in a Brownian Setting (2006)
Vol. 1875: J. Pitman, J. Picard, Combinatorial Stochastic Processes. Ecole d'Eté de Probabilités de Saint-Flour XXXII-2002. Editor: J. Picard (2006)
Vol. 1876: H. Herrlich, The Axiom of Choice. (2006)

Recent Reprints and New Editions

Vol. 1200: V. D. Milman, G. Schechtman (Eds.), Asymptotic Theory of Finite Dimensional Normed Spaces. 1986. – Corrected Second Printing (2001)
Vol. 1471: M. Courtieu, A.A. Panchishkin, Non-Archimedean L-Functions and Arithmetical Siegel Modular Forms. – Second Edition (2003)
Vol. 1618: G. Pisier, Similarity Problems and Completely Bounded Maps. 1995 – Second, Expanded Edition (2001)
Vol. 1629: J.D. Moore, Lectures on Seiberg-Witten Invariants. 1997 – Second Edition (2001)
Vol. 1638: P. Vanhaecke, Integrable Systems in the realm of Algebraic Geometry. 1996 – Second Edition (2001)
Vol. 1702: J. Ma, J. Yong, Forward-Backward Stochastic Differential Equations and their Applications. 1999. – Corrected 3rd printing (2005)

4. Manuscripts should in general be submitted in English. Final manuscripts should contain at least 100 pages of mathematical text and should always include
   - a general table of contents;
   - an informative introduction, with adequate motivation and perhaps some historical remarks: it should be accessible to a reader not intimately familiar with the topic treated;
   - a global subject index: as a rule this is genuinely helpful for the reader.

   Lecture Notes volumes are, as a rule, printed digitally from the authors' files. We strongly recommend that all contributions in a volume be written in the same LaTeX version, preferably LaTeX2e. To ensure best results, authors are asked to use the LaTeX2e style files available from Springer's web-server at

   ftp://ftp.springer.de/pub/tex/latex/mathegl/mono.zip (for monographs) and
   ftp://ftp.springer.de/pub/tex/latex/mathegl/mult.zip (for summer schools/tutorials).

   Additional technical instructions, if necessary, are available on request from:

   lnm@springer-sbm.com.

5. Careful preparation of the manuscripts will help keep production time short besides ensuring satisfactory appearance of the finished book in print and online. After acceptance of the manuscript authors will be asked to prepare the final LaTeX source files (and also the corresponding dvi-, pdf- or zipped ps-file) together with the final printout made from these files. The LaTeX source files are essential for producing the full-text online version of the book. For the existing online volumes of LNM see:
   http://www.springerlink.com/openurl.asp?genre=journal&issn=0075-8434.

   The actual production of a Lecture Notes volume takes approximately 8 weeks.

6. Volume editors receive a total of 50 free copies of their volume to be shared with the authors, but no royalties. They and the authors are entitled to a discount of 33.3 % on the price of Springer books purchased for their personal use, if ordering directly from Springer.

7. Commitment to publish is made by letter of intent rather than by signing a formal contract. Springer-Verlag secures the copyright for each volume. Authors are free to reuse material contained in their LNM volumes in later publications: A brief written (or e-mail) request for formal permission is sufficient.

Addresses:

Professor J.-M. Morel, CMLA,
École Normale Supérieure de Cachan,
61 Avenue du Président Wilson, 94235 Cachan Cedex, France
E-mail: Jean-Michel.Morel@cmla.ens-cachan.fr

Professor F. Takens, Mathematisch Instituut,
Rijksuniversiteit Groningen, Postbus 800,
9700 AV Groningen, The Netherlands
E-mail: F.Takens@math.rug.nl

Professor B. Teissier, Institut Mathématique de Jussieu,
UMR 7586 du CNRS, Équipe "Géométrie et Dynamique",
175 rue du Chevaleret, 75013 Paris, France
E-mail: teissier@math.jussieu.fr

For the "Mathematical Biosciences Subseries" of LNM :
Professor P. K. Maini, Center for Mathematical Biology,
Mathematical Institute, 24-29 St Giles,
Oxford OX1 3LP, UK
E-mail : maini@maths.ox.ac.uk

Springer, Mathematics Editorial I, Tiergartenstr. 17,
69121 Heidelberg, Germany,
Tel.: +49 (6221) 487-8410
Fax: +49 (6221) 487-8355
E-mail: lnm@springer-sbm.com